W9-CHH-326

Immunobiology and Pathogenesis of Persistent Virus Infections

Immunobiology and Pathogenesis of Persistent Virus Infections

Edited by
Carlos Lopez
Division of Viral Diseases
Centers for Disease Control
Atlanta, Georgia

American Society for Microbiology
Washington, D.C.

1913 I Street, N.W.
Washington, DC 20006

Library of Congress Cataloging-in-Publication Data

Immunobiology and pathogenesis of persistent virus infections.

First International Symposium on Immunobiology and Pathogenesis of Persistent Virus Infections, held in Atlanta, Ga., Apr. 1987.
Includes index.
1. Virus diseases—Immunological aspects—Congresses. 2. Virus diseases—Congresses. I. Lopez, Carlos, 1942– . II. International Symposium on Immunobiology and Pathogenesis of Persistent Virus Infections (1st : 1987 : Atlanta, Ga.) [DNLM: 1. Immunologic Deficiency Syndromes—immunology—congresses. 2. Virus Diseases—immunology—congresses. 3. Viruses—pathogenicity—congresses. WC 500 I33 1987]

RC114.5.I46 1988 616.9'25079 88-10369

ISBN 1-55581-000-4

Printed in the United States of America

CONTENTS

I. BASIC CONCEPTS

II. ARENAVIRUS INFECTIONS

III. HUMAN IMMUNODEFICIENCY VIRUS INFECTIONS

IV. EPSTEIN-BARR VIRUS INFECTIONS

CONTRIBUTORS

Rafi Ahmed • Department of Microbiology and Immunology, University of California-Los Angeles School of Medicine, Los Angeles, CA 90024

Anthony C. Allison • Syntex Research, 3401 Hillview Avenue, Palo Alto, CA 94304

Birgitta Asjö • Department of Virology, Karolinska Institute, 10401 Stockholm, Sweden

David H. L. Bishop • Natural Environment Research Council Institute of Virology, Mansfield Road, Oxford OX1 3SR, United Kingdom

M. J. Buchmeier • Department of Immunology, Research Institute of Scripps Clinic, 10666 North Torrey Pines Road, La Jolla, CA 92037

James W. Curran • AIDS Program, Center for Infectious Diseases, Centers for Disease Control, Atlanta, GA 30333

Eva-Maria Fenyö • Department of Virology, Karolinska Institute, 10401 Stockholm, Sweden

John R. Gebhard • Department of Immunology, Scripps Clinic and Research Foundation, 10666 North Torrey Pines Road, La Jolla, CA 92037

Magnus Gidlund • Department of Immunology, Karolinska Institute, 10401 Stockholm, Sweden

Leon Gordis • Department of Epidemiology, School of Hygiene and Public Health, 615 N. Wolfe Street, Johns Hopkins University, Baltimore, MD 21205

Martin S. Hirsch • Massachusetts General Hospital and Harvard Medical School, Boston, MA 02114

Scott D. Holmberg • AIDS Program, Center for Infectious Diseases, Centers for Disease Control, Building 6-285, G 22, Atlanta, GA 30333

Fritz Lehmann-Grube • Heinrich-Pette-Institut für Experimentelle Virologie und Immunologie an der Universität Hamburg, Martinistrasse 52, 2000 Hamburg 20, Federal Republic of Germany

Hanna Lewicki • Department of Immunology, Scripps Clinic and Research Foundation, 10666 North Torrey Pines Road, La Jolla, CA 92037

Carlos Lopez • Viral Exanthems and Herpesvirus Branch, Division of Viral Diseases, Center for Infectious Diseases, Centers for Disease Control, Atlanta, GA 30333

Karin Lundin • Department of Immunology, Karolinska Institute, 10401 Stockholm, Sweden

Shunji Matsuda • Department of Immunology, Karolinska Institute, 10401 Stockholm, Sweden

J. Steven McDougal • Immunology Branch, Division of Host Factors, Center for Infectious Diseases, Centers for Disease Control, Atlanta, GA 30333

Kim W. McIntyre • Department of Pathology, University of Massachusetts Medical School, 55 Lake Avenue North, Worcester, MA 01655

C. A. Mims • Department of Microbiology, United Medical and Dental School, Guy's Campus, London SE1 9RT, United Kingdom

Janet K. A. Nicholson • Division of Host Factors, Center for Infectious Diseases, Centers for Disease Control, Atlanta, GA 30333

Kenneth Nilsson • Department of Pathology, University Hospital, 75185 Uppsala, Sweden

Michael B. A. Oldstone • Department of Immunology, Scripps Clinic and Research Foundation, 10666 North Torrey Pines Road, La Jolla, CA 92037

B. S. Parekh • Department of Immunology, Research Institute of Scripps Clinic, 10666 North Torrey Pines Road, La Jolla, CA 92037

Sandra E. Pike • Laboratory of Molecular Immunology, Division of Biochemistry and Biophysics, Office of Biologics Research and Review, Food and Drug Administration, Bethesda, MD 20892

Thomas C. Quinn • Laboratory of Immunoregulation, National Institute of Allergy and Infectious Diseases, Bethesda, MD 20892; Johns Hopkins University School of Medicine, Blalock 1111, 600 North Wolfe Street, Baltimore, MD 21205

Urban Ramstedt • Department of Immunology, Karolinska Institute, 10401 Stockholm, Sweden

A. B. Rickinson • Department of Cancer Studies, University of Birmingham, Birmingham B15 2TJ, United Kingdom

Kenneth L. Rosenthal • Molecular Virology and Immunology Program, Department of Pathology, McMaster University Health Sciences Centre, Hamilton, Ontario, Canada L8N 3Z5

Peter J. Southern • Department of Immunology, Scripps Clinic and Research Foundation, 10666 North Torrey Pines Road, La Jolla, CA 92037

Stephen E. Straus • Medical Virology Section, Laboratory of Clinical Investigation, National Institute of Allergy and Infectious Diseases, Building 10, Room 11N113, Bethesda, MD 20892

Antoinette Tishon • Department of Immunology, Scripps Clinic and Research Foundation, 10666 North Torrey Pines Road, La Jolla, CA 92037

Giovanna Tosato • Laboratory of Molecular Immunology, Division of Biochemistry and Biophysics, Office of Biologics Research and Review, Food and Drug Administration, Bethesda, MD 20892

E. L. Weber • Department of Immunology, Research Institute of Scripps Clinic, 10666 North Torrey Pines Road, La Jolla, CA 92037

Raymond M. Welsh • Department of Pathology, University of Massachusetts Medical School, 55 Lake Avenue North, Worcester, MA 01655

J. Lindsay Whitton • Department of Immunology, Scripps Clinic and Research Foundation, 10666 North Torrey Pines Road, La Jolla, CA 92037

Hans Wigzell • Department of Immunology, Karolinska Institute, 10401 Stockholm, Sweden

Flossie Wong-Staal • Laboratory of Tumor Cell Biology, National Cancer Institute, Building 37, Room 6A09, Bethesda, MD 20892

K. E. Wright • Department of Immunology, Research Institute of Scripps Clinic, 10666 North Torrey Pines Road, La Jolla, CA 92037

Hyekyung Yang • Department of Pathology, University of Massachusetts Medical School, 55 Lake Avenue North, Worcester, MA 01655

R. M. Zinkernagel • Institut für Pathologie, Universitätsspital Zürich, CH-8091 Zurich, Switzerland

PREFACE

Disease usually results from inconclusive negotiations for symbiosis, an overstepping of the line by one side or the other, a biologic misinterpretation of border.

Lewis Thomas
(*The Lives of a Cell*)

Many of the infectious diseases of humanity have been conquered, and we are left with chronic "degenerative" illnesses heading the list of major causes of morbidity and mortality. Heart disease, cancer, chronic neurological diseases, and chronic musculoskeletal disorders currently top the roll of important human illnesses, while acquired immunodeficiency syndrome (AIDS) creeps up that list. Although it has become fashionable to ascribe many of these maladies to environmental influences, scientific insight strongly suggests that some, if not most, chronic disorders may be the direct or indirect consequences of viral infections. Viruses can act directly via persistent infections or indirectly by triggering pathophysiologic mechanisms.

The viruses which have been found to establish persistent or recrudescing infection are not unique with respect to the type or size of genomes or the characteristics of their replicative cycle; rather, they interact with the host in ways which are different from viruses unable to establish such infections. Immune reactivity usually fails to clear the infection and often causes far more damage than the virus itself. Because of this, we need a thorough understanding of the immunobiology and pathogenesis of persistent infections in order to develop new ways of controlling them.

This volume begins with five chapters which develop some of the basic concepts needed for an understanding of the immunobiology and pathogenesis of persistent virus infections. A section comprising seven chapters focuses on arenavirus infections. Lymphocytic choriomeningitis virus, an arenavirus infection of mice, has been the preeminent model for study of the mechanisms involved in persistence. The next seven chapters focus on human immunodeficiency virus infection, an important and rapidly growing problem. The final three chapters are on persistent infections with Epstein-Barr virus, a human herpesvirus.

I would like to thank the contributors to this volume for providing not only the facts, as they know them, but also the studied understanding which only a major contributor to the field can offer.

Carlos Lopez
Centers for Disease Control
Atlanta, Georgia

Present address:
Eli Lilly & Co.
Indianapolis, Indiana

ACKNOWLEDGMENTS

The First International Symposium on Immunobiology and Pathogenesis of Persistent Virus Infections, which formed the basis for this volume, was supported by the generous contributions of Biokit, Inc., Bristol-Myers Co., Burroughs Wellcome Co., Cancer Research Institute, The Carter Center of Emory University, Center for Infectious Diseases (Centers for Disease Control), Cetus Corp., Connaught Research Institute, E. I. du Pont de Nemours & Co., Eli Lilly and Co., Hoffmann-La Roche Inc., Hyclone Laboratories, Inc., Institut Merieux, Merck Sharp & Dohme Research Laboratories, Miles Laboratories, Inc., Ortho Pharmaceutical (Canada) Ltd., Ortho Pharmaceutical Corp., The Rockefeller Foundation, Schering Corp., E.R. Squibb and Sons, Inc., and Viratek.

Part I.

BASIC CONCEPTS

Chapter 1

Immunobiology and Pathogenesis of Persistent Virus Infections

C. A. Mims

INTRODUCTION

Nowadays, with smallpox gone and poliomyelitis and maybe influenza viruses held at bay, persistent virus infections are the ones that concern us all. There are several reasons why persistent viral infections matter. (i) They can be activated to cause acute episodes of disease. Under natural circumstances, reactivation of certain viruses occurs during pregnancy and old age. Reactivation at other times, for instance during unrelated infections, might also be expected to occur. So far, the only example is with herpes simplex virus during fevers or at certain stages of the menstrual cycle. (ii) They are sometimes associated with immunopathological disease or with other chronic disease, especially of the central nervous system. (iii) They are sometimes associated with neoplasms. (iv) They enable the virus to persist in the community or in the host species.

This last point is important, because it explains why there are such things as persistent viral infections. If there were no selective advantage for viruses that persist, persistence, which is not easily established, would be a relatively rare phenomenon.

One hundred thousand years ago, when humans lived in bands of 20 to 30 individuals in extended family groups that rarely met other groups, there were definite restrictions on the type of virus that could infect the human species. Poliomyelitis virus, measles virus, and rhinoviruses, for instance, could not have maintained themselves. After infecting all members in the group, these viruses, having nowhere else to go and not being very stable in the environment, would have perished. The only

C. A. Mims • Department of Microbiology, United Medical and Dental School, Guy's Campus, London SE1 9RT, United Kingdom.

viruses that stood a chance were either those that infected humans temporarily but persisted in an animal or arthropod cycle independent of humans or those that infected and persisted in humans.

Some years ago, Black et al. (6) studied members of isolated Indian communities in the Amazon basin, some of whom had had no previous contact with other Indians, let alone other races. I am not certain just how easily blood samples were obtained at an early stage, but when antibody tests were done it was clear that only the persistent viruses—the herpesvirus group, polyomavirus, adenovirus, etc., plus some of the arthropod-borne viruses with alternative vertebrate hosts—infected these small communities. There were no antibodies to measles, poliomyelitis, or influenza viruses.

What is the mechanism by which persistent viruses avoid the fate of other viruses, i.e., elimination by antibodies, immune cells, interferons, and other host defense systems?

It turns out that we have no shortage of possible mechanisms, but unfortunately most of them remain no more than possibilities. Let us examine some of them.

PRIVILEGED SITES

There are some locations in the body where immune forces are less efficiently expressed, so that persistence is favored (Fig. 1). In each case the surface of the infected cell faces the outside world, for instance, a feather follicle (Marek's disease virus) or the lumen of a salivary gland (Epstein-Barr virus [EBV] and rabies virus) or mammary gland (mammary tumor virus) or a kidney tubule (polyomaviruses).

Many viruses mature only from the luminal surfaces of epithelial cells rather than from lateral or basal surfaces (54), a topographical restriction of budding that also reflects biochemical gradients in the cell. If there are no destructive or inflammatory results of the infection, blood-borne sensitized lymphocytes or antibodies cannot easily reach the site and eliminate infection. One striking example is provided by the papillomaviruses, which infect epithelial cells in the basal layers of the epidermis. Viral replication, as is so often the case, depends on differentiation of cells. Basal and parabasal cells contain viral DNA, and it is only as cells move up to the surface and become squamified and keratinized that viral antigens and finally fully infectious particles are produced, ready to be shed to the exterior. We must remember that skin shedding, with or without wart viruses, makes a notable contribution to the environment; if you run a finger along a fairly clean surface in a hospital, it will pick up a fine white film of material, 70 to 80% of which is keratin from human skin.

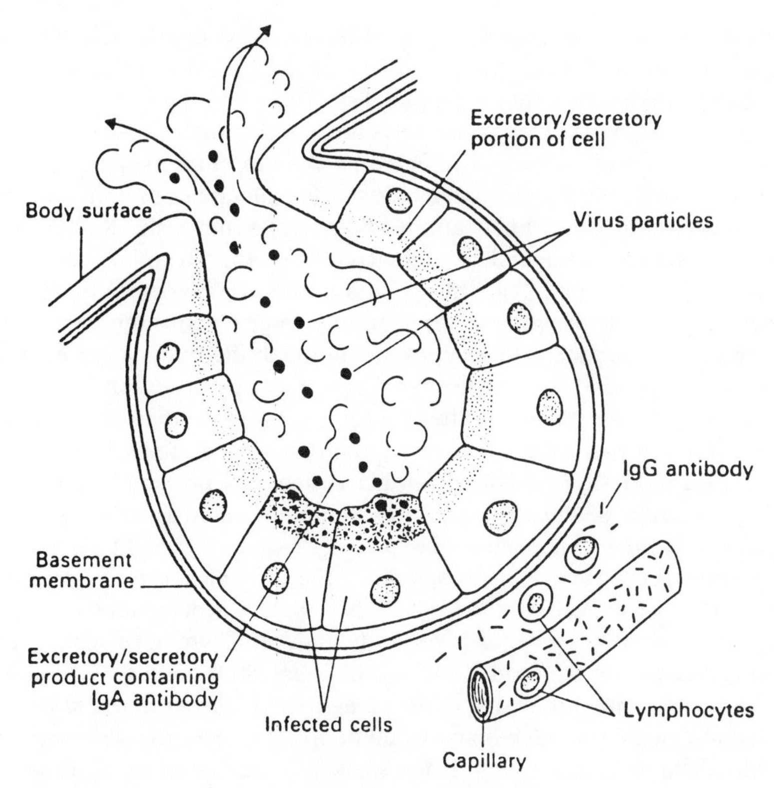

Figure 1. Privileged sites. These include the following: mammary glands, salivary glands, kidney tubules, feather follicles, and bile ducts.

It is no wonder that wart viruses are one of the most successful of human parasites. They have remained with our species, I am sure, during its evolution, yet nearly all the time cause little or no inconvenience. Many infectious wart virus foci in the skin are so well behaved that they are invisible. Lesions on the cervix are often visible as whitish flattened areas of dysplasia only after application of 5% acetic acid, and lesions on the penis, although infectious, may be so small as to be invisible to the naked eye. The types of venereal warts that could at one time be regarded as mere ornamental appendages, but now appear to have a close and sinister connection with carcinoma of the cervix (and penis and vulva), can be regarded as an aberrant handful of the 46 or more better-behaved types that infect humans. These warts have taken advantage of the tremendous possibilities of spread by the sexual route in crowded communities of promiscuous humans who have unfortunately been enjoying barrier-free mucosal contacts for the past 15 to 20 years.

EBV provides another example of a virus whose persistence may be geared to a dependence on cell differentiation. For EBV to survive and be transmitted so well in humans, it must be shed from the body, preferably for long periods. The virus is thus present in saliva and in fluid collected from the parotid gland duct, and it infects epithelial cells. Recent work (58) shows that basal and parabasal cells in the stratified epithelium express C3d receptors. The viruses bind to their receptors on these cells, infect them (with productive replication taking place only as the cells differentiate), lose the C3d receptors, and reach the body surface. If this is the basic persistence strategy of EBV, perhaps infection of B cells, which also happen to have C3d receptors, is no more than an accidental aberration. On the other hand, it may be that infection of these immune cells, with induction of polyclonal activation and possible interference with generation of effective antiviral immune responses, represents an additional strategy for persistence.

Can certain organs or tissues (e.g., the central nervous system) be regarded as privileged sites? Perhaps not, but in a persistently infected site, especially if persistence involves latency, this would be made easier if cells have a long life span. Neurons and oligodendrocytes, like some of the basal cells on epithelial surfaces, long-lived cells in lymphoid tissues, and stem cells in the bone marrow, come into this category. Perhaps this is one of the reasons why epithelial surfaces and central nervous system and lymphoid tissues are favored sites for persistence. For instance, could stem cells in the bone marrow harbor persistent cytomegaloviruses? Unfortunately, we know too little about the nature of stem cells in these parts of the body. Which of these cells are conserved and long lived, not destined to be lost by differentiation, with subsequent death or shedding from the body?

FAILURE OF IMMUNE RESPONSE

There are many examples of the failure of protective immune responses in persistent virus infections (Table 1). Let us now survey the mechanisms by which this occurs.

Primary Failure of Immune Response to Virus in which Critical Viral Antigens or Epitopes Are Poor Immunogens

If the host fails to make an effective immune response, whether this involves neutralizing antibodies, antibody-dependent cell-mediated cytotoxicity, cytotoxic T cells, etc., then persistence is favored. Inadequate

Table 1. Failure of Protective Immune Response in Persistent Virus Infections

Phenomenon	Examples	References
Failure in neutralizing antibody production (low titer of failure to neutralize)	Thymic necrosis virus (mice)	13
	LCMV (mice)	44
	Aleutian disease virus (mink)	49, 59
	African swine fever virus (pigs)	64
	Caprine retrovirus (field strains) (goat)	40
	Bovine virus diarrhea	19
	Mucosal disease (cattle, sheep)	8
	Hepatitis B virus (humans)	
	HIV	65
Delay in production of neutralizing antibody	Cytomegalovirus (humans, mice)	
	Visna virus (laboratory strains)	
	LDV	55
Failure of protective cell-mediated immune response	Congenital rubella rubivirus (humans)	18, 63
	Congenital or neonatal LCM virus (mice)	2
	Congenital cytomegalovirus (humans)	23
	LDV (mice)	

antibody responses would include low-titer responses, responses of poor affinity, or responses directed against irrelevant (nonprotective) epitopes. A delayed immune response, as well as a weak immune response, would tend to favor persistence. Each infection is a race between virus replication and spread on the one hand and the generation of an effective antiviral response on the other; a few days' delay in this response may enable a virus to invade key tissues and initiate persistent infection.

Specific immune responses are controlled by genes linked to those controlling the class 2 major histocompatibility antigens. It seems inevitable that since the genome of viruses, particularly RNA viruses, is continuously varying, viral polypeptides which evoke less powerful immune responses will tend to be produced and will be selected for during virus evolution. In other words, viruses are constantly probing the immunological repertoire of the host species, ready to exploit weaknesses. Possible mechanisms would include weak interaction of a key viral antigen or epitope with class 2 molecules on antigen-presenting cells. Studies of these interactions are just beginning to be made (3). Individuals differ, of course, in their responsiveness to a given antigen or epitope, and there is no doubt that a host species with a great reserve of variability in response is at an advantage. That is to say, polymorphism in the class 2 genes is a good thing.

One has to consider the possibility that a poor immune response is due to similarities between viral and host antigens and that the host is reluctant to generate autoimmune responses even though they may be antiviral. Does molecular mimicry play a part in the failure of immune responses to persistent viruses? Earlier reports of cross-reactivity between viral and host antigens have been greatly extended by computer searching for shared sequences (15). Many viral proteins share six or more consecutive amino acids with the acetylcholine receptor, the insulin receptor, or the encephalitogenic decapeptide of myelin basic protein, and others share sequences with epitopes in the peripheral nerve P2 protein that induces experimental allergic neuritis. Monoclonal antibodies have been shown to bind to viral and to host epitopes, and small peptides consisting of the shared sequences can induce both antiviral and antihost antibodies. The potential is great when one remembers that only a very small number of polypeptides from a small number of viruses were tested. However, autoimmune responses do not necessarily lead to immunopathological results. Also, it is not clear whether viral polypeptides are more likely to have sequences shared with the host than are unselected foreign polypeptides. At present, we do not know whether cross-reactive immune responses have any significance. Sequence homologies could also be important when a viral product mimics a hormone, neurotransmitter, or soluble immune mediator, interfering with the action of one of these substances by competitive binding to the appropriate receptor on cells. For instance, there is a report of a 6-amino-acid sequence shared between the envelope protein of human immunodeficiency virus type 1 (HIV-1) and the cell-binding region of interleukin-2, providing a mechanism for interference with the immune function of interleukin-2 (51). So far, however, there seems little actual evidence that molecular mimicry by viruses exerts a restraining influence on host immune responses or is important in persistent virus infections.

Active Interference of Infecting Virus with Immune Response to Itself

Many persistent viruses invade and replicate in the immune system. Usually, the exact cells invaded are not known; even when they are known, productive infection often depends on the state of differentiation of the host cell, e.g., visna virus infection in monocytes and macrophages (20, 40) or herpes simplex virus infection in T cells (7). The infected cells may die or merely show alterations in function. Actual growth of virus in the responding cells themselves may not be essential. Immunosuppression caused by human cytomegalovirus and measles virus, for instance, involves depression of numerous immune functions of peripheral blood mononuclear cells, which is demonstrable in vitro. However, only a very small proportion of cells are infected, and replication is restricted to expression of immediate early antigens (16, 52). Replicating viruses can produce immunosuppressive molecules, such as the retroviral envelope protein p15E (36, 47, 62). Such molecules could affect the function of uninfected immune cells (46) or the function of immune mediators, but nothing is known of their significance in vivo.

Replication in immune cells can be regarded as a viral strategy, i.e., a method of reducing, delaying, or otherwise weakening the protective immune response to the infecting virus. A broad, antigen-nonspecific immunosuppression would be advantageous for the virus as long as it was only temporary, and this occurs in many acute virus infections (14). Possible mechanisms include production of suppressor factors or cells, interference with antigen presentation, effects on production or action of interleukin-1 or interleukin-2 (41, 53; S. Blackett and C. A. Mims, Arch. Virol., in press), or antigen competition. There would be an advantageous dampening of responses to the infecting virus, giving it time to spread and establish infection in tissues. Even if the virus were nonpersistent, this suppression would enable the pathogenic steps involved in a long incubation period to be accomplished with less interference from the host.

If, however, the general immunosuppression was more severe and long lasting, life-threatening secondary infections would make the host less satisfactory for a persistent virus. HIV, for instance, infects CD4-positive (helper) T cells, which, as a direct or indirect result of this, are severely depleted in numbers. Macrophages, dendritic cells, and Langerhans cells are also infected, and defects in antigen handling and presentation (33) make a separate contribution to the serious immunologic deficit in the resulting disease, acquired immunodeficiency syndrome. Looked at from this point of view, HIV has overshot the mark, as it were, severely damaging the immune system rather than more gently and discreetly modifying immune responses for its own purposes.

Viral immunosuppression is more interesting when it is antigen specific, so that responses to the infecting virus are depressed while responses to unrelated antigens are only temporarily or less severely affected. Lactic dehydrogenase virus (LDV) has long been known to cause immunosuppression to unrelated antigens (55) which lasts for an undetermined period. Depressed antibody and cell-mediated responses are associated with infection and depletion of antigen-presenting, Ia-positive macrophages (25, 26, 28). Other types of Ia-positive cells do not appear to be infected. Protective immune responses to LDV itself are depressed, with a late and feeble production of neutralizing antibodies. Most of the antiviral antibody, however, is directed against the envelope glycoprotein VP3. Delayed hypersensitivity responses to LDV are actively suppressed (27), and so far there have been no reports of cytotoxic

T-cell responses. The mice remain infected and remain well throughout their lives. Circulating complexes of viral antigen plus nonneutralizing antibody are deposited in glomeruli without causing glomerulonephritis, and the more general immunosuppression is not long-lasting or severe enough to allow secondary infections to cause problems. LDV is a better-behaved parasite than HIV, with its more gentle yet supremely successful effect on the host suggesting that it is an ancient infection of murine species, whereas HIV gives all the signs of being a recent human infection with an unbalanced and excessive impact on the host. Left to itself, HIV would presumably settle down to a more stable, less pathogenic state in the human species after having done an unacceptable amount of damage. As with myxomatosis in the Australian rabbit, there would be selection for genetically resistant individuals and for less virulent strains of virus.

Lymphocytic choriomeningitis virus (LCMV), under natural circumstances, establishes its lifelong association with the mouse after infection in utero. It infects immune cells in the infant, and there is a complete failure to produce antiviral neutralizing antibodies or cytotoxic T cells. Responses to unrelated antigens are normal. The severe immunopathological consequences of virus carriage that occur in strains of laboratory mice (glomerulonephritis and renal failure) are not seen in colonies of naturally infected wild mice (34). In certain other persistent virus infections in which immune cells are infected, such as Aleutian disease in mink (49) and mouse thymic virus in infant mice (13), there is an initial broad immunosuppression, together with a failure to develop protective virus-specific responses. Neutralizing antibodies are not detected in either infection, and there is no lasting depression of responses to unrelated antigens. Little is known about cell-mediated immune responses.

Induction of Antibodies That Interfere with Antiviral Responses

The failure to produce protective neutralizing antibodies is often seen against a background of intense B-cell activity. In LDV infection, there is polyclonal activation of B cells, with a 20-fold increase in total immunoglobulin G (IgG) (43). This is seen 7 days after infection, and much of it appears not to be antiviral (38). The demonstration that much of this antibody is IgG2a, rather than the IgG1 antibody produced against most proteins (12), is less surprising in light of the report that most antiviral antibodies are IgG2a (11). However, it highlights our ignorance about the exact mode of action of antiviral antibodies in vivo. Sometimes antiviral antibodies are formed which do not neutralize, but can block, presumably by steric hindrance, the action of neutralizing antibodies (42, 61) or which block the action of cytotoxic lymphocytes. Mink infected with Aleutian disease virus form very large amounts of antiviral antibody, which binds to antigen and ultimately may be the cause of lethal immunopathological injury, but fails to neutralize the virus. Recent work (59) shows that this antibody can indeed neutralize the virus if host phospholipids are first removed from virus particles. Masking of neutralization epitopes by host phospholipids (or other nonspecific molecules) provides an interesting mechanism for an ineffective antiviral response, and this phenomenon seems to occur also with certain nonpersistent viruses (35).

Enhancing antibodies have been described for a variety of viruses (50); they promote infection in susceptible cells if these cells express Fc receptors. They can also enhance virulence in vivo, as in, for instance, the neurovirulence of yellow fever virus. Some progress has been made in defining the conditions for neutralization or enhancement in terms of epitope specificity and antibody affinity. Unfortunately, despite the variety of antiviral actions displayed by antibodies in vitro, we have little evidence about their relative importance in vivo.

Let us now examine the possible mechanisms for the antigen-specific immunosuppression referred to above.

Preferential Deletion of Responding T Cells

Many viruses selectively infect differentiating mitotic T cells, and selective infection of those that are responding to the infecting virus would take place. Effects especially on the virus-specific response would be seen against a background of less marked but more general immunosuppression. HIV would come into this category, since it more successfully infects the activated T cells that in the initial stages of infection are responding to its own antigens.

Clonal Deletion of T Cells

During clonal deletion of T cells, the virus uses the immunologically specific antigen recognition site on the T-cell receptor as the virus receptor. This gives selective infection, with possible destruction or loss of function in precisely the T cells that would otherwise have initiated antiviral responses. This is a remarkable and specific strategy for a persistent virus and presumably happens with viruses that are capable of replicating in T cells, whether or not there is a specific virus receptor on these cells. It has been suggested as a mechanism for the failure of neutralizing antibody and cell-mediated responses in mice persistently infected with LCMV, but there are so far no studies in which its importance has been demonstrated.

Tolerization of Responding Lymphocytes

Both B and T cells can be made tolerant after exposure, without accessory cells, to large amounts of the antigen to which they are reactive. B cells are susceptible to tolerization during their differentiation, and it is possible that this accounts for the failure of hepatitis B carriers to produce antibody to hepatitis B surface antigen (HBsAg). Concentrations of HBsAg that are present in the plasma of carrier individuals ($\geq$100 ng/ml) greatly depress anti-HBs antibody production by immune B cells (4). B-cell tolerization is reversible, but with T cells it appears to be longer lasting. When cloned human T helper cells specifically reactive with the influenza virus hemagglutinin were incubated with moderately high concentrations of hemagglutinin, they became unresponsive, failing to proliferate on further exposure to this antigen, although they continued to multiply (32) and remained responsive to interleukin-2. This lasted for at least 7 days and involved the loss of T3 molecules from the antigen-specific receptor on the T cells. There was a requirement for class 2 major histocompatibility molecules on the T cells (31).

Tolerization of B or T cells would be expected when viruses replicate in lymphoid tissues, with local production of viral antigens in the immediate vicinity of specifically reactive cells. The importance of this persistence strategy, however, is not known.

Failure of Interferon Response

Persistence will be favored when a virus fails to induce significant amounts of alpha or beta interferon or when it is relatively resistant to their action. This is distinct from production of gamma interferon by sensitized T lymphocytes. Some persistent viruses, such as LCMV, LDV, and Aleutian disease virus, induce little or no interferon in the naturally infected host. Interferon is not detected during the chronic stage of LDV infection (17), only small amounts are present in LCMV carrier mice (9, 56), and it is not present in mink infected with Aleutian disease virus (66). LCMV is highly susceptible to interferon, but LDV and Aleutian disease virus are relatively resistant to its action (60, 66). As another example, many adenoviruses are insensitive to interferon, the viral VAI RNA preventing activation of interferon-induced eIF-2α-kinase (29). RNAs from EBV and from encephalomyocarditis virus have a similar action.

ANTIGENIC VARIATION

In a few persistent virus infections, antigenic variants appear in persistently infected individuals. Immune responses to the original infecting virus, even though delayed or weak, may encourage the selective replication of the antigenic variant. The virus thus escapes elimination by the immune system. Antigenic variation in goats infected with visna virus (10, 57), in horses infected with equine infectious anemia virus (30, 48), and possibly in pigs infected with African swine fever virus (24) and humans infected with HIV (22) has been described.

I have surveyed some of the basic concepts in the pathogenesis of persistent virus infections. Persistent viruses display other phenomena that may play a part in their persistence. These phenomena include the induction on cells infected with certain herpesviruses, or even on the virion itself (5), of virus-encoded Fc receptors which can bind antibody nonspecifically and thus block the action of specific antibodies or immune cells (1); they also include the capping and shedding of antigens from measles virus-infected cells by antiviral antibody (45).

Another interesting possible persistence strategy was recently described for human cytomegalovirus. Virus particles shed in urine (and probably saliva, etc.) become coated with beta-2-microglobulin, which not only renders them relatively insusceptible to the action of neutralizing antibody (37) but also, intriguingly, increases their infectivity (21).

The most reliable strategy for a persistent virus is to hide in the cell in the form of nucleic acid sequences, without forming antigens on the cell surface. It thus enjoys an almost complete freedom from immune defenses. One might expect a general movement of persistent viruses in this direction. Some viruses rely on being reactivated and shed from the host for their survival in the host species; these viruses must persist as complete genomes, as is the case for EBV, herpes simplex virus, polyomaviruses, etc. However, once a virus has become integrated into the genome of the host species, production of infectious virus becomes less necessary, and there would be a tendency for the viral genome to be reduced in size. Even a short sequence of DNA can be a highly successful parasite as long as it is regularly replicated, is not recognized as foreign and excised, and is not harmful (39). Its only justification, as with all forms of life, is survival. The status of parasitic DNA is not clear, but it seems inevitable that this ideal form of parasitism has been reached during the course of evolution by many viruses. The parasitic sequences need not necessarily be related to existing retroviruses, or indeed to other viruses.

Literature Cited

1. **Adler, R., J. C. Glorioso, J. Cossman, and M. Levine.** 1978. Possible role of Fc receptors on cells infected and transformed by herpes virus: escape from immune lysis. *Infect. Immun.* **21**:442–447.

2. **Ahmed, R., A. Salmi, L. D. Butler, J. M. Chiller, and M. B. Oldstone.** 1984. Selection of genetic variants of lymphocytic choriomeningitis virus in spleens of persistently infected mice. Role in suppression of cytotoxic T lymphocyte response and viral persistence. *J. Exp. Med.* **60:**521–540.
3. **Allen, A. M., G. R. Matsueda, R. J. Evans, J. B. Dunbar, G. R. Marshall, and E. R. Unanue.** 1987. Identification of the T-cell and Ia contact residues of a T-cell antigenic epitope. *Nature* (London) **327:**713–715.
4. **Barnaba, V., G. Valesini, M. Levrero, C. Zaccari, A. van Dyke, M. Falco, A. Musca, and F. Balsano.** 1985. Immunoregulation of the in vitro anti-HBs antibody synthesis in chronic HBs Ag carriers and in recently boosted anti-hepatitis B vaccine recipients. *Clin. Exp. Immunol.* **60:**259–266.
5. **Baucke, R. B., and P. G. Spear.** 1979. Membrane proteins specified by herpes simplex virus. V. Identification of an Fc-binding glycoprotein. *J. Virol.* **32:**779–789.
6. **Black, F. L., W. J. Hierholzer, F. D. Pinheiro, et al.** 1974. Evidence for persistence of infectious agents in isolated human populations. *Am. J. Epidemiol.* **100:**230–250.
7. **Braun, R. W., and H. Kirchner.** 1986. T lymphocytes activated by interleukin-2 alone acquire permissiveness for replication of herpes simplex virus. *Eur. J. Immunol.* **16:**709–712.
8. **Brownlie, J., M. C. Clark, and C. J. Howard.** 1984. Experimental production of fatal mucosal disease in cattle. *Vet. Rec.* **114:**535–536.
9. **Bukowski, J. F., C. A. Biron, and R. M. Welsh.** Elevated natural killer cell-mediated cytotoxicity, plasma interferon and tumour cell rejection in mice persistently infected with lymphocytic choriomeningitis virus. *J. Immunol.* **131:**991–996.
10. **Clements, J. E., F. S. Pederse, C. Narayan, and W. A. Haseltine.** 1980. Genomic changes associated with antigenic variations of visna virus during persistent infection. *Proc. Natl. Acad. Sci. USA* **77:**4454–4458.
11. **Coutelier, J. P., J. T. van der Logt, F. W. Heessen, G. Warnier, and J. Van Snick.** 1987. IgG2a restriction of murine antibodies elicited by viral infections. *J. Exp. Med.* **165:**64–69.
12. **Coutelier, J. P., E. van Roost, P. Lambotte, and J. Van Snick.** 1986. The murine antibody response to lactic dehydrogenase-elevating virus. *J. Gen. Virol.* **67:**1099–1108.
13. **Cross, S. S., J. C. Parker, W. P. Rowe, and M. L. Robbins.** 1979. Biology of mouse thymic virus, a herpes virus of mice, and the antigenic relationship to mouse cytomegalovirus. *Infect. Immun.* **26:**1186–1195.
14. **Denman, A. M., T. H. Bacon, and B. K. Pelton.** 1983. Viruses: immuno-suppressive effects. *Philos. Trans. R. Soc. London Ser. B* **303:**137–147.
15. **Dyrberg, T., and M. B. A. Oldstone.** 1986. Peptides as probes to study molecular mimicry and virus-induced autoimmunity. *Curr. Top. Microbiol. Immunol.* **130:**25–37.
16. **Einhorn, L., and A. Ost.** 1984. Cytomegalovirus infection of human blood cells. *J. Infect. Dis.* **149:**207–214.
17. **Evans, R., and V. Riley.** 1968. Circulating interferon in mice infected with lactic dehydrogenase elevating virus. *J. Gen. Virol.* **3:**449–452.
18. **Fucillo, D. A., R. W. Steele, S. A. Hensen, M. M. Vincent, J. R. Hardy, and J. A. Bellanti.** 1974. Impaired cellular immunity to rubella virus in congenital rubella. *Infect. Immun.* **9:**81–84.
19. **Gardiner, A. C., P. F. Nettleton, and R. M. Barlow.** 1983. Virology and immunology of a spontaneous and experimental mucosal disease-like syndrome in sheep recovered from clinical border disease. *J. Comp. Pathol.* **93:**463–469.
20. **Gendelman, H. E., O. Narayan, S. Kennedy-Stoskopf, P. G. Kennedy, Z. Ghotbi, J. E. Clements, J. Stanley, and G. Pezechkpour.** 1986. Tropism of sheep lentiviruses for

monocytes: susceptibility to infection and virus gene expression increase during maturation of monocytes to macrophages. *J. Virol.* **58:**67–74.

21. **Grundy, J. E., J. A. McKeating, P. J. Ward, A. R. Sanderson, and P. D. Griffiths.** 1987. Beta 2 microglobulin enhances the infectivity of cytomegalovirus and when bound to the virus enables class I HLA molecules to be used as a virus receptor. *J. Gen. Virol.* **68:**793–803.
22. **Hahn, B. H., G. M. Shaw, M. E. Taylor, R. R. Redfield, P. D. Markham, S. Z. Salahuddin, F. Wong-Staal, R. C. Gallo, E. S. Parks, and W. P. Parks.** 1986. Genetic variation in HTLVIII/LAV over time in patients with AIDS or at risk from AIDS. *Science* **232:**1548–1553.
23. **Hayward, A. R., M. J. Herberger, J. Groothuis, and M. R. Levin.** 1984. Specific immunity after congenital or neonatal infection with cytomegalovirus or herpes simplex virus. *J. Immunol.* **133:**2469–2473.
24. **Hess, W. R.** 1981. African swine fever: a reassessment. *Adv. Vet. Sci. Comp. Med.* **25:**39–69.
25. **Inada, T., and C. A. Mims.** 1984. Mouse Ia antigens are receptors for lactic dehydrogenase virus. *Nature* (London) **309:**59–61.
26. **Inada, T., and C. A. Mims.** 1985. Pattern of infection and selective loss of Ia positive cells in suckling and adult mice inoculated with lactic dehydrogenase virus. *Arch. Virol.* **86:**151–165.
27. **Inada, T., and C. A. Mims.** 1986. Live lactic dehydrogenase virus (LDV) induces suppressor T cells that inhibit the development of delayed hypersensitivity to LDV. *J. Gen. Virol.* **67:**2103–2112.
28. **Isakov, N., M. Feldman, and S. Segal.** 1982. Acute infection of mice with lactic dehydrogenase virus (LDV) impairs the antigen-presenting capacity of their macrophages. *Cell. Immunol.* **66:**317–332.
29. **Kitajewski, J., R. J. Schneider, B. Safer, S. M. Munemitsu, C. E. Samuel, B. Thimmappaya, and T. Shenk.** 1986. Adenovirus VAI RNA antagonizes the antiviral action of interferon by preventing activation of the interferon-induced eIF-2α-kinase. *Cell* **45:**195–200.
30. **Kono, Y., K. Kobayashi, and Y. Fukunaga.** 1973. Antigenic drift of equine infectious anaemia virus in chronically infected horses. *Arch. Gesamte Virusforsch.* **41:**1–10.
31. **Lamb, J. R., and M. Feldman.** 1984. Essential requirement for major histocompatibility complex recognition in T cell tolerance induction. *Nature* (London) **308:**72–74.
32. **Lamb, J. R., B. J. Skidmore, N. Greene, J. M. Chiller, and M. Feldman.** 1983. Induction of tolerance to influenza virus in immune T lymphocyte clones with synthetic peptides of influenza haemagglutinin. *J. Exp. Med.* **157:**1434–1447.
33. **Lane, H. C., J. M. Depper, W. C. Greene, G. Whalen, T. A. Waldmann, and A. S. Fauci.** 1985. Qualitative analysis of immune function in patients with the acquired immunodeficiency syndrome. *N. Engl. J. Med.* **313:**79–84.
34. **Lehman-Grube, F., L. M. Peralta, M. Bruns, and J. Lohler.** 1982. Persistent infection of mice with the lymphocytic choriomeningitis virus, p. 43. *In* H. Fraenkel-Conrat and R. R. Wagner (ed.), *Comprehensive Virology*, vol. 18. Plenum Publishing Corp., New York.
35. **Lemon, S. M., and L. N. Binn.** 1985. Incomplete neutralisation of hepatitis A virus in vitro due to lipid associated virions. *J. Gen. Virol.* **66:**2501–2505.
36. **Mathes, L. E., R. G. Olsen, L. C. Heberbrand, E. A. Hoover, J. P. Schaller, P. W. Adams, and H. Nichols.** 1979. Immuno-suppressive properties of a virion polypeptide: a 15,000 dalton protein from feline leukaemia virus. *Cancer Res.* **39:**950–955.
37. **McKeating, J. A., P. D. Griffiths, and J. E. Grundy.** 1987. Cytomegalovirus in urine

specimens has host beta 2 microglobulin bound to the viral envelope: a mechanism of evading the host immune response? *J. Gen. Virol.* **68:**785–792.
38. **Michaelides, M. C., and E. S. Simms.** 1980. Immune responses in mice infected with lactic dehydrogenase virus. III. Antibody response to a T-dependent and a T-independent antigen during acute and chronic LDV infection. *Cell. Immunol.* **50:**253–260.
39. **Mims, C. A.** 1981. Vertical transmission of viruses. *Microbiol. Rev.* **45:**267–286.
40. **Narayan, O., D. Sheffer, D. E. Griffin, J. Clements, and J. Hess.** 1984. Lack of neutralizing antibodies to caprine arthritis encephalitis lentivirus in persistently infected goats can be overcome by immunization with inactivated *Mycobacterium tuberculosis. J. Virol.* **49:**349–355.
41. **Nicholas, J. A., M. E. Levely, R. J. Brideau, and A. E. Berger.** 1987. During recovery from cytomegalovirus infection T lymphocyte subsets become selectively responsive to activation and have depressed interleukin-2 (Il-2) secretion and Il-2 receptor expression. *Microb. Pathogenesis* **2:**37–47.
42. **Notkins, A. L., M. Mage, W. K. Ashe, and S. Mahar.** 1968. Neutralisation of sensitised lactic dehydrogenase virus by anti-gamma-globulin. *J. Immunol.* **100:**314–320.
43. **Notkins, A. L., S. E. Mergenhagen, A. A. Rizzo, C. Scheele, and T. A. Waldmann.** 1966. Elevated gamma globulin and increased antibody production in mice infected with lactic dehydrogenase virus. *J. Exp. Med.* **123:**347–364.
44. **Oldstone, M. B. A., and F. J. Dixon.** 1967. Lymphocyte choriomeningitis: production of antibody by "tolerant" infected mice. *Science* **158:**1193–1195.
45. **Oldstone, M. B. A., R. S. Fujinami, and P. W. Lampert.** 1980. Membrane and cytoplasmic changes in virus-infected cells induced by interaction of antiviral antibody with surface viral antigen. *Prog. Med. Virol.* **26:**45–93.
46. **Orosz, C. G., N. E. Zinn, R. G. Olsen, and L. E. Mathes.** 1985. Retrovirus-mediated immunosuppression. II. FeLV-UV alters in vitro murine T lymphocyte behaviour by reversibly impairing lymphokine secretion. *J. Immunol.* **135:**583–590.
47. **Pahwa, S., R. Pahwa, C. Saxinger, R. C. Gallo, and R. A. Good.** 1985. Influence of the human T lymphotropic virus/lymphadenopathy-associated virus on functions of human lymphocytes: evidence for immunosuppression effects and polyclonal B cell activation by banded viral preparations. *Proc. Natl. Acad. Sci. USA* **82:**8198–8202.
48. **Payne, S., B. Parekh, R. C. Montelaro, and C. J. Issel.** 1984. Genomic alterations associated with persistent infection by equine infectious anaemia virus, a retrovirus. *J. Gen. Virol.* **65:**1395–1399.
49. **Porter, D. D., A. E. Larson, and H. G. Porter.** 1969. The pathogenesis of Aleutian disease of mink. I. In vivo viral replication and host antibody response to viral antigen. *J. Exp. Med.* **130:**575–593.
50. **Porterfield, J. S.** 1986. Antibody-dependent enhancement of viral infectivity. *Adv. Virus Res.* **31:**335–355.
51. **Reiher, W. E., III, E. Blalock, and T. K. Brunck.** 1986. Sequence homology between acquired immunodeficiency syndrome virus envelope protein and interleukin-2. *Proc. Natl. Acad. Sci. USA* **83:**9188–9192.
52. **Rice, G. P. A., R. D. Schier, and M. B. A. Oldstone.** 1984. Cytomegalovirus infects human lymphocytes and monocytes—virus expression is restricted to immediate-early gene products. *Proc. Natl. Acad. Sci. USA* **81:**6134–6138.
53. **Rodgers, B. C., D. M. Scott, J. Mundin, and J. G. Sissons.** 1985. Monocyte-derived inhibition of interleukin-1 induced by human cytomegalovirus. *J. Virol.* **55:**527–532.
54. **Rodriguez-Boulan, E. J., and D. D. Sabatini.** 1978. Asymmetric budding of viruses in epithelial monolayers: a model system for study of epithelial polarity. *Proc. Natl. Acad. Sci. USA* **75:**5071–5074.

55. **Rowson, K. E. K., and B. W. J. Mahy.** 1985. Lactic dehydrogenase-elevating virus. *J. Gen. Virol.* **66:**2297.
56. **Saron, M. F., Y. Riviere, A. G. Hovanessian, and J. C. Guillon.** 1982. Chronic production of interferon in carrier mice congenitally infected with lymphocytic choriomeningitis virus. *Virology* **117:**253–256.
57. **Scott, J. V., L. Stowring, A. T. Haase, O. Narayan, and R. Vigne.** 1979. Antigenic variation in visna virus. *Cell* **18:**321–327.
58. **Sixbey, J. W., D. S. Davis, L. S. Young, L. Hutt-Fletcher, T. F. Tedder, and A. B. Rickinson.** 1987. Human epithelial cell expression of an Epstein Barr virus receptor. *J. Gen. Virol.* **68:**805–811.
59. **Stoize, B., and O.-R. Kaaden.** 1987. Apparent lack of neutralizing antibodies in Aleutian disease is due to masking of antigenic sites by phospholipids. *Virology* **158:**174–180.
60. **Stueckemann, J. A., M. Holth, W. J. Swart, K. Kowalchyk, M. S. Smith, A. J. Welstenholme, W. A. Cafruny, and P. G. Plagemann.** 1982. Replication of lactate dehydrogenase elevating virus in macrophages. 2. Mechanism of persistent infection in mice and in cell cultures. *J. Gen. Virol.* **59:**263–272.
61. **Symington, J., A. K. McCann, and M. J. Schlesinger.** 1977. Infectious virus-antibody complexes of Sindbis virus. *Infect. Immun.* **15:**720–725.
62. **Synderman, R., and G. J. Cianciolo.** 1984. Immunosuppressive activity of the retroviral envelope protein p15E and its possible relation to neoplasia. *Immunol. Today* **5:**240–242.
63. **Verder, H., E. Dickmeiss, S. Haahr, E. Kappelgaard, J. Leerboy, A. Mouer-Larsen, H. Nielsen, I. Platz, and C. Koch.** 1986. Late-onset rubella syndrome: coexistence of immune complex disease and defective cytotoxic effector cell function. *Clin. Exp. Immunol.* **63:**367–375.
64. **Wardley, R. C., C. de M. Andrade, D. N. Black, F. L. de Castro Portugal, L. Enjuanes, W. R. Hess, C. Mebus, A. Ordas, and D. Rutili.** 1983. African swine fever: brief review. *Arch. Virol.* **76:**73–90.
65. **Weiss, R. A., P. R. Clapham, R. Cheinsong-Popov, A. G. Dalgleish, C. A. Carne, I. V. Weller, and R. S. Tedder.** 1985. Neutralisation of human T-lymphotropic virus type III be sera of AIDS and AIDS-risk patients. *Nature* (London) **316:**69–72.
66. **Wiedbrauk, D. L., W. J. Hadlow, L. C. Ewalt, and D. L. Lodmell.** 1986. Interferon response in normal and Aleutian disease virus-infected mink. *J. Virol.* **59:**514–517.

Chapter 2

Role of Monocytes, Macrophages, Langerhans Cells, and Follicular Dendritic Cells in Persistent Virus Infections

Anthony C. Allison

INTRODUCTION

Monocytes, macrophages, Langerhans cells, and follicular dendritic cells (FDC) play an important role in persistent virus infections. First, they can support infections by such viruses. Second, these cell types, collectively termed accessory cells, are usually required for presentation of antigens to lymphocytes. This process involves the physical presentation of antigens associated with class II major histocompatibility (MHC) glycoproteins and the production of cytokines that are cofactors in stimulating the proliferation of clones of lymphocytes with receptors for the antigens. Overproduction of such factors can lead to antigen-specific and polyclonal hypergammaglobulinemia, which is observed in some persistent virus infections. Other products of persistently infected macrophages, such as prostaglandins, can have immunosuppressive effects. Abnormal antigen presentation may play a role in the pathogenesis of disorders of immune function, as in the case of infection of FDC by human immunodeficiency virus (HIV) discussed below. Thus, products of accessory cells can positively or negatively influence lymphocyte responses to antigens.

Third, macrophages are major effector cells in antiviral immunity. Their capacity to limit virus infections can be augmented by lymphokines, so that interactions of macrophages and lymphocytes are reciprocal. Morahan et al. (55) have made a distinction between intrinsic and

Anthony C. Allison • Syntex Research, Palo Alto, California 94304.

extrinsic resistance of macrophages to virus infections. Intrinsic resistance is defined as the permissiveness or nonpermissiveness of the macrophages themselves to support virus replication. Extrinsic resistance is defined as the ability of the macrophages to inactivate extracellular virus or reduce production in cells that are normally permissive. Activated macrophages or their products lyse some virus-infected cells: a mechanism that is not only protective but could have immunopathological consequences. Other immunopathological effects of macrophages include the production of prostaglandins and leukotrienes, which increase vascular permeability, and of thromboplastin, which has procoagulant effects. These could be important in the pathogenesis of hemorrhagic fevers and shock syndromes. Related to the pathogenesis of shock syndromes is the enhancement of virus infections by antibodies. First shown in infections of human monocytes by dengue viruses, this phenomenon is now known to be true of several other viruses. Antibody-mediated enhancement of virus infections could well apply to FDC and could provide one explanation of why it is so difficult to elicit protection against HIV.

Relationships of bone marrow-derived precursors, blood monocytes, and tissue macrophages, cells of the Langerhans lineage, and FDC are complex and imperfectly understood. A brief review of available information on this topic provides a background for a discussion of the possible role of various types of accessory cells in persistent virus infections.

TYPES OF ACCESSORY CELLS

Bone Marrow-Derived Monocytes and Macrophages

Monocytes and neutrophils arise from common precursors in bone marrow which are responsive to colony-stimulating factors (CSFs). In the presence of M-CSF, also known as CSF-1, the precursors proliferate and differentiate into monocytes. When stimulated by G-CSF, they proliferate and differentiate into neutrophils, and when stimulated by GM-CSF, they proliferate and differentiate into both cell lineages. Human and murine CSFs have been cloned, expressed, and sequenced, and their major biological effects have been defined (51).

Studies with labeled cells and cells bearing genetic markers after irradiation and bone marrow reconstitution have shown that monocytes are precursors of pulmonary alveolar macrophages (1, 17). However, there is also evidence that macrophages can multiply in tissues, so that not all are of hematogenous origin (22). These cell types may be heterogeneous in their capacity to support virus replication and in their function as accessory cells or effectors in immune responses. Even the state of differentiation of a single cell lineage may influence these activities. Differentiated sheep monocyte-derived macrophages produce more visna virus than immature precursors do (26). Human alveolar macrophages, presumably terminally differentiated, produce less interleukin-1 (IL-1) than do peripheral blood monocytes from the same individuals stimulated with lipopolysaccharide in culture (94). Often, viruses replicate in subsets of macrophages (52), and a distinction has seldom been made between macrophages and other types of accessory cell.

The Langerhans Cell Lineage

Labeling with tritiated thymidine or uridine has shown that cells of the Langerhans lineage originate in the bone marrow, reside in the skin (and presumably other tissues) for a relatively short period (a few days to 1 week), and migrate through afferent lymphatics to the paracortical thymus-dependent areas of lymph nodes of the drainage chain (7). These cells are termed veiled cells when in afferent lymph and interdigitating cells when in the lymph nodes, because they have cytoplasmic extensions among T lymphocytes.

Langerhans cells in the skin, and malignant cells of the same lineage (histocytosis X cells), express class II MHC antigens and CD4 (72). Restricted to cells of this lineage are CD6 and strong membrane ATPase reactions. Class II MHC expression in these cells persists in culture, does not require stimulation by alpha interferon (IFN-α) or GM-CSF, and may be constitutive.

Langerhans cells can be isolated from human skin by rosetting procedures and maintained in culture. They efficiently present antigens, including herpesvirus antigens, to elicit immune responses in histocompatible T lymphocytes (14). Veiled cells can be recovered from lymph without proteinase modification by cannulating afferent lymphatics of experimental animals. These cells efficiently present antigens associated with their surfaces, for example, contact-sensitizing chemicals and myelin basic protein, to elicit T-lymphocyte-dependent immune responses (43, 44). Dendritic cells isolated from lymphoid tissues (76) have similar properties and may be of the same lineage. Because of possible confusion with FDC, which have a different location and different properties, the term Langerhans cells is used in this review.

FDC

As their name implies, FDC are found in lymphoid follicles, where their branching cytoplasmic extensions are closely related to B lymphocytes. FDC express CD4, class II MHC, and complement receptors CR1, CR2, and CR3 (67, 97). The CD23, p45 antigen, detected by monoclonal antibody R4/23, seems to be confined to FDC (59), and another monoclonal antibody (2BF11) shows a similar specificity (40).

The origin of FDC and their relationship to other cell types is unknown, but mouse reconstitution experiments suggest that they are not derived from precursors in bone marrow (37). Isolating FDC without damaging them has been difficult and has impaired analysis of their immune functions or capacity to support virus replication in culture. Immune complexes activating complement injected into mice become localized on FDC, and this process appears to be required for the generation of B-lymphocyte memory, i.e., proliferation of clones of B lymphocytes responding to the antigen (42). The high-affinity complement receptor (CR1) expressed on FDC is presumably involved in the localization of the complexes, and the use of emulsions bearing C3b for targeting of vaccines to FDC is discussed below.

Antigens can be transferred to B lymphocytes and through them to T lymphocytes. Depletion of B lymphocytes by repeatedly giving antibodies against the μ chain of surface membrane immunoglobulin to mice markedly depresses the capacity of the mice to mount T-lymphocyte responses to antigens following subcutaneous immunization (68). Hence, it seems likely that B lymphocytes are major links in the chain of antigen presentation in lymph nodes.

Antibody responses against viral antigens that have been investigated require T-lymphocyte help (18, 90). Helper T lymphocytes, stimulated by antigen, produce IFN-γ and GM-CSF (57), both of which can activate macrophages. The cytotoxic subset of T lymphocytes lyses virus-infected cells in a genetically restricted fashion and, when stimulated by virus-infected histocompatible cells, releases IFN-γ (56). T lymphocytes have been shown to lyse cytomegalovirus-infected human monocytes in HLA-DR-restricted fashion (47). Expression of DR on monocytes and macrophages allows recognition of antigens in association with either class I or class II MHC glycoproteins. For all these reasons, T-cell responses to virus antigens are required to elicit protective immunity, and that requires intact accessory cell function.

VIRUS REPLICATION IN MONOCYTES AND MACROPHAGES

Inherited Differences in the Capacity of Macrophages To Support Virus Replication

Many viruses replicate in human monocytes and in macrophages of other species (1). The outcome of the infection can be lysis of the cells, fusion to form giant cells, or persistent productive or nonproductive infection. Often, the capacity of macrophages to support a particular virus infection parallels that of other cell types of the same species or strain of animal.

In some well-studied cases, inherited differences in the capacity to support virus replication are specific for cells of macrophage lineage. This is true of murine hepatitis viruses (coronaviruses) in different strains of mice, as originally shown by Bang for murine hepatitis virus type 2 (8) and extended in my laboratory for murine hepatitis virus type 3 (88). The latter virus is lethal for most strains of mice, and in peritoneal macrophages there is permissive infection with giant-cell formation. A/J mice and their macrophages are relatively resistant to murine hepatitis virus type 3 infection, and persistent infection of macrophages occurs. This is associated with severely deficient immune responses to unrelated antigens (91), which will be discussed below.

Inherited differences in the capacity of mouse macrophages to support the replication of influenza virus have been localized to a gene termed *Mx*. In this case, IFN-α and IFN-β have antiviral effects only in macrophages from mice bearing the *Mx* resistance gene (52).

Effect of Age on Macrophage Resistance

Newborn humans and experimental animals are often more susceptible to virus infections than are older hosts. Neonatal herpes simplex virus (HSV) infections in children (usually HSV-2) are severe and often fatal, and neonatal HSV-1 infections in mice are also lethal. Peritoneal macrophages from very young mice support HSV production, whereas in macrophages from adult mice there is incomplete replication (77). Transfer of macrophages from adult to young mice increases resistance to HSV infection (35), and injections of anti-macrophage serum or silica, which is cytotoxic for macrophages but not other cell types, decrease the resistance of adult mice to intraperitoneal HSV infection (98). Macrophages of older animals produce substantial amounts of IFN, whereas those of young animals do not (33). Phytohemagglutinin-stimulated leukocytes from newborn children produce very little IFN-γ, and reconstitution experiments with adult cells show that the functionally immature monocytes are primarily responsible for this defect (80).

Newborn mice were also found to be highly susceptible to intraperitoneal infection by coxsackievirus type B-3. Resistance could be conferred by transfer of syngeneic adult peritoneal macrophages and small amounts of immune serum, but not by either one alone (65). Antibodies presumably facilitate the uptake of virus by macrophages, but the macrophages of newborn mice are unable to abolish the infectivity of the virus.

Role of Cytokines in Macrophage Resistance

In both of the situations just discussed, inherited and age-related differences in susceptibility to virus infections, cytokines, in particular IFNs, participate. The role of IFNs in resistance to infections is a large and complex subject, but a few points relevant to macrophages can be summarized.

IFN-α

The IFN-α family of molecules is produced by monocytes and macrophages exposed to many viruses. Human monocyte-derived macrophages are reported to produce different IFN-αs with distinct antiviral spectra (12). IFN decreases the replication of rubella virus in human monocytes (86), and anti-IFN-α/β augments the replication of a human respiratory coronavirus in macrophages (61). Anti-IFN increases the permissiveness of mouse peritoneal macrophages for vesicular stomatitis and encephalomyocarditis viruses (11). Anti-IFN enhances the susceptibility of mice to HSV-1 (28).

HSV-1 provides an example of the complexity of host responses to infection. Effects of anti-lymphocyte serum (35) and cell transfer experiments (66) show that T lymphocytes play an important role in immunity to HSV, and the evidence just presented suggests that macrophages and IFN contribute to resistance. The IFN may exert its effects through activating natural killer cells, which have also been implicated in host defense against the virus (48). Thus, many components of resistance act in concert, their relative importance depending on the virus, strain, genetic configuration of the host, and other factors.

IFN-γ

Production of IFN-γ by cloned T lymphocytes has defined the lymphocyte subsets with this capacity and the conditions under which they are activated. A subset of helper T lymphocytes (T_H1) stimulated by antigen produces IFN-γ, whereas another subset (T_H2) does not (57). Clones of cytotoxic T lymphocytes recognizing influenza virus nucleoprotein produce IFN-γ in the presence of syngeneic target cells bearing the nucleoprotein antigen (56). We showed that IFN-γ increases the resistance of mouse peritoneal macrophages to influenza virus (89). IFN-γ also increases the expression of class II MHC glycoproteins on macrophages (75), which could augment their capacity to function as accessory cells in immune responses and also as targets for class II MHC-restricted cytotoxic lymphocytes.

Infection of sheep and goat macrophages with visna virus does not induce production of IFN, but the infected cells produce a factor that induces the production of IFN by lymphocytes (41). Kennedy et al. (41) suggest that this IFN induces class II MHC antigens in macrophages, facilitating a lymphoproliferative response and restricting virus replication.

GM-CSF

GM-CSF is produced by several cell types including helper T lymphocytes activated by antigen (57). Not only is it a stimulator of monocyte precursor replication, but it is also emerging as a major activator of monocytes and macrophages, for example, inducing the expression of class II MHC and augmenting microbicidal capacity.

IL-1

Monocytes and macrophages activated by various agents, including immune complexes and C3b, produce IL-1β and some IL-1α. IL-1 is required for helper functions of T lymphocytes and has various proinflammatory effects, including recruitment of leukocytes and production of prostaglandins (A. C. Allison and S. W. Lee, *Agents Actions*, in press).

TNF-α

Tumor necrosis factor-α (TNF-α) is produced by monocytes and macrophages activated by viruses and in other ways. TNF-α lyses certain virus-infected cells, and this activity is increased by IFN-γ (95, 96). In this way, TNF-α and TNF-β (which is produced by T lymphocytes) can exert antiviral activity.

IFN-β_2

The cytokine IFN-β_2 is different from other IFNs in several respects (34, 74). Its constitutive expression in human fibroblasts is increased by other cytokines such as TNF-α and IL-1β. IFN-β_2 is identical with B-cell stimulatory factor-2, which induces the differentiation of B lymphocytes into immunoglobulin-secreting cells, and so has an important role in immunity. In addition, IFN-β_2 can increase resistance to viruses of certain cell types under certain conditions (74). IFN-β_2 is produced by human monocytes and some T-cell lines, as well as other cell types (74).

Enhancement by Antibodies of Virus Infections in Monocytes and Macrophages

The phenomenon of antibody-mediated enhancement of virus infections in monocytes and macrophages was discovered when Scott Halstead and Edward O'Rourke were spending a sabbatical in my laboratory. Dengue virus was shown to replicate in human peripheral blood monocytes (33), and nonneutralizing antibody (against a heterologous serotype) or highly diluted antibody (against the homologous serotype) increased this replication (32). $F(ab)_2$ fragments of human immunoglobulin G antibodies retained neutralizing capacity, but had no enhancing activity, which could be reconstituted with rabbit anti-Fab before exposure to monocytes (23, 31). Aggregated γ-globulin inhibited the enhancement. Trypsinization of monocytes blocked infection with virus but not with virus-antibody complexes. These findings suggest that dengue virus can enter monocytes by a protease-sensitive receptor or, when complexed with nonneutralizing antibody, through Fcγ receptors. Enhancement also occurs in vivo: nonneutralizing antibody considerably augmented dengue virus viremia in monkeys (31).

Epidemiological evidence has been obtained that antibody-mediated enhancement plays a role in the pathogenesis of dengue hemorrhagic fever and shock syndrome (31). Antibody-mediated enhancement has also been demonstrated with West Nile, yellow fever, Murray Valley encephalitis, Japanese encephalitis, Kuijin, Getah, and rabbitpox virus infections in mammalian or avian leukocytes (31).

An important question is whether antibody enhances infections of macrophages by a taxonomically diverse group of persistent viruses (Table 1). These viruses have common features, including persistence, the production of poor-quality neutralizing antibodies, and circulation of infectious virus-antibody complexes. If acute infections occur, these are often hemorrhagic fever syndromes.

The pathogenesis of these syndromes is complex and incompletely understood. Two main factors seem to be involved. One is increased vascular permeability, which is dramatically manifested in children with dengue shock syndrome. We have found (in collaboration with S. Kliks) that monocytes with enhanced dengue virus infections produce factors increasing vascular permeability. T lymphocytes activated by IL-2 may also increase vascular permeability; a complication of the therapeutic use of IL-2 is fluid retention from what is presumed to be a capillary leak syndrome (69).

The multiple localized hemorrhages in hemorrhagic fever seem to be due to thrombosis, endothelial injury, thrombolysis, and subsequent hemorrhage. We showed that immune complexes activate monocyte thromboplastin, a potent procoagulant (64), and the same appears to be true for monocytes with enhanced dengue virus infections. Immune complexes and complement may also activate endothelial cell thromboplastin.

Since FDC have receptors for Fcγ and complement components, it is likely that antibody can also enhance infections by these cells. This is

Table 1. Viruses Persistently Infecting Monocytes, Macrophages, or Other Accessory Cells

Virus group	Virus	Reference
Herpesvirus	Murine cytomegalovirus	16
Retrovirus—oncovirus	Friend leukemia virus	83
	Leukemia viruses	79
Retrovirus—lentivirus	Visna-maedi virus	60
	Equine infectious anemia virus	21
	HIV	25, 36
Coronavirus	Mouse hepatitis virus	8, 88
	Feline infectious peritonitis virus	39
Togavirus	Lactate dehydrogenase virus	38
Arenavirus	Lymphocytic choriomenigitis virus	82
	Hemorrhagic fever viruses	45
Paramyxovirus	Sendai virus	54
Parvovirus	Aleutian mink disease virus	63
Iridovirus	African swine fever virus	93

relevant to the discussion of HIV replication in FDC as well as in monocytes and macrophages.

Persistent Infections of Monocytes and Macrophages by Lentiviruses

HIV has been shown to resemble morphologically and share DNA base sequence homology with lentiviruses (2, 27). The lentiviruses are exogenous, nononcogenic retroviruses producing persistent infections and progressive pathological changes. Lentiviruses include visna-maedi virus of sheep, caprine arthritis-encephalitis virus of goats, equine infectious anemia virus, and bovine visnalike virus.

All of the lentiviruses that have been investigated produce persistent infections of monocytes and macrophages. Infection of sheep with visna virus is restricted to cells of this lineage, and productive infection is greater in differentiated cells than in precursors (26). No infection of lymphocytes has been observed (41).

Equine infectious anemia virus proliferates in monocytes and macrophages (21). Periodic episodes of fever are attributed to the emergence of antigenic variants (71), with increasing viremia and production of endogenous pyrogens.

Role of Macrophages in HIV Infections of the Lungs and Brain

The original cell type in which HIV was shown to replicate was the $CD4^+$ T lymphocyte. However, several groups of authors have reported that peripheral blood monocytes (24, 30, 36) and cells of macrophage type from lymph nodes (30), lungs (20, 62, 70), and brains (75) can support the replication of HIV. The conclusion that this is true in vivo has been reached from demonstration of virus-specific mRNA by in situ hybridization, demonstration of viral antigens by immunocytochemistry, and demonstration of characteristic virus particles by electron microscopy.

In lungs of patients with acquired immunodeficiency syndrome (AIDS), expression of viral mRNA is rare (20). The expression is much greater in patients with lymphocytic interstitial pneumonitis, a condition observed in a small minority of adult HIV-infected persons but in about one-half of children with the infection. Lymphocytic interstitial pneumonitis can occur without severe immunodeficiency and secondary *Pneumocystis carinii* or cytomegalovirus infections, and it may have an immunopathological basis. Plata et al. (62) have shown that adherent cells obtained by bronchoalveolar lavage from patients with lymphocytic interstitial pneumonitis are lysed in a genetically restricted fashion by $T8^+$ lymphocytes. Cells bearing HIV antigens and syngeneic class I MHC antigens as a result of transfection were also lysed by the lymphocytes of the patients, defining the presumed target of the cytotoxic cells. Such cytotoxicity might limit viral replication in pulmonary cells, and production of lymphokines could be a factor in the increased vascular permeability and other manifestations of inflammation in lymphocytic interstitial pneumonitis. The parallel with the pathogenesis of pneumonitis in visna virus infection (60) is evident.

Lentiviruses such as visna-maedi and caprine arthritis-encephalitis viruses produce progressive infections of the central nervous system. Cells of the monocyte-macrophage lineage carry virus to the central nervous system and are generally believed to play a role in immunopathological events leading to dysfunction (60).

A generalized encephalopathy associated with progressive dementia is also produced by HIV infection. Studies by in situ hybridization, cell culture, and immunofluorescence of tissues from HIV- and simian immunodeficiency virus-infected humans and monkeys have shown viruses predominantly in cells of macrophage lineage, including giant cells and glio-mesenchymal cell nodules (25, 78, 92). Infected macrophages are observed in close opposition to glial cells and neurons, and macrophage-derived cytokines may play a role in the pathogenesis of functional disorders.

HIV isolated from infected macrophages is reported to replicate more readily in monocytes than is virus isolated from lymphocytes of the same patient (25). Hence, different isolates of HIV may show preferential tropism for one cell type or the other.

Although yields of HIV from macrophage-type cells are relatively low, there is little or no cytopathic effect and virus production can continue over a long period. Cells of the monocyte-macrophage lineage could be a source of virus to continuously infect $T4^+$ lymphocytes, in which the virus has cytopathic effects. In view of evidence that antibodies can enhance virus infections of monocytes (see above), such enhancement may also occur in HIV infection and make neutralization more difficult. This possibility should be carefully investigated; the phenomenon is important enough to require thorough documentation before it is generally accepted.

REPLICATION OF HIV IN FDC

In 1984, Armstrong and Horne (6) reported the demonstration by electron microscopy of retroviruslike particles in association with expanded FDC in hyperplastic lymph node follicles of homosexual males with various categories of AIDS. This observation has been repeatedly confirmed, and FDC in hyperplastic follicles have been shown to contain HIV antigens in humans and simian immunodeficiency virus antigens in monkeys (81). The occurrence of budding profiles among FDC suggests that they are replicating viruses and not just trapping virions and immune complexes (19). Replication of HIV is, in fact, more prominent in FDC than in lymphocytes, macrophages, or interdigitating cells in lymph nodes.

These findings have interesting implications. It seems that FDC may support the replication of HIV better than monocytes, macrophages, or Langerhans cells. Cells with dendritic appearance bearing HLA-DR have been observed in vaginal and cervical epithelia (15). These were thought to be Langerhans cells, but some might be FDC, in which case they could be infected after sexual contact. Cameron et al. (19) point out that FDC might initially be stimulated by HIV infection, which could account for excessive B-cell activation. High antibody levels are found in asymptomatic patients with an intact FDC network and low virion load. In AIDS patients with a high virion load, FDC showed degenerative changes (19). These degenerative changes may be secondary to virus multiplication in

FDC or, conceivably, to lytic effects of $CD8^+$ T lymphocytes or other effector cells, as discussed below. In view of the importance of FDC in antigen presentation, this activity could be defective in AIDS patients with degenerate FDC and could be a factor in the pathogenesis of immunodeficiency, in addition to the well-known lytic effects of HIV on $CD4^+$ T lymphocytes.

Since FDC express CD4 (67), this could be a receptor for HIV. However, the possibility of infection of FDC by nonneutralizing complexes of HIV and antibody through receptors for Fcγ or complement deserves investigation. If this mechanism operates, it could be a factor in the persistence of the virus and the difficulty of eliciting protective immunity. It would also imply that procedures designed to block HIV attachment to CD4 are unlikely to prevent persistent infection.

Murine leukemia retroviruses can also replicate in FDC (79), and so these cells may support persistent infections with other viruses. Bovine visnalike virus produces enlargement of subcutaneous lymph nodes, which display prominent follicles; this lymphoproliferative reaction is accompanied by a persistent lymphocytosis (87). It would be interesting to ascertain whether this lentivirus proliferates in FDC and whether degenerative changes and immunodeficiency can eventually occur.

In Aleutian disease of mink, which is produced by a persistent parvovirus infection, there is lymph node hyperplasia with follicular expansion, plasmacytosis, and hypergammaglobulinemia (63). FDC could be stimulated by viral infection itself or by the immune complexes which are formed in this disease.

HIV IN EPIDERMAL LANGERHANS CELLS

Skin biopsies of 40 HIV-seropositive persons were examined for the viral core protein antigen p17 by using a monoclonal antibody (85). In samples from seven persons, viral antigen was demonstrated in epidermal dendritic cells expressing CD6. Electron micrographs of one of the biopsies showed particles typical of HIV in cells with the appearance of Langerhans cells. These findings suggest that HIV may replicate in Langerhans cells, at least in some persons. It has been suggested that depletion of Langerhans cells may contribute to immunodeficiency in patients infected with HIV (13). However, since viral antigen is observed in Langerhans cells in only a minority of patients and is found in FDC rather than interdigitating cells in lymph nodes, the role of Langerhans cells in supporting the replication of HIV and the pathogenesis of HIV remains to be clarified.

IMMUNOSUPPRESSIVE EFFECTS OF VIRUS-INFECTED MACROPHAGES: ROLE OF PROSTAGLANDINS

Monocytes or macrophages infected with cytomegalovirus, influenza virus, Sendai virus, or poliomyelitis virus have been reported to suppress lymphocytic responses to antigens or mitogens (58). Lactate dehydrogenase virus infection interferes with antigen presentation (38). The mechanisms by which these in vitro effects are mediated are unknown.

Many virus infections produce immunosuppression (5). Usually this is transient, resolving after the acute infection is over. However, sometimes persistent virus infections are accompanied by long-term immunosuppression. We studied one example, persistent macrophage infection of A/J mice by the coronavirus murine hepatitis virus type 3. The mice were found to have profoundly suppressed responses to sheep erythrocytes, as shown by Jerne plaque assays (91). Because prostaglandins of the E type are characteristic macrophage products and suppress immune responses by several mechanisms, Lahmy and Virelizier (46) studied the effect of the cyclooxygenase inhibitor indomethacin. This drug was found to restore antibody responses in the persistently virus-infected mice to nearly normal levels.

Whether prostaglandins play any role in the pathogenesis of immunodeficiency in HIV infections is unknown. A preliminary report suggests that CD4/CD8 ratios and responses to mitogens in AIDS-related complex and AIDS patients can sometimes be increased by indomethacin (29). Further systematic studies of prostaglandin production in HIV infections and effects of cyclooxygenase inhibitors are required. Since diarrhea is caused by prostaglandins and is a problem in AIDS, especially in Africa, it would be interesting to know whether cyclooxygenase inhibitors could have therapeutic effects on this condition, as well as on the HIV-associated pneumonitis discussed above.

ROLE OF MACROPHAGE-MEDIATED CYTOTOXICITY IN VIRUS INFECTIONS

T lymphocytes activate macrophages so that they have increased capacity to lyse virus-infected cells, and low levels of antibody can increase macrophage-mediated cytotoxicity (49). Products of activated macrophages, e.g., TNF-α, can lyse some cells infected with viruses (95, 96). This might not only be part of a defense mechanism; it could also contribute to pathogenesis. Macrophage-mediated cytotoxicity could be important in diseases such as virus hepatitis. A model is the priming of macrophages and their recruitment into the liver by *Corynebacterium parvum* and subsequent activation by lipopolysaccharide, which damages hepatocytes (24). Analogous mechanisms could occur with hepatitis B virus (HBV). Infection with HBV itself does not appear to lyse hepatocytes, but could render these cells more susceptible to the lytic effects of macrophage-derived mediators. There is a need for improved vaccines that extend protection against viruses of this type—not only the acute damaging effects of the viruses, but also persistent infections, which have long-term sequelae.

AN ADJUVANT FORMULATION TARGETING VACCINES TO FDC AND ITS POSSIBLE ROLE IN THE PREVENTION OF LATENCY

An exciting development is the ability to express virus antigens by recombinant DNA technology in a form able to elicit protective immunity. We have previously (4) summarized reasons why we believe that recombinant protein vaccines are preferable to peptides, which often elicit immune responses to linear determinants, whereas responses to conformational determinants are usually required. Moreover, genetically controlled differences in the ability to respond well to peptides (e.g., of hepatitis B surface antigen [HBsAg] [53]) frequently occur and are observed in humans (84). To protect a high proportion of individuals in genetically diverse populations, it is desirable that a vaccine have several epitopes (as in a recombinant protein) rather than a small number (as in a peptide) (4). HBsAg is the first recombinant vaccine to be shown to be efficacious in humans (73) and to be authorized for use in the United States.

Recombinant proteins do not elicit protective immunity unless they are administered with adjuvants. The only adjuvants authorized for use in humans are aluminum hydroxide and phosphate, which have limitations. We have developed an adjuvant formulation (SAF-1) with an efficacy comparable to that of Freund complete adjuvant in eliciting cell-mediated immunity and antibodies of protective isotypes in experimental animals (3, 4). One component of SAF-1 is the threonyl analog of muramyl depeptide, which induces the formation of IL-1 by accessory cells. A second component is an emulsion of squalane and a pluronic block copolymer, L121. Antigens become associated with the surfaces of the lipid spherules in the emulsion, as do activated complement components. The latter association facilitates targeting of the antigen to FDC, which is required for generation of B-lymphocyte memory. SAF-1, unlike lipopolysaccharide and several other adjuvants, elicits the formation in the mouse of antibodies of isotypes such as immunoglobulin G2a, which activate complement and strongly bind Fcγ receptors on monocytes and macrophages. Studies with monoclonal antibodies show that antibodies of these isotypes are more protective against infections and tumors than are those of other isotypes having the same variable regions (4).

Our adjuvant formulation can be used with inactivated viruses or recombinant envelope proteins to prevent some persistent infections. Recombinant glycoprotein D of HSV-2 administered in SAF-1 to guinea pigs markedly reduced lesions following genital challenge and prevented mortality and infection of the dorsal root ganglion (4), a site of latency in this disease. With inactivated feline leukemia virus (4) and simian immunodeficiency virus (50), good protection against viremia and the late manifestations of the disease were obtained. Unfortunately, a first attempt to immunize rhesus monkeys against simian immunodeficiency virus by using inactivated virus in SAF-1 did not prevent viremia and lethal effects (4).

In Southeast Asia, perinatal infections with HBV are often acquired from mothers and produce persistent infections, with cirrhosis and primary hepatocellular carcinoma as late manifestations (9). Taiwanese children can be protected against perinatal HBV infection by use of antibodies and conventional immunization (10). Cell-mediated and humoral immune responses against internal antigens (HBcAg and HBeAg) can be protective (58). Because of the desirability of having many epitopes in a vaccine, we propose as an ideal vaccine HBsAg (with pre-S1), HBcAg, and HBeAg in SAF-1. This might prevent persistent HBV infections, as well as the late manifestations of the infections, in millions of children in China, Africa, and other countries where HBV is prevalent.

Persistent infections with Epstein-Barr virus are associated with nasopharyngeal carcinoma and Burkitt's lymphoma in China and Africa. A surface antigen of Epstein-Barr virus in SAF-1 can elicit strong immune responses in experimental animals and might provide a vaccine that attenuates primary infections with the virus and decreases the probability of late sequelae (4). Thus, the possibility of preventing at least some common and important persistent infections by vaccination seems realistic.

Literature Cited

1. **Alblas, A. B., and R. van Furth.** 1979. Origin, kinetics, and characteristics of pulmonary macrophages in the normal steady state. *J. Exp. Med.* **149:**1504–1509.
2. **Alizon, M., and L. Montagnier.** 1986. Lymphodenopathy/AIDS virus: genetic organization and relationship to animal lentiviruses. *Anticancer Res.* **6**(Part B)**:**403–411.
3. **Allison, A. C., and N. E. Byars.** 1986. An adjuvant formulation that selectively elicits the formation of antibodies of protective isotypes and of cell-mediated immunity. *J. Immunol. Methods* **95:**157–165.

4. **Allison, A. C., and N. E. Byars.** 1987. Vaccine technology: developmental strategies and adjuvants for increased efficiency. *Bio/Technology* **5:**1041–1045.
5. **Allison, A. C., and J.-L. Virelizier.** 1975. Effects of viruses on immune responses. *Adv. Nephrol.* **5:**115–125.
6. **Armstrong, J. A., and R. Horne.** 1984. Follicular dendritic cells and virus like particles in AIDS related lymphadenopathy. *Lancet* **ii:**370–372.
7. **Balfour, B. M., H. A. Drexhage, E. W. A. Kamperdijk, and E. C. M. Hoefsmit.** 1981. Antigen-presenting cells, including Langerhans cells, veiled cells and interdigitating cells. *CIBA Found. Symp.* **84:**281–293.
8. **Bang, F. B.** 1978. Genetics of resistance of animals to viruses. I. Introduction and studies in mice. *Adv. Virus Res.* **23:**270–348.
9. **Beasley, R. P., and L. Y. Hwang.** 1984. Hepatocellular carcinoma and hepatitis B virus. *Semin. Liver Dis.* **4:**1131–1153.
10. **Beasley, R. P., L. Y. Hwang, G. C. Lee, C. C. Lan, C. H. Roan, F. Y. Hwang, and C. L. Chen.** 1983. Prevention of perinatally transmitted hepatitis B virus infections with hepatitis B immune globulin and hepatitis B vaccine. *Lancet* **ii:**1099–1101.
11. **Belardelli, F., F. Vignaux, E. Proietti, and I. Gresser.** 1984. Injection of mice with antibody to interferon renders peritoneal macrophages permissive for vesicular stomatitis virus and encephalomyocarditis virus. *Proc. Natl. Acad. Sci. USA* **81:**602–606.
12. **Bell, D. M., N. J. Roberts, Jr., and C. B. Hall.** 1983. Different antiviral spectra of human macrophage interferon activities. *Nature* (London) **305:**319–321.
13. **Belsito, D. V., and G. J. Thorbecke.** 1984. Reduced Ia-positive Langerhans cells in AIDS. *N. Engl. J. Med.* **311:**857–858.
14. **Bjercke, S., J. Elg, L. Braathen, and E. Thorsby.** 1984. Enriched epidermal Langerhans cells are potent antigen-presenting cells for T cells. *J. Invest. Dermatol.* **83:**286–289.
15. **Bjercke, S., H. Scott, L. R. Breathen, and E. Thorsby.** 1983. HLA-DR-expressing Langerhans-like cells in vaginal and cervical epithelium. *Acta Obstet. Gynecol. Scand.* **62:**585–589.
16. **Brautigam, A. R., F. J. Dutko, L. B. Olding, and M. B. A. Oldstone.** 1979. Pathogenesis of murine cytomegalovirus infection: the macrophage as a permissive cell for cytomegalovirus infection, replication, and latency. *J. Gen. Virol.* **59:**345–356.
17. **Brunstetter, M.-A., J. A. Hardie, R. Schiff, J. P. Lewis, and C. E. Cross.** 1971. The origin of pulmonary alveolar macrophages. *Arch. Intern. Med.* **127:**1054–1058.
18. **Burns, W. H., and A. C. Allison.** 1975. Virus infections and the immune responses they elicit, p. 480–506. *In* M. Sela (ed.), *The Antigens*, vol. 3. Academic Press, Inc., New York.
19. **Cameron, P. V., R. L. Dawkins, J. A. Armstrong, and E. Bonifacio.** 1987. Western blot profiles, lymph node ultrastructure and viral expression in HIV-infected patients: a correlative study. *Clin. Exp. Immunol.* **68:**465–478.
20. **Chayt, K. J., M. E. Harper, L. M. Marselle, E. B. Lewin, R. M. Rose, J. M. Oleske, L. G. Epstein, F. Wong-Staal, and R. C. Gallo.** 1986. Detection of HTLV-III RNA in lungs of patients with AIDS and pulmonary involvement. *J. Am. Med. Assoc.* **256:**2356–2359.
21. **Cheevers, W. P., and T. C. McGuire.** 1985. Equine infectious anemia: immunopathogenesis and persistence. *Rev. Infect. Dis.* **7:**83–88.
22. **Daems, W. T., and J. M. de Bakker.** 1982. Do resident macrophages proliferate? *Immunobiology* **161:**204–211.
23. **Daughaday, C. C., W. E. Brandt, J. M. McCown, and P. K. Russell.** 1981. Evidence for two mechanisms of dengue virus infection of adherent human monocytes: trypsin-

sensitive virus receptors and trypsin-resistant immune complex receptors. *Infect. Immun.* **32**:469–473.

24. **Ferluga, J., and A. C. Allison.** 1978. Role of mononuclear infiltrating cells in pathogenesis of hepatitis. *Lancet* **ii**:610–611.
25. **Gartner, P., P. Markovits, D. M. Markovitz, M. H. Kaplan, R. C. Gallo, and M. Popovic.** 1986. The role of mononuclear phagocytes in HTLV-III/LAV infection. *Science* **233**:215–219.
26. **Gendelman, H. E., O. Narayan, S. Kennedy-Stotskoff, P. G. E. Kennedy, Z. Ghobi, J. E. Clements, J. Stanley, and G. Pezeshkpour.** 1986. Tropism of sheep lentiviruses for monocytes: susceptibility to infection and virus gene expression increase during maturation of monocytes to macrophages. *J. Virol.* **58**:67–74.
27. **Gonda, M. A., M. J. Braun, J. E. Clements, J. M. Pyper, F. Wong-Staal, R. C. Gallo, and R. V. Gilden.** 1986. Human T-cell lymphotropic virus type III shares sequence homology with a family of pathogenic lentiviruses. *Proc. Natl. Acad. Sci. USA* **83**:4007–4011.
28. **Gresser, I., M. G. Tovey, C. Maury, and M.-T. Bandu.** 1976. Role of interferon in the pathogenesis of virus diseases in mice as demonstrated by the use of anti-interferon serum. *J. Exp. Med.* **144**:1316–1323.
29. **Grieco, M. H., M. M. Reddy, M. L. Moriarty, K. K. Ahiya, S. Bellomo, H. Holtz, J. Dobro, J. W. Keslak, E. Cohen, and P. C. T. Dickson.** 1986. Study of *in vivo* immunomodulation by indomethacin in acquired immune deficiency syndrome and AIDS-related complex. *Int. J. Immunother.* **II**:295–300.
30. **Gyorkey, F., J. L. Melnick, J. G. Sinkovics, and P. Gyorkey.** 1985. Retrovirus resembling HTLV in macrophages of patients with AIDS. *Lancet* **i**:106.
31. **Halstead, S. B.** 1982. Immune enhancement of viral infections. *Prog. Allergy* **31**:301–364.
32. **Halstead, S. B., and E. J. O'Rourke.** 1977. Dengue viruses and mononuclear phagocytes. I. Infection enhancement by non-neutralizing antibody. *J. Exp. Med.* **146**:201–217.
33. **Halstead, S. B., E. J. O'Rourke, and A. C. Allison.** 1977. Dengue viruses and mononuclear phagocytes. II. Identity of blood and tissue leukocytes. *J. Exp. Med.* **146**:218–229.
34. **Herano, T., T. Toga, K. Yasukawa, K. Nakajima, N. Nakano, T. Takasuki, M. Shimizu, A. Murashima, S. Tsunasokawa, F. Sakijama, and T. Kishimoto.** 1987. Human B-cell differentiation factor defined by an anti-peptide antibody and its role in autoantibody production. *Proc. Natl. Acad. Sci. USA* **84**:228–231.
35. **Hirsch, M. S., B. Zisman, and A. C. Allison.** 1970. Macrophages and age-dependent resistance to herpes simplex virus in mice. *J. Immunol.* **104**:1160–1165.
36. **Ho, D. D., T. R. Rota, and M. S. Hirsch.** 1986. Infection of monocyte/macrophages by human T lymphotropic virus type III. *J. Clin. Invest.* **77**:1712–1715.
37. **Humphrey, J. H., D. Grennan, and V. Sundaram.** 1984. The origin of follicular dendritic cells in the mouse and the mechanism of trapping of immune complexes on them. *Eur. J. Immunol.* **14**:859–863.
38. **Isakov, N., M. Feldman, and S. Segal.** 1982. Acute virus infection of mice with lactate dehydrogenase virus (LDV) impairs the antigen-presenting capacity of their macrophages. *Cell. Immunol.* **66**:317–332.
39. **Jacobse-Geels, H. E. L., and M. C. Horzinek.** 1983. Expression of feline infectious peritonitis coronavirus antigens on the surface of feline macrophage-like cells. *J. Gen. Virol.* **64**:1859–1866.
40. **Johnson, G. D., D. L. Hardie, N. R. Ling, and I. C. M. MacLennan.** 1986. Human

follicular dendritic cells (FDC): a study with monoclonal antibodies (MoAb). *Clin. Exp. Immunol.* **64:**205–213.

41. **Kennedy, P. G. E., O. Narayan, Z. Ghotbe, J. Hopkins, H. E. Gendelman, and J. E. Clements.** 1985. Persistent expression of Ia antigen and viral genome in visna-maedi virus-induced inflammatory cells. Possible role of lentivirus-induced interferon. *J. Exp. Med.* **162:**1970–1982.
42. **Klaus, G. G., J. H. Humphrey, A. Kunkl, and D. W. Dongworth.** 1980. The follicular dendritic cell: its role in antigen presentation in the generation of immunological memory. *Immunol. Rev.* **53:**3–28.
43. **Knight, S. C., J. Krejci, M. Malkovsky, V. Colizzi, A. Gautam, and G. L. Asherson.** 1985. The role of dendritic cells in the initiation of immune responses to contact sensitizers. 1. *In vivo* exposure to antigen. *Cell. Immunol.* **94:**427–434.
44. **Knight, S. C., J. Mertin, A. Stackpole, and J. Clarke.** 1983. Induction of immune responses *in vivo* with small numbers of veiled (dendritic) cells. *Proc. Natl. Acad. Sci. USA* **80:**6032–6035.
45. **Laguens, M., J. G. Chambo, and R. P. Laguens.** 1983. In vivo replication of pathogenic and attenuated strains of junin virus in different cell populations in lymphoid tissue. *Infect. Immun.* **39:**955–959.
46. **Lahmy, C., and J.-L. Virelizier.** 1981. Prostaglandins as probable mediators of the suppression of antibody production by mouse hepatitis virus infection. *Ann. Inst. Pasteur Immunol.* **132C:**101–105.
47. **Lindsley, M. D., D. J. Torpey, and C. R. Rinaldo.** 1984. Lymphocyte-mediated cytotoxicity of cytomegalovirus-infected human monocytes restricted by the HLA-DR locus, p. 429–433. *In* S. Plotkin (ed.), *Pathogenesis and Prevention of Human Cytomegalovirus Disease.* Alan R. Liss, Inc., New York.
48. **Lopez, C.** 1984. Natural resistance mechanisms against herpesvirus in health and disease, p. 45–70. *In* B. T. Rouse and C. Lopez (ed.), *Immunobiology of Herpes Simplex Virus Infection.* CRC Press, Inc., Boca Raton, Fla.
49. **Macfarlan, R. I., W. H. Burns, and D. O. White.** 1977. Two cytotoxic cells in the peritoneal cavity of virus-infected cells: antibody-dependent macrophages and non-specific killer cells. *J. Immunol.* **119:**1569–1574.
50. **Marx, P. A., N. C. Pedersen, N. W. Lerche, K. G. Osborn, L. J. Lowenstine, A. A. Lackner, D. H. Maul, H.-S. Kwang, J. D. Kluge, C. P. Zaiss, V. Sharpe, A. D. Spinner, A. C. Allison, and M. B. Gardner.** 1986. Prevention of simian acquired immune deficiency syndrome with a Formalin-inactivated type D retrovirus vaccine. *J. Virol.* **60:**431–435.
51. **Metcalf, D.** 1987. The molecular control of normal and leukaemic granulocytes and macrophages. *Proc. R. Soc. Lond. Ser. B.* **230:**389–423.
52. **Meyer, T., and M. A. Horisburger.** 1984. Combined action of mouse α and β interferons in influenza virus-infected macrophages carrying the resistance gene Mx. *J. Virol.* **49:**709–716.
53. **Milich, D. R., and F. V. Chisari.** 1982. Genetic regulation of the immune response to hepatitis B virus surface antigen (HBsAg). 1. H2 restriction of the murine humoral immune response to the a and d determinants of HBsAg. *J. Immunol.* **129:**320–325.
54. **Mills, J.** 1979. Effects of Sendai virus infection on function of cultured mouse alveolar macrophages. *Am. Rev. Respir. Dis.* **120:**1239–1244.
55. **Morahan, P., J. R. Connor, and K. R. Leary.** 1985. Viruses and the versatile macrophage. *Br. Med. Bull.* **41:**15–21.
56. **Morris, A. G., Y.-L. Lin, and B. A. Askonas.** 1982. Immune interferon release when a

cloned cytotoxic T-cell line meets its correct influenza-infected target cell. *Nature* (London) **295:**150–152.
57. **Mossman, T. R., H. Cherwinski, M. W. Bond, M. A. Giedlin, and R. L. Coffman.** 1986. Two types of murine helper T cell clone. 1. Definition according to profiles of lymphokine activities and secreted proteins. *J. Immunol.* **136:**2348–2357.
58. **Murray, K., S. A. Bruce, P. A. Wingfield, P. van Eerd, A. de Reus, and H. Schellekens.** 1987. Protective immunization against hepatitis B with an internal antigen of the virus. *J. Med. Virol.* **23:**101–107.
59. **Naiem, M., J. Gerdes, Z. Abdulaziz, H. Stein, and D. Y. Mason.** 1983. Production of a monoclonal antibody reactive with human dendritic reticulum cells and its use in the immunohistological analysis of lymphoid tissue. *J. Clin. Pathol.* **36:**167–175.
60. **Narayan, O., and L. C. Cork.** 1985. Lentiviral diseases of sheep and goats: chronic pneumonia, leukoencephalitis and arthritis. *Rev. Infect. Dis.* **7:**89–98.
61. **Patterson, S., and M. R. MacNaughton.** 1982. Replication of human respiratory coronavirus strain 229E in human macrophages. *J. Gen. Virol.* **60:**307–314.
62. **Plata, F., B. Autran, L. P. Martins, S. Wain-Hobson, M. Raphael, C. Mayand, M. Denis, J.-M. Guillon, and P. Debre.** 1987. AIDS virus-specific T lymphocytes in lung disorders. *Nature* (London) **328:**348–351.
63. **Porter, D. D., A. E. Larsen, and H. E. Porter.** 1972. The pathogenesis of Aleutian disease of mink. II. Enhancement of tissue lesions following the administration of a killed virus vaccine or passive antibody. *J. Immunol.* **109:**1–7.
64. **Prydz, H., T. Lyberg, P. Deteix, and A. C. Allison.** 1979. *In vitro* stimulation of tissue thromboplastin (factor III) activity in human monocytes by immune complexes and lectins. *Thromb. Res.* **15:**465–474.
65. **Rager-Zisman, B., and A. C. Allison.** 1973. The role of antibody and host cells in the resistance of mice against infection by Coxsackie B-3 virus. *J. Gen. Virol.* **19:**329–338.
66. **Rager-Zisman, B., and A. C. Allison.** 1976. Mechanism of immunologic resistance to herpes simplex virus (HSV-1) infection. *J. Immunol.* **116:**35–40.
67. **Reynes, M., J. P. Aubert, J. H. Cohen, J. Audoin, V. Tricollet, J. Diebald, and M. D. Kazatchkine.** 1985. Human follicular dendritic cells express CR1, CR2 and CR3 complement receptor antigens. *J. Immunol.* **135:**2687–2694.
68. **Ron, Y., and J. Sprent.** 1987. T cell priming *in vivo*: a major role for B cells in presenting antigen to T cells in lymph nodes. *J. Immunol.* **138:**2848–2856.
69. **Rosenberg, S. A., and M. T. Lotze.** 1986. Cancer immunotherapy using interleukin-2 and interleukin-2 activated lymphocytes. *Annu. Rev. Immunol.* **4:**681–709.
70. **Salahuddin, S. Z., R. M. Rose, J. E. Groopman, P. D. Markham, and R. C. Gallo.** 1986. Human T-lymphotropic virus type III infection of human alveolar macrophages. *Blood* **68:**281–284.
71. **Salinovich, O., S. L. Payne, R. C. Montelero, K. A. Hussain, C. J. Issel, and K. L. Schnorr.** 1986. Rapid emergence of novel antigenic and genetic variants of equine infectious anemia virus during persistent infection. *J. Virol.* **57:**71–80.
72. **Schmitt, D., M. Foure, C. Dambuyant-Dezutter, and J. Thivolet.** 1984. The semi-quantitative distribution of T4 and T6 surface antigens on human Langerhans cells. *Br. J. Dermatol.* **111:**655–661.
73. **Scolnick, E. M., A. A. McLean, D. J. West, W. J. McAleer, W. J. Miller, and E. B. Bunyak.** 1984. Clinical evaluation in healthy adults of a hepatitis B vaccine made by recombinant DNA. *J. Am. Med. Assoc.* **251:**2812–2815.
74. **Seghal, P. B., L. T. May, I. Tamm, and J. Vilcek.** 1987. Human β_2 interferon and B-cell differentiation factor BSF-2 are identical. *Science* **235:**731–732.
75. **Steeg, P. S., R. N. Moore, H. M. Johnson, and J. J. Oppenheim.** 1982. Regulation of

murine macrophage Ia antigen expression by a lymphokine with immune interferon activity. *J. Exp. Med.* **156:**1780–1793.

76. **Steinman, R. M., and M. C. Nussenzweig.** 1980. Dendritic cells: features and functions. *Immunol. Rev.* **53:**127–147.
77. **Stevens, J. G., and M. L. Cook.** 1971. Restriction of herpes simplex virus by macrophages. An analysis of the cell virus interaction. *J. Exp. Med.* **133:**19–38.
78. **Stoler, M. H., T. A. Eskin, S. Benn, R. C. Angerer, and L. M. Angerer.** 1986. Human T-cell lymphotropic virus type III infection of the central nervous system. *J. Am. Med. Assoc.* **256:**2360–2364.
79. **Swartzenduber, D. C., B. I. Ma, and W. H. Murphy.** Detection of virus particles in lymphatic tissue germinal centers, p. 203–210. *In* L. Fiore-Donati and M. G. Hanna (ed.), *Lymphatic Tissue and Germinal Centers in Immune Response*. Plenum Publishing Corp., New York.
80. **Taylor, S., and Y. J. Bryson.** 1985. Impaired production of gamma-interferon by newborn cells *in vitro* is due to a functionally immature macrophage. *J. Immunol.* **134:**1493–1497.
81. **Tenner-Racź, K., P. Racź, M. Bofill, A. Schulz-Meyer, M. Dietrich, P. Kern, J. Weber, A. J. Pinching, F. Veronese-Dinarzo, M. Popovic, D. Klatzmann, J. C. Gluckmann, and A. Janossy.** 1986. HTLVIII/LAV viral antigens in lymph nodes of homosexual men with persistent generalized lymphodenopathy and AIDS. *Am. J. Pathol.* **123:**9–15.
82. **Thomsen, A. R., and M. Volkert.** 1983. Studies on the role of mononuclear phagocytes in resistance to acute lymphocytic choriomeningitis virus infection. *Scand. J. Immunol.* **18:**271–277.
83. **Toniolo, A., D. Matteucci, M. P. Pistillo, Z. Goni, and M. Bendinelli.** 1980. Early replication of Friend leukemia viruses in spleen macrophages. *J. Gen. Virol.* **49:**203–208.
84. **Townsend, A. R. M., J. Rothbard, F. M. Gotch, G. Bahadur, D. Wraith, and A. J. McMichael.** 1986. The epitopes of influenza nucleoprotein recognized by cytotoxic T-lymphocytes can be defined by short synthetic peptides. *Cell* **44:**959–968.
85. **Tschachler, E., V. Groh, M. Popovic, D. Mann, K. Konrad, B. Safai, L. Eron, F. D. Veronese, K. Wolff, and G. Stingl.** 1987. Epidermal Langerhans cells—a target for HTLV-III/LAV infection. *J. Invest. Dermatol.* **88:**233–237.
86. **van der Logt, J. T. M., A. M. van Loon, and J. van der Veen.** 1980. Replication of rubella virus in human mononuclear blood cells. *Infect. Immun.* **27:**309–314.
87. **van der Maaten, M. J., A. D. Boothe, and C. L. Seger.** 1972. Isolation of a virus from cattle with persistent lymphocytosis. *J. Natl. Cancer Inst.* **49:**1649–1657.
88. **Virelizier, J. L., and A. C. Allison.** 1976. Correlation of persistent mouse hepatitis virus (MHV-3) infection with its effect on mouse macrophage cultures. *Arch. Virol.* **50:**279–285.
89. **Virelizier, J. L., A. C. Allison, and E. de Maeyer.** 1977. Production by mixed lymphocyte cultures of a type II interferon able to protect macrophages against virus infections. *Infect. Immun.* **17:**282–285.
90. **Virelizier, J.-L., A. C. Allison, and G. C. Schild.** 1974. Antibody responses to antigenic determinants of influenza virus hemagglutinin. 1. Thymus dependence of antibody formation and thymus independence of immunological memory. *J. Exp. Med.* **140:**1559–1570.
91. **Virelizier, J.-L., A.-M. Virelizier, and A. C. Allison.** 1976. The role of circulating interferon in the modifications of immune responsiveness by mouse hepatitis virus (MHV-3). *J. Immunol.* **117:**748–753.
92. **Ward, J. M., T. J. O'Leary, G. B. Baskin, R. Benveniste, C. A. Harris, P. L. Nara, and**

R. H. Rhodes. 1987. Immunohistochemical localization of human and simian immunodeficiency viral antigens in fixed tissue sections. *Am. J. Pathol.* **127:**199–205.
93. **Wardley, R. C., F. Hamilton, and P. J. Wilkinson.** 1979. The replication of virulent and attenuated strains of African swine fever virus in porcine macrophages. *Arch. Virol.* **61:**217–225.
94. **Wewers, M. D., S. I. Rennard, A. J. Hance, P. B. Bitteman, and R. B. Crystal.** 1984. Normal human macrophages obtained by bronchoalveolar lavage have a limited capacity to release interleukin-1. *J. Clin. Invest.* **74:**2208–2218.
95. **Wong, G. H. W., and D. V. Goeddel.** 1986. Tumour necrosis factors α and β inhibit virus replication and synergize with interferons. *Nature* (London) **323:**819–822.
96. **Wong, G. H. W., and D. V. Goeddel.** 1987. The antiviral effects of TNF, p. 273–277. *In* K. Cantell and H. Schellekens (ed.), *The Biology of the Interferon System*. Martinus Nijhoff, Boston.
97. **Wood, G. S., R. R. Turner, R. A. Schiurba, L. Eng, and R. A. Warnke.** 1985. Human dendritic cells and macrophages. In situ immunophenotypic definition of subsets that exhibit specific morphologic and microenvironmental characteristics. *Am. J. Pathol.* **119:**73–82.
98. **Zisman, B., M. S. Hirsch, and A. C. Allison.** 1970. Selective effects of anti-macrophage serum, silica and anti-lymphocyte serum on pathogenesis of herpes virus infection of young adult mice. *J. Immunol.* **104:**1155–1159.

Chapter 3

Role of Natural Killer Cells in Resistance to Virus Infections

Carlos Lopez

INTRODUCTION

Defense of a host against an invading virus pathogen depends on many different cell populations and cytokines acting in concert to limit and clear the infection (1). The natural defense system of the infected host constitutes the first barrier of active defense against that agent. These mechanisms act nonspecifically and require no prior exposure to the invading microorganism to be active. In recent years, natural killer (NK) cells were shown to be an important natural defense mechanism in controlling certain virus infections. This chapter will focus on the role of NK cells in resistance to virus infections as demonstrated first in animal studies and second in studies of human effector cells.

Since natural defense systems are immediately available for response against an invading virus, they are thought to play an especially important role in controlling primary infections. The fact that most primary virus infections in humans are asymptomatic would suggest that natural defense systems control those infections before lesions or systemic disease develops. These mechanisms should play a pivotal role in the control of virus infections before antigen-specific immune responses such as neutralizing antibody, cytotoxic T cells, and delayed-type hypersensitivity develop (48). The mechanisms responsible for resistance against primary virus infections have received considerable attention during the past few years. Animal models, especially those involving inbred strains of mice, have been used to explore various aspects of natural and adaptive immunity (5). These models allow for manipulations that cannot be

---

Carlos Lopez • Viral Exanthems and Herpesvirus Branch, Division of Viral Diseases, Center for Infectious Diseases, Centers for Disease Control, Atlanta, Georgia 30333.

carried out with outbred mice or humans. One must recognize, however, that these animal models have possible disadvantages. For example, if a virus is not indigenous to the animals under study, the diseases caused by the virus and the host defense mechanisms that control those infections may not be the same as those found in humans. Therefore, studies with mouse models must be corroborated by studies with humans.

MOUSE STUDIES

Genetic Resistance to HSV-1 in the Mouse

Inbred strains of mice were evaluated for resistance against an intraperitoneal infection with a virulent strain of herpes simplex virus type 1 (HSV-1) (34). Resistant C57BL/6 mice were found to survive a challenge with 10^6 PFU of HSV-1, whereas the susceptible A/J mice died after challenge with as little as 10^2 PFU. Susceptible strains of mice demonstrated hind-leg paralysis 5 to 8 days after inoculation and died shortly thereafter. Although the viruses tested varied greatly in their virulence for mice, a lower concentration of each virus was needed to kill the susceptible mice than the moderately susceptible mice, and none of the virus strains killed resistant mice (34, 37). These observations have been confirmed by a series of studies by Kirchner and co-workers (30, 31, 61), as well as by other groups (14, 51).

Genetic studies indicated that resistance is a dominant trait controlled by two or more genes (36). Also, we conducted experiments to determine whether immune response genes within the major histocompatibility region of the mouse (*H-2*) influence resistance to HSV-1. In these studies, the *H-2* of susceptible strains failed to diminish the resistance of resistant mice and the *H-2* of resistant mice failed to transfer resistance to susceptible mice. Therefore, genetic resistance to HSV-1 could not be linked to the major histocompatibility genes of the mouse. More recent studies of male and female progeny indicate that genetic resistance in this model is not sex linked (C. Lopez, manuscript in preparation).

Three observations suggest that genetic resistance to HSV-1 is mediated by natural defense systems rather than adaptive immunity. First, mice challenged with HSV-1 usually died from infection long before an adaptive immune response could be detected (34). Second, athymic nude (*nu/nu*) mice, deficient in T cells and unable to generate cell-mediated immune responses, have been found to be about as resistant to HSV-1 as were the controls (35, 60). Third, strontium-89 treatment of mice has been shown to suppress the natural defense mechanisms required for resistance to allogeneic marrow and to suppress genetic resistance to HSV-1 (6, 7, 39). Since ^{89}Sr treatment of mice does not significantly suppress adaptive immune mechanisms, suppression of the natural defense systems is most probably responsible for the increased susceptibility found in treated mice.

Genetic resistance to lethal infection with HSV-2 has also been demonstrated (2; K. E. Schneweis and V. Saftig, personal communication). In addition, a model of genetic resistance to HSV-1-induced latent infections of the peripheral nervous system in mice has been established (27). In both of these models, C57BL/6 or C57BL/10 series mice were found to be resistant, and resistance was shown to be a dominant trait. These results suggest that the resistance mechanisms operative in intraperitoneally challenged mice may also be responsible for resistance to lethal HSV-2 infections and the establishment of latency by HSV-1.

Genetic Resistance to Murine CMV

The studies of Chalmer et al. (15) demonstrated that the host genotype strongly influences both resistance to the lethal effects of murine cytomegalovirus (CMV) infection and the characteristics of the virus-induced pathogenesis. *H-2* genes and non-*H-2* genes were found to influence resistance to intraperitoneal challenge with murine CMV.

In the murine CMV model, both natural defense mechanisms and adaptive immunity appear to be required for resistance to infection. Evidence supporting a requirement for natural defense systems includes the observations that many susceptible mice succumb to virus infection before a cell-mediated immune response can be detected (15) and that abrogation of natural defense systems in genetically resistant mice results in greatly diminished resistance of those mice (12, 41, 50). Adaptive immune mechanisms were also required, however, since athymic mice, with a deficiency in their ability to generate a cell-mediated immune response, were shown to be less resistant to murine CMV infection than the appropriate controls were (52). Athymic mice died late after challenge with the virus, indicating that the role of the adaptive immune response was to clear virus late in infection.

NK Cells in Resistance to Virus Infections

NK cells were first described as effector cells capable of lysing certain tumor targets (23, 29). NK cells were detected in nonsensitized individuals and therefore were thought to play a role as a first line of defense against arising neoplasia. Although most NK cell studies have been carried out with tumor targets, it has become clear in the last several years that these effectors can also lyse virus-infected cells.

The earliest indications that NK cells might play an active role against virus infections came from investigations of virus-specific T-cell cytotoxicity. Nonspecific lytic activity, peaking just before specific cytotoxic T-cell activity, has been reported for several virus infections, including those with lymphocytic choriomeningitis virus, ectromelia orthopoxvirus, and Semliki Forest virus (8, 43, 47). Although this nonspecific cytotoxicity was originally thought to be due to T cells or macrophages, it was probably due to the augmented NK cell activity usually found in mice shortly after virus infection. In parallel studies, NK cells were described to be the effector cells that gave the high background lysis in studies of tumor immunity (25, 26).

Role of NK Cells in HSV-1 Infections

The characteristics of genetic resistance to HSV-1 in the mouse were similar to those of genetic resistance to bone marrow allographs (39). Since NK cells were shown to mediate resistance against allogeneic marrow (28), studies were undertaken to determine whether NK cells played a role in genetic resistance to HSV-1 in the mouse.

^{89}Sr treatment of mice abrogated genetic resistance to HSV-1 (39) without suppressing adaptive immune responses, but markedly depressed NK cell function (7). When challenged with HSV-1, ^{89}Sr-treated mice demonstrated persistent infections of visceral tissues. Virus was also able to migrate to and replicate in cerebrospinal tissues. Since ^{89}Sr treatment of mice selectively inhibits NK cell function, these findings support an important role for NK cells in genetic resistance to HSV-1.

Rager-Zisman et al. (46) have also provided evidence for an important role for NK cells in resistance to fatal HSV-1 infections in mice. Mice were rendered susceptible to HSV-1 infection by treatment with cyclophosphamide 1 day before challenge. Subpopulations of normal (nonsensitized) spleen cells were then used to attempt to transfer resistance. NK cells from normal animals transferred resistance, whereas spleen cells from NK-cell-deficient beige mice failed. Thus, NK cells in the normal spleen cells of mice mediated resistance to HSV-1.

Role of NK cells in Resistance to Murine CMV

Several observations suggest that the NK cell response is responsible, in part, for genetic resistance to murine CMV. Mice shown to be genetically susceptible to murine CMV were found to have low NK cell activity, whereas genetically resistant animals had high NK cell activity in response to infection (3). In addition, mice bearing the beige mutation, which has been associated with a deficiency of NK cell function, were shown to be much more susceptible to murine CMV infection than appropriate controls were (50). Also, suppression of NK cell function in adult mice by treatment with antibody to asialo GM_1 (13) or ^{89}Sr (41) resulted in mice that were far more susceptible to murine CMV.

Adoptive transfer experiments have also been used to show that NK cells are required for resistance to murine CMV (11). Adult splenic leukocytes were shown to protect the normally susceptible newborn mice from fatal murine CMV infection by reducing virus concentrations in their spleens and other organs (11). The leukocyte subpopulation responsible for this function was shown to have the NK cell phenotype. In this system, NK cells were shown to function within the first 3 days after infection and were not effective 6 days after challenge. Protection was not mediated by interferon (IFN) production and appears to be a cytotoxic function of NK cells.

Role of NK Cells in Other Murine Virus Infections

The role of NK cells in resistance to other murine virus infections has not been well explored. Murine hepatitis virus (MHV) infection requires T-cell, macrophage, and NK cell functions for resistance to be manifest (33). Depletion of NK cell function by treatment of mice with asialo GM_1 markedly enhanced viral replication in challenged mice (12). In addition, adoptive transfer experiments showed that NK cells were required for resistance (33).

Intratracheal infection of mice with influenza virus has been shown to result in a three- to fourfold increase of NK cell activity confined to the lungs (40). This increased NK cell activity in the lungs appears to be necessary for resistance to infection, since diminution of NK cell function by treatment of mice with anti-asialo GM_1 resulted in significantly increased morbidity and mortality due to pulmonary influenza infection (53). These findings suggest that a highly localized NK cell response in the lungs is required for resistance to infection with this virus.

Surprisingly few studies have been carried out to determine the role of NK cells in resistance to retrovirus infections. However, acute retrovirus infections have been shown to induce elevated NK cell activity in mice (22), and depressed NK cell function in beige mice or in anti-asialo GM_1-treated mice has been associated with increased replication of Friend leukemia virus (58). Thus, these findings support the concept that retrovirus infections and retrovirus-induced tumors may also be regulated by NK cells.

For some viruses, clear evidence shows that NK cells do not contribute to resistance. Even though some of the earliest studies demonstrating that virus infections could boost NK cell function in vivo were carried out with lymphocytic choriomeningitis virus, these effector cells appear to play no significant role in resistance to the virus infection or in the immunopathology associated with infection (12, 59). The elevated levels of NK cell activity closely follow the levels of IFN found in the challenged host. Diminished NK cell function, as found in beige mice or in anti-asialo GM_1-treated mice, was not associated with higher levels of virus replication in organs or greater mortality due to the virus infection. In similar studies, augmented NK cell function failed to suppress Sindbis virus (24) or Pichinde virus (21) infection.

HUMAN STUDIES

The earliest evidence that NK cells might play a role in resistance against virus infections in humans was the observation that human effector cells could preferentially lyse virus-infected cells in culture. Diamond et al. (17) first reported that effector cells from nonimmune donors could lyse CMV-infected fibroblasts more efficiently than they lysed uninfected cells. The activity was dependent on the dose of effector cells used, was found to be independent of antibody to CMV, and was mediated by effectors that were not T cells or macrophages. Several studies have now documented preferential lysis of fibroblasts infected with HSV-1, influenza virus, paramyxovirus, and measles virus (16, 44, 49).

Characteristics of the NK Cells That Lyse Virus-Infected Targets

Several studies have now shown that NK cells are capable of lysing virus-infected targets. Most of these studies have characterized the effectors as NK cells that are using certain properties including the absence of cell surface markers expressed on mature T cells, B cells, or macrophages; the lack of a requirement for histocompatibility between target cells and donor lymphoid cells; and the lack of a requirement for presensitization to virus antigens.

Since we were interested in the possible role of NK cells in resistance to herpesvirus infections in humans, we developed an assay involving the use of HSV-1-infected fibroblasts as targets [NK(HSV-Fs)] (16). Although the effector cells that lyse HSV-Fs targets are similar to cells that lyse the commonly used K562 erythroleukemia targets [NK(K562)] in the properties which define them as NK cells, they differ from them in certain other characteristics (18).

Heterogeneity of NK Cells

Both the NK(K562) and the NK(HSV-Fs) effectors have been found to express CD16, a cell surface determinant detected by monoclonal antibodies and found on all NK effector cells (18). In contrast, monoclonal antibodies to other cell surface determinants differentiated among the subpopulations of effector cells that mediated lysis of these two targets. Thus, NK(K562) effectors were found to be positive for CD1, CD2, and CD4 cell surface markers, whereas NK(HSV-Fs) effectors were negative.

Results from studies of certain groups of patients have provided the most compelling evidence for heterogeneity of human NK effector cells and have also provided the first evidence that these activities are under independent regulation in vivo (18, 42). Patients with Wiskott-Aldrich syndrome were consistently shown to have low levels of NK(HSV-Fs) activity, but many of these patients demonstrated normal NK(K562) function. Conversely, patients with consistently normal levels of NK(HSV-Fs) activity but low levels of NK(K562) function have been found (42). These observations provide strong evidence for the existence of at least two subpopulations of effector cells which mediate lysis of the targets and indicate that they are independently regulated.

More recent studies provide new evidence for a difference between the effectors that lyse K562 targets and those that lyse HSV-Fs. Thus, lysis of HSV-Fs targets has been shown to require an accessory cell function, whereas lysis of K562 targets has not (N. Feldman, S. Curl, and P. Fitzgerald-Bocarsly, *Fed. Proc.* **46**:483, 1987). The presence of an HLA-DR^+ subpopulation of cells was necessary for the $CD16^+$ effector cells to lyse HSV-Fs targets. HLA-DR^+ cells are not required for K-562 target cell lysis. The accessory cell function has also been shown to be required for lysis of CMV-infected fibroblasts (4). Thus, at least two different subpopulations of NK effector cells can be distinguished by the requirement for an accessory cell function.

Preferential Lysis of HSV-Infected Fibroblasts

Peripheral blood mononuclear cells from both seropositive and seronegative individuals are capable of lysing HSV-Fs cells in a dose-dependent manner (16). Uninfected fibroblasts are usually poor targets for NK effector cells. Many studies have been carried out to try to determine what the virus infection does to fibroblasts to make them better targets for NK cells.

Role of interferon in NK lysis of HSV-Fs. In some of the earliest reports describing the preferential lysis of virus-infected cells, Trinchieri et al. (56) and Santoli et al. (49) described the production of IFN during NK cell assays with virus-infected targets. These investigators also noted that IFN greatly augmented NK cell function with infected and uninfected targets. Because of this, they suggested that the IFN might be responsible for preferential lysis of infected rather than uninfected targets. In more recent studies, we have found that although IFN-α is produced during the NK(HSV-Fs) assay, the presence of the IFN cannot account for the preferential lysis of the virus-infected targets (20). No correlation was found between the levels of cytotoxicity obtained in the NK assay and the amount of IFN-α generated during that same assay. In addition, anti-IFN-α antibody, at levels sufficient to neutralize all the antiviral activity generated during the assay, failed to reduce NK cell cytotoxicity. Further evidence that IFN-α generation and NK cell cytotoxic activity segregate independently was derived from a study of five patients who had severe combined immunodeficiency disease and who were found to have normal levels of NK(HSV-Fs) activity despite their ability to produce only very low levels of IFN-α during the cytotoxicity assay (42). Thus, IFN-α is generated during the NK(HSV-Fs) assay, but is not responsible for the selective lysis of those target cells. It might still play a role in vivo, however, by recruiting and activating NK cells and increasing NK cell activity at the site of the virus infection.

Possible destabilization of infected fibroblasts by HSV. Still another possible explanation for the increased susceptibility of HSV-infected targets to NK cell lysis is that the virus destabilizes the fibroblast membrane and impairs its ability to repair itself, leaving the infected target cell inherently more susceptible to lysis than the intact uninfected cell. Several experimental approaches have been taken to evaluate this. For example, both infected and uninfected cells have been subjected to hypotonic conditions and found to be equally sensitive to osmotic shock (P. Fitzgerald-Bocarsly and C. Lopez, manuscript in preparation). Also, infected and uninfected targets have been found to be equally susceptible to lysis by cytotoxic T cells directed at major histocompatibility determinants. Infected cells did not appear to be less stable than uninfected cells and appeared to be lysed by NK effector cells as a result of recognition of viral antigens or virus-induced changes to target cells.

Role of HSV replication and HSV gene products in induction of lysis. Replication of HSV-1 in fibroblasts is required for the target cells to be preferentially lysed (Fitzgerald-Bocarsly and Lopez, in preparation). Therefore, UV inactivation of HSV-1 before infection resulted in the nonpreferential lysis of targets, although virus antigens had been adsorbed to the surfaces of those cells.

In HSV-1-infected targets, virus gene expression proceeds as a coordinately regulated and sequentially ordered cascade: immediate early (alpha) gene products are expressed first, followed by early (beta) gene products and then by late (gamma) gene products (25). To determine whether the gamma gene products, which make up most of the structural proteins of the virus, are required for the induction of NK cells, we carried out cytotoxicity assays in the presence of phosphonoacetic acid, an inhibitor of HSV-1 DNA polymerase and gamma gene product expression. Since treated and untreated HSV-Fs targets were killed equally well, it appears that the gamma gene products of HSV-1, and, thus, the viral glycoproteins, are not required for the induction of NK cells (Fitzgerald-Bocarsly and Lopez, in preparation).

We also conducted experiments to determine whether beta gene products were required for induction of NK cells. A temperature-sensitive mutant of HSV-1, tsLB2, fails to make both beta and gamma gene products at nonpermissive temperatures (25). Target cells infected with tsLB2 were lysed even more efficiently at nonpermissive temperatures than were the targets infected with wild-type HSV-1 (Fitzgerald-Bocarsly and Lopez, in preparation). These results suggest that neigher a beta nor a gamma gene product induces lysis by NK cells and that an alpha gene product is probably responsible. Similar results were obtained by Borysiewicz et al. (10) with CMV-infected targets.

Target binding and triggering of NK cell-mediated lysis. Using the single-cell assay, we were able to show that human effectors enriched for NK cells bound HSV-Fs and Fs targets equally well (Fitzgerald-Bocarsly and Lopez, in preparation). Although conjugate formation was similar, only the HSV-Fs–effector cell conjugates were efficiently lysed. Thus, both infected and uninfected fibroblasts expressed structures recognized and used for conjugate formation by NK effector cells, but only the virus-infected targets were capable of triggering NK cell lytic function.

Role of NK Cells in Resistance to Virus Infections in Humans

Two approaches have been taken to try to document a role for NK effector cells in resistance to virus infections in humans. First, in vitro assays have been devised to evaluate the ability of NK effector cells to inhibit the replication of the virus in question. Second, studies of patients unusually susceptible to severe disease caused by virus infections have been evaluated for NK cell function to determine whether a deficiency might correlate with susceptibility to infection.

Role of NK Cells in Resistance to HSV Infections in Humans

Fitzgerald et al. (19) first showed that NK effector cells were capable of reducing the amount of HSV-1 produced by infected cells in a dose-dependent manner. The reduction of virus yield was mediated by NK effector cells as determined by cell surface marker studies. Although IFN-α was generated during these assays, it was shown not to be responsible for reducing HSV-1 replication. In other studies, Leibson et al. (32) showed that CD16-positive NK effector cells suppressed HSV replication in vitro. This was a long-term assay, and IFN-α and IFN-γ, produced during culture, were required for maximum effect.

Two groups of patients known to be unusually susceptible to HSV infections were shown to have NK(HSV-Fs) activity significantly below normal (38). Peripheral blood mononuclear cells from premature infants have been found to have responses greater than 3 standard deviations below the normal mean. Most cord blood mononuclear cells have also been found to have NK cell function significantly below the normal mean. Most cord blood mononuclear cells have also been found to have NK cell function significantly below normal. These observations correlate with the increased susceptibility of newborns to HSV infection. We have also found similarly low NK cell activity in patients with Wiskott-Aldrich syndrome, a primary immunodeficiency in which HSV infections are often fatal (38, 42). Thus, the unusually severe manifestations of HSV infections found in these two groups of individuals were clearly associated with a marked deficiency in NK(HSV-Fs) activity.

Since newborns and patients with Wiskott-Aldrich syndrome are known to have other deficiencies that might account for their susceptibility to HSV infections, we studied a group of individuals suffering from unusually severe HSV infections but with no known underlying primary or secondary cellular immunodeficiency. Members of this group were also found to have NK(HSV-Fs) responses greater than 2 standard deviations below the normal mean, suggesting that the low NK cell activity might be responsible for the unusually severe nature of their infections (38).

Role of NK Cells in Resistance to CMV Infections in Humans

As with HSV infections, CMV infections are most severe in patients with diminished NK cell function, i.e., congenital infections in newborns or in immunosuppressed adults. For example, Quinnan et al. (45) showed that bone marrow transplant recipients suffering from severe CMV infections had abnormally low levels of NK cell activity against CMV-infected fibroblasts. As with the studies correlating deficient NK(HSV-Fs) activity with severe disease caused by HSV infections, however, newborns and bone marrow transplant recipients demonstrate other immunologic deficiencies that might contribute to their susceptibility to CMV infections.

Role of NK Cells in Resistance to EBV Infections in Humans

NK cells have been shown to inhibit the outgrowth of autologous Epstein-Barr virus (EBV)-infected B lymphocytes (26). Effectors enriched for NK cells by Percoll density gradient fractionation were capable of inhibiting the outgrowth of EBV-transformed B lymphocytes. The responsible effector cells had the cell surface phenotype of NK cells and were active during the first 8 to 10 days of culture. In parallel studies, effector cells from both seronegative and seropositive donors were shown to mediate high levels of lysis against lymphoblastoid cells stimulated to productive EBV infection (9). Thus, cytotoxic effectors may limit the outgrowth of EBV-transformed cells by lysing the target cells vegetatively infected with EBV.

Sullivan et al. (54) found low NK cell activity in patients with X-linked lymphoproliferative syndrome, which is associated with unusually severe disease following EBV infection. More recently, however, Sullivan et al. suggested that the deficient NK cell activity in these patients was a response to infection rather than a predisposing factor for it (55). Virelizier et al. (57) also found low NK cell activity in a group of patients unusually susceptible to severe disease caused by EBV. Thus, it is possible that deficiencies of NK cell function are associated with unusual susceptibility to EBV infection.

CONCLUSIONS

NK cells constitute an important segment of the host natural resistance mechanism. These nonspecific systems clearly play a decisive role early during many virus infections. NK cells are immediately available to respond to an invading viral pathogen and can inhibit local replication of the virus and perhaps systemic spread of the infection. Failure of NK cells to suppress virus infection at the site of entry would theoretically result in widely disseminated infection long before an adaptive immune response could be induced and be capable of clearing that infection. For certain virus infections (it is interesting that herpesviruses have been studied most extensively), a rapidly growing literature suggests that NK cells are probably a necessary component of host defense. These cells probably constitute one of the first barriers of host defense against those virus infections.

Study of the basic biology of NK effector cells has led to a better understanding of their heterogeneity, the cellular interactions that result in lysis of target cells, and the target cell structures that bind the effectors, as well as the determinants that trigger NK cell-mediated lysis of targets. A more precise understanding of the basic biology of these effector cells and the interactions required for the normal response is a necessary first step toward developing new modalities of treatment that either augment these defense mechanisms or replace components required in the response.

Many of the studies that have documented a role for NK effector cells in resistance to virus infections have been carried out with animal models. Although studies with animal models have been very instructive about the role that NK cells might play in resistance to virus infections, observations made with such models must be confirmed by parallel studies with humans.

Literature Cited

1. **Allison, A. C.** 1974. Interaction of antibodies, complement components and various cell types in immunity against the virus and pyogenic bacteria. *Transplant. Rev.* **19:**3–55.
2. **Armerding, D., and H. Rossiter.** 1981. Induction of natural killer cells by herpes simplex virus type II in resistant and sensitive inbred mouse strains. *Immunobiology* **158:**369–379.
3. **Bancroft, G. J., G. R. Shellam, and J. E. Chalmers.** 1981. Genetic influences on the augmentation of natural killer (NK) cells during murine cytomegalovirus infection. Correlation with patterns of resistance. *J. Immunol.* **126:**988–994.
4. **Bandyopadhyay, S., B. Perussia, G. Trinchieri, D. S. Miller, and S. T. Starr.** 1986. Requirement for HLA-DR^{+} accessory cells in natural killing of cytomegalovirus-infected fibroblasts. *J. Exp. Med.* **164:**180–195.
5. **Bang, F. B.** 1978. Genetics of resistance of animals to viruses: introduction and studies in mice. *Adv. Virus Res.* **23:**269–347.
6. **Bennett, M.** 1973. Prevention of marrow allograft rejection with radioactive strontium: evidence for marrow dependent effector cells. *J. Immunol.* **110:**510–516.
7. **Bennett, M., E. E. Baker, J. W. Eascott, V. Kumar, and D. Yonkosky.** 1976. Selective elimination of marrow precursors with the bone-seeking isotope ^{89}Sr: implications for hemopoesis, lymphopoesis, viral leukemogenesis and infection. *RES J. Reticuloendothel. Soc.* **20:**71–87.
8. **Blanden, R. V., and J. Gardner.** 1976. The cell-mediated immune response to ectromelia virus infection. I. Kinetics and characteristics of the primary effector T cell response *in vivo*. *Cell. Immunol.* **22:**271–282.
9. **Blazar, B., M. Patarroyo, E. Klein, and G. Klein.** 1980. Increased sensitivity of human lymphoid lines to natural killer cells after induction of the Epstein-Barr viral cycle by super infection or sodium butyrate. *J. Exp. Med.* **151:**614–627.
10. **Borysiewicz, L., V. Rodger, S. Morris, S. Graham, and J. Sissons.** 1985. Lyses of human cytomegalovirus-infected fibroblasts by natural killer cells: demonstration of an interferon-independent component requiring expression of early viral proteins and characterization of effector cells. *J. Immunol.* **134:**2695–2701.
11. **Bukowski, J. F., J. F. Warner, G. Dennert, and R. M. Welsh.** 1985. Adoptive transfer studies demonstrating the antiviral effects of NK cells *in vivo*. *J. Exp. Med.* **161:**40–52.
12. **Bukowski, J. F., B. A. Woda, S. Habu, K. Okumura, and R. M. Welsh.** 1983. Natural killer cell depletion enhances virus synthesis and virus-induced hepatitis *in vivo*. *J. Immunol.* **131:**1531–1538.
13. **Bukowski, J. F., B. A. Woda, and R. M. Welsh.** 1984. Pathogenesis of murine cytomegalovirus infection in natural killer cell-depleted mice. *J. Virol.* **52:**119–128.
14. **Caspary, L., B. Schindling, S. Dundarov, and D. Falke.** 1980. Infections of susceptible and resistant mouse strains with herpes simplex virus type I and II. *Arch. Virol.* **65:**219–227.
15. **Chalmer, J. E., J. S. MacKenzie, and N. F. Stanley.** 1977. Resistance to murine cytomegalovirus linked to the major histocompatibility complex of the mouse. *J. Gen. Virol.* **37:**107–112.

16. **Ching, C., and C. Lopez.** 1979. Natural killing of herpes simplex virus type 1-infected target cells: normal human responses and influence of antiviral antibody. *Infect. Immun.* **26:**49–55.
17. **Diamond, R. D., R. Keller, G. Lee, and D. Finkel.** 1977. Lysis of cytomegalovirus-infected human fibroblasts and transformed human cells by peripheral blood lymphoid cells from normal human donors. *Proc. Soc. Exp. Biol. Med.* **154:**259–263.
18. **Fitzgerald, P. A., R. Evans, D. Kirkpatrick, and C. Lopez.** 1983. Heterogenity of human NK cells: comparison of the effectors that lyse HSV-1 infected fibroblasts and K562 erythroleukemia targets. *J. Immunol.* **130:**1663–1667.
19. **Fitzgerald, P. A., N. Mendelsohn, and C. Lopez.** 1985. Human natural killer cells limit replication of herpes simplex virus type-1 *in vitro*. *J. Immunol.* **134:**2666–2672.
20. **Fitzgerald, P. A., P. von Wussow, and C. Lopez.** 1982. Role of interferon in natural kill of HSV-1 infected fibroblasts. *J. Immunol.* **129:**819–823.
21. **Gee, S. R., D. A. Clark, and W. E. Rawls.** 1979. Differences between Syrian hamster strains in natural killer cell activity induced by infection with pichinde virus. *J. Immunol.* **123:**2618–2626.
22. **Herberman, R. B., M. E. Nunn, H. T. Holden, S. Staal, and J. Y. Djeu.** 1977. Augmentation of natural cytotoxic reactivity of mouse lymphoid cells against syngeneic and allogeneic target cells. *Int. J. Cancer* **19:**555–564.
23. **Herberman, R. V., M. E. Nunn, and D. H. Lavrin.** 1975. Natural cytotoxic reactivity of mouse lymphoid cells against syngeneic and allogeneic tumors. I. Distribution of reactivity and specificity. *Int. J. Cancer* **16:**216–229.
24. **Hirsch, R. L.** 1980. Natural killer cells appear to play no role in the recovery of mice from Sindbis virus infection. *Immunology* **43:**81–89.
25. **Honess, R., and D. Watson.** 1977. Unity and diversity in the herpes viruses. *J. Gen. Virol.* **37:**15–37.
26. **Kaplan, J., and T. C. Shope.** 1985. Natural killer cells inhibit outgrowth of autologous Epstein-Barr virus-infected B lymphocytes. *Nat. Immun. Cell Growth Regul.* **4:**40–47.
27. **Kastrukoff, L. F., A. S. Lau, and M. L. Putterman.** 1986. Genetics of natural resistance to herpes simplex virus type I latent infection of the peripheral nervous system in mice. *J. Gen. Virol.* **67:**613–621.
28. **Kiessling, R., P. S. Hochman, O. Haller, G. M. Shearer, H. Wigzell, and G. Cudkowicz.** 1977. Evidence for a similar or common mechanism for natural killer activity and resistance to hemopoietic grafts. *Eur. J. Immunol.* **7:**655–663.
29. **Kiessling, R., E. Klein, and H. Wigzell.** 1975. "Natural" killer cells in the mouse. I. Cytotoxic cells with specificity for mouse Moloney leukemia cells. Specificity and distribution according to genotype. *Eur. J. Immunol.* **5:**112–117.
30. **Kirchner, H., H. M. Hirt, D. L. Rosenstreich, and S. E. Mergenhagen.** 1978. Resistance of C3H/HeJ mice to lethal challenge with herpes simplex virus. *Proc. Soc. Exp. Biol. Med.* **157:**29–32.
31. **Kirchner, H., M. Kochen, H. M. Hirt, and K. Munk.** 1978. Immunological studies of HSV infection of resistant and susceptible inbred strains of mice. *Z. Immunitaetsforsch.* **154:**147–154.
32. **Leibson, P. J., N. Hunter-Laszlo, and A. R. Heyward.** 1986. Intervention of herpes simplex virus type 1 replication in fibroblast culture by human blood mononuclear cells. *J. Virol.* **57:**976–982.
33. **Levy-Leblond, E., and J. M. Dupuy.** 1977. Neonatal susceptibility to MHV_3 infection in mice. I. Transfer of resistance. *J. Immunol.* **118:**1219–1222.
34. **Lopez, C.** 1975. Genetics of natural resistance to herpes virus infections in mice. *Nature* (London) **258:**152–153.

35. **Lopez, C.** 1978. Immunological nature of genetic resistance of mice to herpes simplex virus-type I infection, p. 775–778. *In* G. De The, W. Henle, and F. Rapp (ed.), *Oncongenesis and Herpesviruses III*. International Agency for Research on Cancer, Lyon, France.
36. **Lopez, C.** 1980. Resistance to HSV-1 in the mouse is governed by two major, independently segregating non-H-2 loci. *Immunogenetics* **11:**87–92.
37. **Lopez, C.** 1981. Resistance to herpes simplex virus-type 1 (HSV-1), p. 15–24. *In* O. Haller (ed.), *Natural Resistance to Tumors and Viruses*. Springer-Verlag, New York.
38. **Lopez, C., D. Kirkpatrick, S. Read, P. A. Fitzgerald, J. Pitt, S. Pahwa, C. Y. Ching, and E. M. Smithwick.** 1983. Correlation between low natural kill of HSV-1-infected fibroblasts, NK(HSV-1), and susceptibility to herpes virus infections. *J. Infect. Dis.* **147:**1030–1035.
39. **Lopez, C., R. Ryshke, and M. Bennett.** 1980. Marrow-dependent cells depleted by ^{89}Sr mediate genetic resistance to herpes simplex virus type 1 infections in mice. *Infect. Immun.* **28:**1028–1032.
40. **Mann, D. W., G. Sonnenfeld, and J. Stein-Streilein.** 1985. Pulmonary compartmentalization of interferon and natural killer cell activity. *Proc. Soc. Exp. Biol. Med.* **180:**214–230.
41. **Masuda, A., and M. Bennett.** 1981. Murine cytomegalovirus stimulates natural killer cell function but kills genetically resistant mice treated with radioactive strontium. *Infect. Immun.* **34:**970–974.
42. **Messina, C., D. Kirkpatrick, P. Fitzgerald, R. O'Reilly, F. Siegal, C. Cunningham-Rundles, N. Blaese, J. Oleske, S. Pahwa, and C. Lopez.** 1986. Natural killer cell function and interferon generation in patients with primary immunodeficiencies. *Clin. Immunol. Immunopathol.* **39:**394–404.
43. **Pfizenmaier, K., H. Trostmann, M. Rollinghoff, and H. Wagner.** 1975. Temporary presence of self-reactive cytotoxic T-lymphocytes during murine lymphocytic choriomeningitis. *Nature* (London) **258:**238–240.
44. **Piontek, G. E., R. Weltzin, and W. A. F. Tompkins.** 1980. Enhanced cytotoxicity of mouse natural killer cells for vaccinia and herpes virus-infected targets. *RES J. Reticuloendothel. Soc.* **27:**175–188.
45. **Quinnan, G. V., N. Kirmani, A. H. Brook, H. J. Manishewitz, L. Jackson, G. Moreschi, G. W. Santos, R. Saral, and W. H. Burns.** 1982. HLA-restricted T-lymphocyte and non T-lymphocyte cytotoxic responses correlate with recovery from cytomegalovirus infection in bone-marrow transplantation. *N. Engl. J. Med.* **307:**7–13.
46. **Rager-Zisman, B., P.-C. Quan, M. Rosner, J. R. Moller, and B. R. Bloom.** 1987. Role of NK cells in protection of mice against herpes simplex virus-I infection. *J. Immunol.* **138:**884–888.
47. **Rodda, S. J., and D. O. White.** 1976. Cytotoxic macrophages: a rapid, non-specific response to viral infection. *J. Immunol.* **117:**2067–2072.
48. **Rouse, B. T.** 1984. Cell-mediated immune mechanisms, p. 107–120. *In* B. T. Rouse and C. Lopez (ed.), *Immunobiology of Herpes Simplex Virus Infection*. CRC Press, Inc., Boca Raton, Fla.
49. **Santoli, D., G. Trinchieri, and F. S. Lief.** 1978. Cell-mediated cytotoxicity against virus infected target cells in humans. I. Characterization of the effector lymphocyte. *J. Immunol.* **121:**526–531.
50. **Shellam, G. R., J. E. Allan, J. M. Papadimitriou, and G. J. Bancroft.** 1981. Increase susceptibility to cytomegalovirus infection in beige mutant mice. *Proc. Natl. Acad. Sci. USA* **78:**5104–5108.

51. **Shellam, G. R., and J. P. Flexman.** 1986. Genetically determined resistance to murine cytomegalovirus and herpes simplex virus in newborn mice. *J. Virol.* **58:**152–156.
52. **Starr, S. E., and A. C. Allison.** 1977. Role of T-lymphocytes in recovery from murine cytomegalovirus infection. *Infect. Immun.* **17:**458–462.
53. **Stein-Streilein, J., and J. Guffee.** 1986. *In vivo* treatment of mice and hamsters with antibodies to asialo GM_1 increases morbidity and mortality to pulmonary influenza infection. *J. Immunol.* **136:**1435–1441.
54. **Sullivan, J. L., B. Byron, F. E. Brewster, and D. T. Purtillo.** 1980. Deficient natural killer cell activity in X-linked lymphoproliferative syndrome. *Science* **210:**543–545.
55. **Sullivan, J. L., K. S. Byron, F. E. Brewster, S. M. Baker, and H. A. Ochs.** 1983. X-linked lymphoproliferative syndrome. *J. Clin. Invest.* **71:**1765–1778.
56. **Trinchieri, G., D. Santoli, and H. Koprowski.** 1978. Spontaneous cell-mediated cytotoxicity in humans: role of interferon and immunoglobulins. *J. Immunol.* **120:**1849–1855.
57. **Virelizier, J. L., M. Lipinski, T. Tursz, and C. Griscelli.** 1979. Defects of immune interferon secretion and natural killer activity in patients with immunological disorders. *Lancet* **ii:**696–697.
58. **Welsh, R. M.** 1986. Regulation of virus infections by natural killer cells. A review. *Nat. Immun. Cell Growth Regul.* **5:**169–199.
59. **Welsh, R. M., and R. W. Kiessling.** 1980. Natural killer cell response to lymphocytic choriomeningitis virus in beige mice. *Scand. J. Immunol.* **11:**363–367.
60. **Zawatzky, R., J. Hilfenhaus, and H. Kirchner.** 1979. Resistance of nude mice to herpes simplex virus and correlation with production of interferon. *Cell. Immunol.* **47:**424–428.
61. **Zawatzky, R., J. Hilfenhaus, F. Marucci, and H. Kirchner.** 1981. Experimental infection of inbred mice with herpes simplex virus type I. 1. Investigation of humoral and cellular immunology and of interferon induction. *J. Gen. Virol.* **43:**31–38.

Chapter 4

Immunopathology by Antiviral T-Cell Immunity

R. M. Zinkernagel

INTRODUCTION

The function of the immune system is to maintain homeostasis in higher vertebrates; its two main arms, cellular and humoral immunity, very efficiently defend the host against cytopathic agents of acute infections. The main targets of T cells are intracellular agents, whereas those of antibodies are primarily extracellular ones (10, 18). However, this immune defense is not equally efficient against all viruses and bacteria, and many other pathogens, such as protozoa, metazoa, or tumor cells, may escape, since they are more or less out of the reach of efficient immune surveillance.

T-cell immunity is at its best when dealing with intracellular agents of acute infection, such as cytopathic viruses. This T-cell activity is easily measurable in a classic ^{51}Cr-release cytotoxicity assay in vitro (2, 16). T-cell-mediated lysis of virus-infected target cells is virus specific, since only target cells infected with the proper virus are lysed. However, in addition, lysis of virus-infected cells depends on T-cell and target cell sharing classic transplantation antigens. Many experiments over the past 12 years have clearly documented that the following general rules govern lymphocyte-lymphocyte and lymphocyte-somatic cell interactions (11, 16). The first rule is that T cells recognize self-transplantation antigens together with foreign antigenic determinants exclusively on cell surfaces. The second rule is that T-cell specificity for self-transplantation antigens (i) is specific for polymorphic determinants; (ii) is selected for during differentiation in the thymus; (iii) determines the effector function of T

R. M. Zinkernagel • Institut für Pathologie, Universitätsspital Zürich, CH-8091 Zurich, Switzerland.

cells (cytotoxic T cells recognize class I major histocompatibility gene products, i.e., the classic transplantation antigens, whereas differentiation-promoting T cells [helper or delayed-type hypersensitivity T cells] are specific for those of class II); and (iv) regulates T-cell responsiveness (i.e., the quality and quantity of cytotoxic T-cell response are regulated by class I major histocompatibility gene products, whereas the quality and quantity of differentiation-promoting T cells are regulated by those of class II).

ROLE OF CYTOTOXIC T CELLS

One may question the physiological role of cytolytic T cells: why should T cells mediate cell and tissue destruction to combat intracellular infectious agents (2, 10)? There is good evidence that cytotoxic T cells destroy virus-infected cells before viral progeny are assembled, thus eliminating virus during the eclipse phase of virus replication. For cytopathic viruses, virus elimination via immunological host cell destruction is an efficient way to prevent virus spread and the resulting more extensive virus-mediated cell and tissue damage (16). This immunological defense mechanism becomes less attractive for combating noncytopathic viruses, because host cells are not destroyed by virus but only by the T-cell immune response. Because T cells apparently cannot distinguish cytopathic from noncytopathic viruses, immune-system-mediated cell and tissue damage results, in the latter infections, in immunopathology.

LCM: A T-CELL-MEDIATED IMMUNOPATHOLOGY

Examples of infections with noncytopathic viruses are lymphocytic choriomeningitis (LCM) in mice (5, 8, 9) and hepatitis B in humans (1, 6). LCM in mice develops after intracerebral injection of LCM virus (LCMV) only in immunocompetent mice. Mice lacking T cells or mice immunosuppressed by irradiation or cytostatic drugs do not develop inflammatory reactions and hence do not develop LCM; however, they fail to eliminate virus and as a result become LCMV carriers (8, 11, 16). LCM has been carefully analyzed and has been clearly shown to be T-cell mediated (3, 5). Lethal LCM apparently depends upon preferential recruiting of effector T cells to the acutely infected leptomeninges. This notion is supported by the finding that high doses of LCMV simultaneously injected intracerebrally and intravenously often do not cause fatal LCM, because effector T cells are recruited to infected cells all over the organism and therefore are diluted out.

In both LCM and hepatitis B, the severity of disease depends upon the balance between virus spread and immune response. Efficient T-cell-mediated immune response leads to rapid elimination of the virus, limited cell and tissue damage, and therefore limited disease. Absence of an immune response results in unchecked growth of virus and in a virus carrier state. Slow and insufficient immune responsiveness allows extensive spread of virus accompanied by chronic T-cell-mediated tissue destruction, a classic immunopathological conflict situation.

Since major transplantation antigens are recognized as self by T cells, define their function, and regulate their responsiveness, they may drastically influence the balance between virus spread and immune-system-mediated tissue damage. Among many other variables characteristic of the virus or the host, severity of disease has also been shown to be determined by major transplantation antigens in both hepatitis B virus infections in humans (4, 6) and LCMV infections in mice (19).

FACTORS INFLUENCING SUSCEPTIBILITY TO LCM

Various LCMV isolates indistinguishable by conventional serology have been tested and were found to vary greatly with respect to the disease which they induced (12). The reasons for these differences are unclear, but it seems that, included among possibilities such as susceptibility to interferons, macrophages, and natural killer cells, etc., their antigenic quality also appears to vary. At least two types of host factors regulate the susceptibility of mice to LCM. First, there is a very important general genetic influence that is independent of the major histocompatibility complex (*H-2*) (17, 19). Depending upon the genetic background, the two LCMV strains used induce four possible disease susceptibility patterns, i.e., susceptibility to both, to neither, or to one or the other of the LCMV strains. Second, in one inbred mouse strain tested in detail (B10), the major transplantation antigens encoded by the *H-2D* region determine whether B10 mice are susceptible to one of the LCMV isolates (19). Since the *H-2D* class I transplantation antigens mediate cytotoxic T-cell function and regulate antiviral cytotoxic T cells, it obviously became important to evaluate whether these *H-2D*-regulated differences in susceptibility to LCM correlated with T-cell-mediated immunopathology caused by LCMV-specific cytotoxic T cells. The result showed that high and early cytotoxic LCMV-specific and *H-2D*-restricted T-cell activity correlated positively with susceptibility to LCM.

HEPATITIS, A T-CELL-MEDIATED DISEASE CAUSED BY LCMV

The model infection with LCMV in mice documents how delicate the balance between infectious agents and the host immune response may be;

depending on the balance between the two, the aspect of protection of the host by immunity or damage to the host by immunopathology will dominate. A model for immunologically T-cell-mediated hepatitis was established by using mice infected with LCMV (17). The severity of hepatitis was monitored histologically and by determining changes in levels of aminotransferases and alkaline phosphatase in serum. The kinetics of histological disease manifestations, increases of liver enzyme levels in the serum, and cytotoxic T-cell activities in livers and spleens were all related and were dependent upon the following parameters: (i) the LCMV isolate, (ii) the virus dose, (iii) the route of infection, (iv) the general genetic background of the murine host (of the strains tested, Swiss mice and strain A mice were more susceptible than C57BL or CBA mice, whereas BALB/c and DBA/2 mice were the least susceptible), (v) the degree of immunocompetence of the murine host (T-cell-deficient *nu/nu* mice never developed hepatitis, whereas *nu*/+ or +/+ mice always did), and (vi) the local cytotoxic T-cell activity (mononuclear cells isolated from livers during the period of overt hepatitis were two to five times more active than equal numbers of spleen cells).

Thus, LCMV-induced hepatitis in mice is an immunopathologically mediated disease caused by T-cell-mediated destruction of infected liver cells. Overall, this disease parallels many aspects of acute viral hepatitis in humans, which is caused by hepatitis B virus.

IMMUNE SUPPRESSION BY ANTI-LCMV T CELLS

Mice infected with LCMV and then superinfected with vesicular stomatitis virus after a few days exhibited a severe impairment of their capacity to produce immunoglobulin or immunoglobulin G antibodies to vesicular stomatitis virus (unpublished observations). This effect of LCMV depended upon the LCMV isolate, the relative times of LCMV infection and vesicular stomatitis virus superinfection, the dose of LCMV, the strains of mice used, and the T-cell immunocompetence of the mice. Similarly, LCMV-infected mice were found to be much more susceptible to tumor growth (fibrosarcoma and melanoma [M. Kohler, unpublished observations]) than uninfected control mice were. Although the results have not yet been analyzed in detail, they suggest that LCMV causes a severe immune suppression in these mice. These findings are particularly relevant to our understanding of acquired immunodeficiency, which is caused mainly by the human immunodeficiency viruses (7, 13–15). Recent evidence suggests that as in LCMV infections, cytotoxic T cells may actually be instrumental in controlling human immunodeficiency virus replication, but may also be responsible for destroying lymphocytes and macrophages and thus causing immune suppression (15). The example of human immunodeficiency virus infections in humans may therefore in a way represent the ultimate perversity in the balance between host immune system and infectious agent: noncytopathic viruses infect lymphocytes and macrophages, the essential partners of an immune response, which are then destroyed by the immune response; i.e., the virus infection forces the immune system to destroy itself.

Literature Cited

1. **Bianchi, L.** 1981. The immunopathology of acute type B hepatitis. *Springer Semin. Immunopathol.* **3:**421–438.
2. **Cerottini, J. C., and K. T. Brunner.** 1974. Cell-mediated cytotoxicity, allograft rejection and tumor immunity. *Adv. Immunol.* **18:**67–132.
3. **Cole, G. A., N. Nathanson, and R. A. Prendergast.** 1972. Requirement for thetabearing cells in lymphocytic choriomeningitis virus-induced central nervous system disease. *Nature* (London) **238:**335–337.
4. **Dausset, J., and A. Svejgaard.** 1977. Introduction, p. 9–11. *In* J. Dausset and A. Svejgaard (ed.), *HLA and Disease*. Munksgaard, Copenhagen.
5. **Doherty, P. C., and R. M. Zinkernagel.** 1974. T cell-mediated immunopathology in viral infection. *Transplant. Rev.* **19:**89–120.
6. **Eddleston, A. L. W. F., and R. Williams.** 1978. HLA and liver disease. *Br. Med. Bull.* **34:**295–300.
7. **Folks, T., D. M. Powell, M. M. Lightfoote, S. Benn, M. A. Martin, and A. S. Fauci.** 1986. Induction of HTLV-III/LAV from a nonvirus-producing T-cell line: implications for latency. *Science* **231:**600–602.
8. **Hotchin, J.** 1971. Persistent and slow virus infections. *Monogr. Virol.* **3:**1–211.
9. **Lehmann-Grube, F.** 1971. Lymphocytic choriomeningitis virus. *Virol. Monogr.* **10:**1–173.
10. **Mims, C. A.** 1982. *Pathogenesis of Infectious Disease*, 2nd ed. Academic Press, Inc. (London), Ltd., London.
11. **Möller, G.** 1984. T cell receptor and genes. *Immunol. Rev.* **81:**1–167.
12. **Oldstone, M. B. A.** 1986. Distortion of cell functions by noncytotoxic viruses. *Hosp. Pract.* **21:**82–92.
13. **Ruscetti, F. W., J. A. Mikovits, V. S. Kalyanaraman, R. Overton, H. Stevenson, K. Stromberg, R. B. Herberman, W. L. Farrar, and J. R. Ortaldo.** 1986. Analysis of effector mechanisms against HTLV-I- and HTLV-III/LAV-infected lymphoid cells. *J. Immunol.* **136:**3619–3624.
14. **Swain, S. L., and P. R. Panfili.** 1979. Helper cells activated by allogeneic H-2K or H-2D differences have a Ly phenotype distinct from those responsive to I differences. *J. Immunol.* **122:**383–391.
15. **Walker, C. M., D. J. Moody, D. P. Stites, and J. A. Levy.** 1986. $CD8^+$ lymphocytes can control HIV infection in vitro by suppressing virus replication. *Science* **234:**1563–1566.
16. **Zinkernagel, R. M., and P. C. Doherty.** 1979. MHC-restricted cytotoxic T cells: studies on the biological role of polymorphic major transplantation antigens determining T cell restriction-specificity, function and responsiveness. *Adv. Immunol.* **27:**52–142.
17. **Zinkernagel, R. M., E. Haenseler, T. P. Leist, A. Cerny, H. Hengartner, and A. Althage.** 1986. T cell mediated hepatitis in mice infected with lymphocytic choriomeningitis virus. *J. Exp. Med.* **164:**1075–1092.

18. **Zinkernagel, R. M., H. Hengartner, and L. Stitz.** 1985. On the role of viruses in the evolution of immune responses. *Br. Med. Bull.* **41**:92–97.
19. **Zinkernagel, R. M., C. J. Pfau, H. Hengartner, and A. Althage.** 1985. Susceptibility to murine lymphocytic choriomeningitis maps to class I MHC genes—a model for MHC/disease associations. *Nature* (London) **316**:814–817.

Chapter 5

Epidemiologic Approaches to the Study of Chronic Diseases

Leon Gordis

INTRODUCTION

Epidemiology plays a major role in the study of chronic diseases. It is used to investigate the causes of, and the risk factors for, chronic diseases and to quantify the relative contributions of environmental and genetic factors to the risk of human disease. It is also used to study the natural history of chronic disease and to evaluate new forms of therapy and prevention. In addition, it is used to determine the community burden of these diseases and to develop projections of future incidence and prevalence.

This chapter will focus on several issues: first, the interpretation of epidemiologic information in regard to causal inferences; second, some of the specific methodologic approaches used by the epidemiologist; and third, a comparison of bench laboratory work and epidemiologic investigations.

The use of epidemiology to determine whether there is a causal association between a given environmental exposure and a specific disease comprises two steps. The first is the demonstration of such an association. The second, if such an association is demonstrated, is the determination of whether the observed association is a causal one. Whether an observed association is causal is often one of the most critical issues in epidemiologic studies.

Leon Gordis • Department of Epidemiology, School of Hygiene and Public Health, Johns Hopkins University, Baltimore, Maryland 21205.

Table 1. Elements of Immunological Proof of Causation[a]

Element	Description
1	Antibody to the agent is regularly absent prior to both the disease and exposure to the agent (i.e., before the incubation period)
2	Antibody to the agent regularly appears during illness
3	Presence of antibody to the agent indicates immunity to the clinical disease associated with primary infection by the agent
4	Absence of antibody to the agent indicates susceptibility to both infection and the disease produced by the agent
5	Antibody to no other agent should be similarly associated with the disease unless it is a cofactor in its production

[a] Adapted from reference 3.

EPIDEMIOLOGY AND DISEASE CAUSATION

For many years, Koch's postulates (or the Koch-Henle postulates) were the hallmark of causal thinking in relation to human disease (see reference 3 and citations therein). They originated in work in infectious diseases and can be briefly summarized as follows: (i) the organism is always found with the disease; (ii) the organism is not found with any other disease; and (iii) the organism, when isolated from an individual with the disease and cultured through several generations, produces the disease in experimental animals.

A number of problems are associated with these postulates, and some were even recognized by Koch at the time he proposed them. He therefore added a postscript to the effect that even when an infectious disease cannot be transmitted to animals, the ''regular'' and ''exclusive'' presence of the organism [postulates (i) and (ii)] proves a causal relationship.

Useful as these postulates have been, they are not completely satisfactory, largely because they were developed only with infectious diseases in mind, and also because they obviously cannot take into account technological advances which took place after they were formulated. In 1976, Evans proposed a series of elements of immunologic proof of causation (Table 1) advocating in effect that as new technology becomes available, it be built into the framework of causal thinking (3).

However, the difficulties with Koch's postulates go beyond subsequent technological advances. First, we know that a given agent can cause different diseases and that a disease can often be produced by different agents. Furthermore, in many cases the agent can no longer be isolated at the time the diagnosis is made. Thus, there are serious limitations with the postulates, even when they are applied to infectious diseases, for which they were originally developed.

When we turn to apparently noninfectious diseases, the problems are even more serious. In the 1950s, when the issue of smoking and lung cancer arose, the Surgeon General appointed an expert committee to assess the evidence. The committee needed criteria for determining whether the relationship of lung cancer to smoking was a causal one. Since it was quite clear that they were not dealing with an agent that could be grown in vitro, a set of criteria different from Koch's postulates was needed. The criteria which they proposed can be summarized as follows (6).

First, an inference of causation will rest on the strength of the association, also called the relative risk. Second, there should be a dose-response relation: if factor A causes disease B, we would expect that the greater the dose of factor A, the greater the outcome in the form of disease B. Third, there should be a temporal relationship: if A causes B, A should occur before B. Fourth, the association should be specific. Fifth, there should be a consistency of association: the association should be consistent from study to study and in different populations. Finally, the relationship should be biologically plausible.

Two of these criteria merit comment. The criterion of specificity has been raised as a problem in discussions of the causation of lung cancer by cigarette smoking. Cigarette manufacturers have claimed that since cigarette smoking has been implicated in emphysema, heart disease, bladder cancer, and cancer of the pancreas, among other diseases, there is no specificity and the relationship should therefore be questioned. These objections are, however, quite weak. Since cigarettes contain many different chemical compounds, it is not surprising that they are linked to many human diseases. Moreover, since so many structural and biochemical characteristics are common to all cells in the human body, it is reasonable to assume that if, for example, a given compound affects DNA in cells of one tissue, it might have other effects on cells of other tissues.

The second criterion meriting comment is that of biologic plausibility. It is certainly reassuring when in epidemiologic studies we observe associations that are biologically plausible, i.e., coherent with existing biological knowledge. However, we should also recognize that at times, epidemiologic evidence can precede biologic plausibility. For example, Jenner's first vaccination against smallpox was based on epidemiologic evidence: observations that milkmaids who developed cowpox did not contract smallpox when later exposed. Jenner had no knowledge of virology. The first descriptions of the congenital rubella syndrome appeared before anything was known about viral teratogenesis. Thus, in certain cases, observational data can precede biologic knowledge, although we would hope that the two would be congruent as often as possible.

Some of the issues that arise in the study of disease causation in human populations are as follows: selection of the study population; selection of a comparison group(s); definition and documentation of exposure and dose; latency of many diseases or effects; definition of and ascertaining outcome; and size of the sample being studied. They will be addressed briefly below.

EPIDEMIOLOGIC APPROACHES TO DISEASE CAUSATION

One approach used for linking a putative causative agent with a disease is epidemiologic surveillance, i.e., the ongoing gathering of data about disease occurrence in a community. Surveillance has been commonly used for infectious diseases, but in recent years it has also been used for chronic diseases. Examples include cancer surveillance by cancer registries and congenital malformation surveillance through congenital malformation registries. By characterizing the individuals who have been identified by the surveillance process as having a condition, it may be possible to link certain environmental factors with an increased risk of the disease in question.

Another approach is that of seroepidemiology, which is used to identify people who have been infected, even in the absence of clinical disease. Many years ago it was recognized that when studying the dynamics of poliomyelitis distribution in a community, the observations should not be limited to children who are clinically ill with poliomyelitis, because these children represent only the tip of the iceberg. Consequently, to understand the dynamics of how a disease is spread in a community, we must have means above and beyond clinical observation for identifying people who have been infected.

One of the first steps taken by epidemiologists is often a descriptive study of the disease in a population. Figure 1 (1) shows a map of the distribution of cases of Burkitt's lymphoma which were identified by Burkitt in Africa. The geographic distribution which he observed led to the hypothesis linking Epstein-Barr virus and malaria with the etiology of Burkitt's lymphoma. Thus, the geographic distribution of a disease can be very useful in suggesting appropriate biologic hypotheses for further testing.

Other types of descriptive evidence can also be very valuable. Clusters of cases may provide clues to risk and etiology. Examples

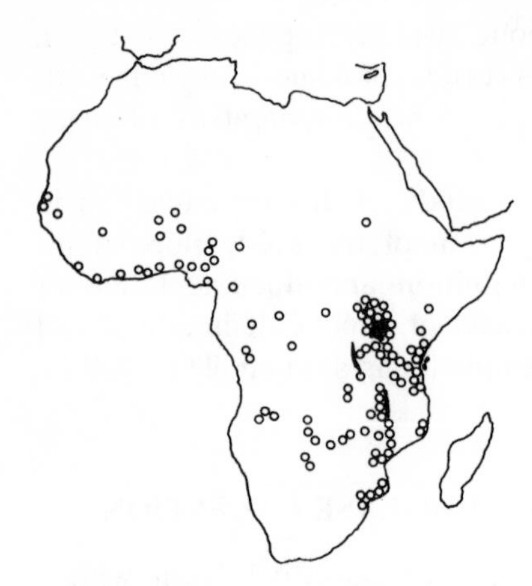

Figure 1. Map of Africa showing the distribution of cases of Burkitt's lymphoma identified by Burkitt (1).

include clusters of cases of infectious diseases relating to care in intensive care units, day care centers, and other settings. Other clusters of interest may include transplant recipients or other patients receiving immunosuppressive therapy who may be particularly susceptible to infectious disease.

How do we proceed to link a chronic disease to an infectious agent? The first step is often to try and isolate an infectious agent from individuals with the disease, although often the agent is no longer recoverable because of a long latent period between the time of the infection and the time when clinical disease is diagnosed. Sometimes, however, relevant serologic evidence can be obtained even when the agent itself cannot be isolated. Thus, for example, streptococci are infrequently isolated from children with rheumatic fever, but serologic tests for antistreptococcal antibodies are almost invariably positive.

Another descriptive approach is to see how disease outbreaks relate in time to outbreaks of infectious agents. For example, early evidence linking rheumatic fever to streptococcal infections was provided by the observation that rheumatic fever outbreaks regularly followed outbreaks of streptococcal infections, which were often milk borne. Many other examples have also shown a parallelism between outbreaks of infectious and chronic diseases.

STUDY DESIGNS

Several study designs are used in epidemiology to link infections agents and specific chronic diseases. Figure 2 shows the schema of the

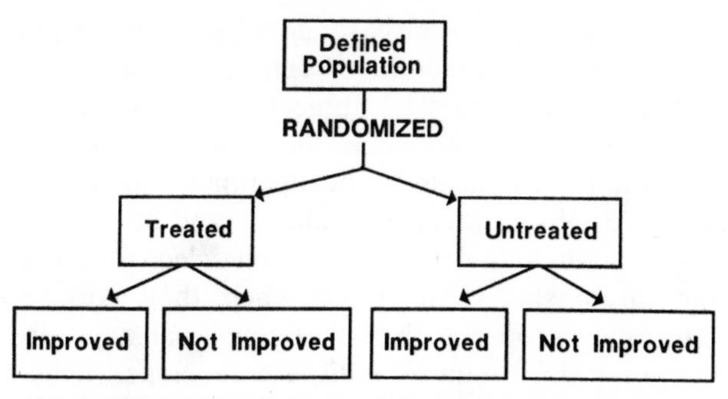

Figure 2. Diagrammatic scheme of a randomized clinical trial. In this example, the trial is being used to evaluate a therapy.

randomized clinical trial as it is used for evaluating a therapy. We identify a study population, randomly assign the subjects to treatment or no treatment (or to different treatments), and monitor them to determine whether their conditions improve. If the treatment is effective, we expect a higher rate of improvement in the treated group than in the untreated group. This approach can also be used to evaluate preventive agents such as vaccines. In theory it could also be used to evaluate disease-producing agents, such as infectious agents, and toxic environmental agents, but practical and ethical reasons preclude the use of a randomized trial for investigating hazardous agents. We therefore use alternatives to approximate this type of study design. These nonrandomized approaches are often termed observational studies.

Prospective Studies

The first type of observational study design is the near relative of the clinical trial—the prospective or cohort study (Fig. 3). In this study design we begin with a population of exposed individuals and compare them with a population of nonexposed individuals. We monitor the subjects in both

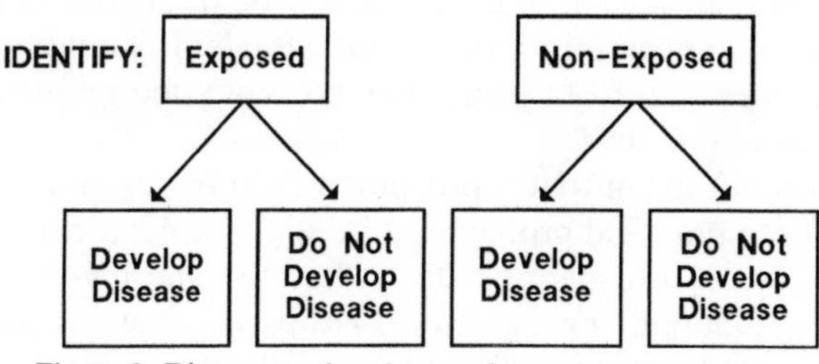

Figure 3. Diagrammatic schema of a prospective study.

groups to ascertain whether disease develops. The major difference between the experimental randomized clinical trial and the prospective study is that in the clinical trial subjects are randomly allocated to exposure or no exposure, whereas in the prospective study there is no random allocation. If we consider an occupational exposure, for example, some people have jobs in a certain industry and others do not. It is a nonrandom allocation. Either individuals self-select for a certain type of work or other circumstances determine where they work. We can then monitor both groups (those with the occupational exposure and those without it) to ascertain whether the disease develops. If the exposure is related to the disease in question, the rate of disease development will be higher in exposed than in nonexposed individuals.

The assumption is that both the exposed and nonexposed groups are similar in all respects except for the exposure and thus in effect come from a single population. In practical terms in real life, this is often not the case. When we begin with an exposed group and a nonexposed group as just described, the exposed and nonexposed groups differ in a number of characteristics other than presence or absence of exposure. Consequently, after identifying an exposed group for study, a major methodologic issue is how to select an appropriate nonexposed group for comparison. Another design option, therefore, is to begin with a defined and as yet unexposed population and monitor it until it self-selects into exposed and nonexposed groups.

A serious practical problem with this approach has to do with the time needed to complete such a study. Let us assume that we want to study the relationship of lung cancer and smoking. To do this we identify a population of children in 1987, and 10 years later they become smokers or remain nonsmokers. If we assume a hypothetical 10-year latent period for development of lung cancer, a 20-year interval is required to observe the study population for development of the disease. Clearly, this poses practical difficulties. First, the subjects in this type of study can outlive the investigators. Second, the National Institutes of Health does not look kindly on proposals requesting funding for 20 years. Finally, many faculty members have at least a passing interest in promotion, but Committees on Appointments and Promotions do not take IOUs. It would be difficult for a faculty member in 1987 to sway such a Committee by promising some very exciting data in 2007.

This does not mean that a prospective study design cannot be used because of these practical problems. The study design described above is called a concurrent prospective study, because the investigator accompanies that population through time. However, there is an alternative approach. Assume that it is 1987 and from school records we can identify

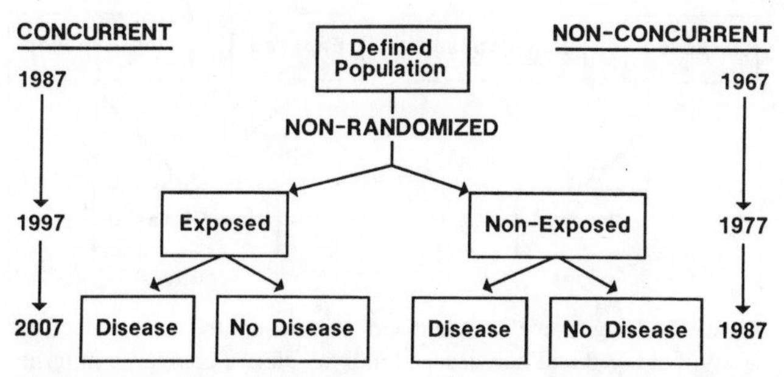

Figure 4. Comparison of concurrent and nonconcurrent prospective studies. Time frames for a hypothetical concurrent and a hypothetical nonconcurrent study conducted in 1987 are shown.

a population of children in 1967 and determine whether they were smokers or nonsmokers in 1977. Now, in 1987, we could ascertain the development of disease without that 20-year wait. This is called a nonconcurrent prospective study. The study design is identical in both a concurrent prospective study and a nonconcurrent prospective study. In both, we are comparing exposed and nonexposed individuals. The only difference is calendar time (Fig. 4).

An example of a nonconcurrent prospective study is the study of the etiology of breast cancer carried out by Cowan et al. (2). Among the known risk factors for breast cancer is a late age at first pregnancy. The question addressed was that of why a late age at first pregnancy is associated with an increased risk of breast cancer. Is it that an early first pregnancy protects the woman or is it that she has both a late age at first pregnancy and an increased risk of breast cancer because of an underlying hormonal problem that prevents an early first pregnancy and also increases the risk for development of breast cancer? Teasing apart these two explanations is very difficult.

Cowan et al. developed an ingenious approach to this problem. In 1977, they obtained records of women who had been treated at the Johns Hopkins Hospital Infertility Clinic from 1948 to 1965. The women had been characterized in great detail in regard to the cause of the infertility. Some had underlying hormonal problems, such as progesterone deficiencies; others had mechanical problems, such as tubal obstruction; and for others the problem was associated with the husband, such as a low sperm count. Cowan et al. were therefore able to determine which women had hormonal abnormalities as the cause of their infertility (the exposed group) and which women had other causes (the nonexposed group). They then monitored the women in both groups and calculated the rate of

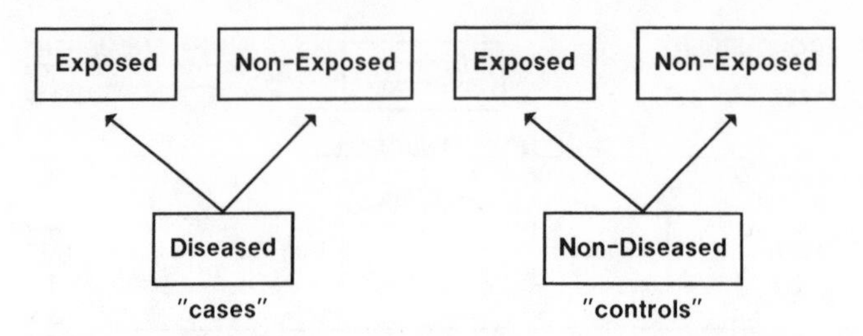

Figure 5. Diagrammatic schema of a case-control (retrospective) study. Diseased and nondiseased individuals are compared in terms of exposure to a possible causal factor.

breast cancer in each group. If the explanation is that an early first pregnancy protects, the breast cancer rate would be the same in both the hormonally deficient and the hormonally normal groups of infertile women. If, on the other hand, late age at first pregnancy and increased risk of breast cancer both result from the hormonal deficiency, the breast cancer rate should be higher in the hormonally deficient group. This nonconcurrent prospective study was conducted in 1977, and it was not necessary to wait 20 years to complete it. Thus, we see that the nonconcurrent prospective study design provides a means of doing a prospective investigation without the tremendous time lag that would be involved in a concurrent study.

Case-Control Studies

An alternative study design is the case-control study (Fig. 5), in which we compare people who have the disease (cases) with people who do not have the disease (controls). We determine from history, medical records, serologies, and other data sources whether the individuals under study had a history of exposure to a particular agent. If the exposure is related to the disease, we would expect a higher rate of exposure in the cases than in the controls. The case-control study is an appealing starting point, because very often we do not have enough information to know what specific exposure would be worth exploring in a prospective study.

In a prospective study, we compare exposed people with nonexposed people and can look at one or multiple outcomes in both groups. In a case-control study, we can explore a number of different exposures which may be related to the disease being studied.

There are other reasons for choosing a case-control design. Attractive as the prospective design is, many of the diseases we study are rare. They may be so infrequent that to obtain enough cases for a prospective

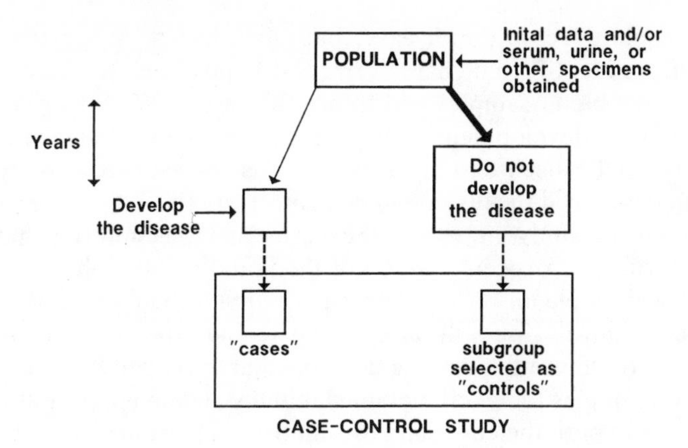

Figure 6. Diagrammatic schema of a nested case-control study.

study, we would have to monitor an enormous population. In a case-control study, we can identify a group of cases and compare them with controls without having to monitor such a large population.

The terminology describing different study designs in the literature is very confusing: for example, a concurrent prospective study is also known as a longitudinal or cohort study, a nonconcurrent prospective study is also known as a historical cohort study, an experimental study is the same as a randomized clinical trial, and a case-control study is the same as a retrospective study. The term retrospective is an unfortunate one, because it implies that the hallmark of a case-control study is to go back in time. That is not so. One can have a prospective study that goes back in time—a nonconcurrent prospective study. The criterion of a case-control study is that we compare cases and controls. In a prospective study, we compare exposed individuals and nonexposed individuals.

One problem in a case-control study is that there may be a significant latent period between the exposure and the development of disease. If we ask the patient about certain infections in the past, the information we obtain may be very limited. If we examine serologies at the time of the illness, we may not be obtaining data relevant to the previous serologic status of the subject. This is a serious problem in case-control studies, and another approach has therefore been used increasingly in recent years and has proven particularly appropriate to the study of infectious diseases. This is called the nested case-control study; we nest a case-control study inside a prospective study.

Figure 6 shows a schema of a nested case-control study. We begin with a population which is monitored over time. During that time interval, a few people develop the disease, but most do not. When the study is initiated, we identify and characterize this population. We may interview them or take blood samples and freeze the serum. We then monitor this population for development of the disease. We can consider the people who develop the disease to be cases and consider a subgroup of those who have not developed the disease to be controls. In this way, we end up with a case-control study in which the cases and the controls have been selected from a prospective study of the initial population.

What do we gain by this approach? First, when we compare these cases and controls, we will have serologic and other information which was obtained before the disease developed in these people. We can thaw the serum samples that were obtained initially and do appropriate tests on samples from both the cases and the controls. The study is much cheaper than a prospective study because we do not have to carry out the antibody tests on the total initial population. In other words, we save the serum samples, and only when the cases develop and we identify controls do we do antibody determinations on those individuals. Thus, it gives the advantages of being relatively economical and of providing premorbid data on the cases and controls which we do not have in many of the usual case-control studies.

An example of this approach is a study done some years ago by Evans and Comstock (4). Johns Hopkins University maintains a Training Center for Public Health Research in Washington County, Md., about 75 miles west of Baltimore. In 1974, a community survey was carried out in which people responding to a Johns Hopkins census were also asked to give serum; about 26,000 people gave serum, which was saved in a serum bank. This population was monitored. Subsequently, Comstock and Evans became interested in the relation of Epstein-Barr virus to Hodgkin's disease. The population is small, but over a period of a few years, two cases of Hodgkin's disease developed. Comstock and Evans thawed the premorbid serum samples from those two patients and from four controls who were matched to each patient and tested the samples for antibodies to various antigens of the Epstein-Barr virus. They found increased antibody levels in the two Hodgkin's disease patients but not in the controls. This is an example of a nested case-control study which also nicely demonstrates the great potential value of serum banks.

Another example would be that of a population of homosexual men being studied longitudinally in several cities (Fig. 7). Initial data and serum, urine, or other specimens are obtained from a population of homosexual men who are being monitored for development of acquired immunodeficiency syndrome (AIDS). The objective is to characterize the risk factors for AIDS development on the basis of initial data. It is

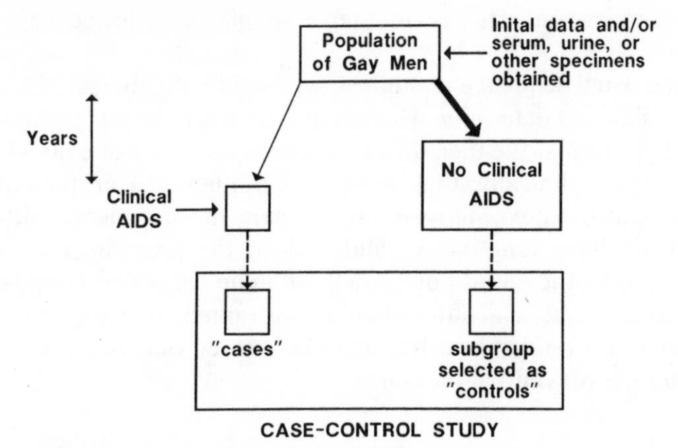

Figure 7. Nested case-control study of a population of homosexual men in terms of the risk factors for the development of AIDS.

possible instead to monitor the population to determine who shows seroconversion and who does not and to determine, among those showing seroconversion, who develops clinical AIDS and who does not (Fig. 8). Thus, we can then study the risk factors for seroconversion within this population and the risk factors for AIDS among those showing seroconversion. This is now a prospective study. However, it is also possible to compare the patients who develop AIDS with the patients who do not develop AIDS in a case-control design by using the original data obtained in this large prospective study to provide the information for the case-

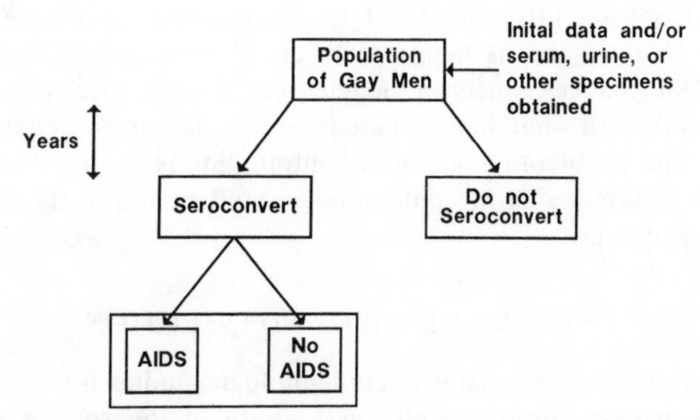

Figure 8. Prospective study of a population of homosexual men (see Fig. 7) to find the relationship between seroconversion and development of clinical AIDS.

control comparison. This is another example of a nested case-control study.

In the usual sequence of studies, we begin with clinical observations at the bedside and determine whether any routinely available data exist to test our hypothesis. We then do a case-control study. For example, it was first observed clinically that most or all lung cancer patients were smokers, and the next step was to compare those patients with people who did not have lung cancer and look at the prevalence of smoking histories. Thus, the natural outgrowth of a clinical observation is often a case-control study, and after that, if warranted, a prospective study. Occasionally, a randomized trial may be carried out, but generally only for evaluating preventive measures.

CONFOUNDING

A major problem in observational epidemiologic studies results from the fact that we have not randomly assigned people to exposure or no exposure as we do in clinical trials. The resulting problem is called confounding. Some years ago, MacMahon et al. reported a case-control study of coffee drinking and cancer of the pancreas (5). They found a strong association between coffee drinking and cancer of the pancreas. One problem which MacMahon et al. addressed in that study was interpreting the meaning of the observed association. Was it really a causal association, such that drinking coffee caused cancer of the pancreas, or could the association be due to some third factor (Fig. 9)?

Cigarette smoking is a known risk factor for cancer of the pancreas. It is rare to find a cigarette smoker who does not drink coffee. There is therefore a close association between cigarette smoking and coffee drinking. Consequently, the third factor involved could be cigarette smoking. That is, the association observed might result not because coffee drinking causes cancer of the pancreas, but because coffee drinking is associated with smoking, which is a risk factor for cancer of the pancreas. The problem of potential confounding is dealt with in several ways both in the design of epidemiologic studies and in the analysis of epidemiologic data.

DEFINING DISEASE AND INFECTION

An important issue arising in epidemiologic studies is the difficulty of defining a disease. For example, in a study of the role of viruses in rheumatoid arthritis, we must define rheumatoid arthritis for study purposes. Different sets of criteria have been developed, and depending

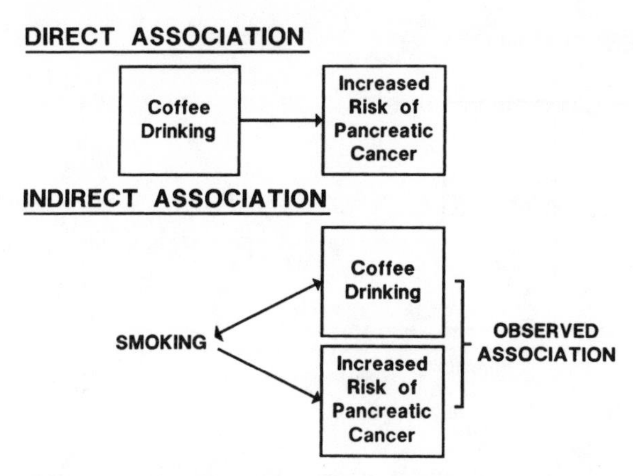

Figure 9. Confounding in association between coffee drinking and cancer of the pancreas. The association may be direct (top) or indirect as a consequence of a common cause (bottom).

on which criteria are used, different individuals will be considered to have rheumatoid arthritis. In general, when we rely on criteria for defining a disease, we do so because we do not understand the pathogenesis of the disease and are forced to resort to criteria.

A related issue is exemplified by the use of the enzyme-linked immunosorbent assay (ELISA) and the Western immunoblot for identifying human immunodeficiency virus infection. We often hear that the ELISA is both highly sensitive and highly specific, perhaps varying with which kit we use. What do we mean by highly sensitive and highly specific? Sensitivity is defined as the ability of the test to correctly identify people who actually have the condition. Specificity is defined as the ability of the test to correctly identify people who do not have the condition. How do these apply to the ELISA? In assessing the ELISA, the Western blot is often used as the measure of truth: the "gold standard" (Fig. 10). The ELISA and the Western blot can each be positive or negative. The four alternatives are then as follows: ELISA and Western blot both positive (a), ELISA positive and Western blot negative (b), ELISA negative and Western blot positive (c), and ELISA and Western blot both negative (d). If the Western blot is positive, the subject is considered to be infected. The sensitivity would be $a/(a+c)$, and the specificity would be $d/(b+d)$ (Fig. 10).

However, the problem with assessing the ELISA is how we get the data needed for calculating its sensitivity and specificity. In general, only

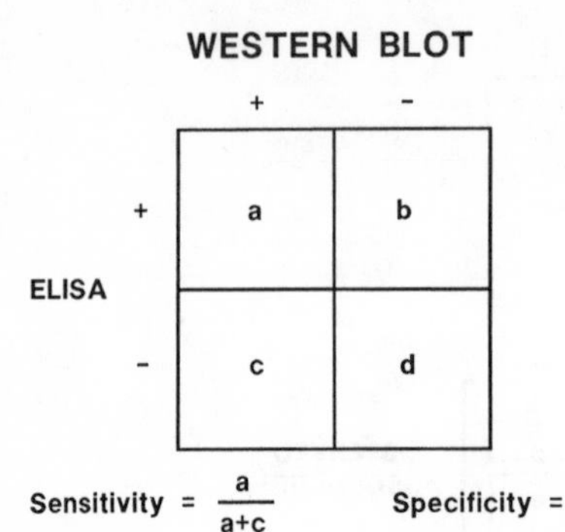

Figure 10. Use of Western blot analysis to confirm ELISA results in terms of sensitivity and specificity.

positive ELISAs are sent for Western blot analysis. We generally do not have information on alternatives c and d, because ELISA-negative samples are not sent for Western blot analysis. We must therefore ask whether there are data that tell us the sensitivity and specificity of the ELISA (compared with the Western blot) are very high. Such data may exist, but they have not been well disseminated in the medical literature. One should therefore be critical and ask for supporting evidence when the statement is made; it is important not only for epidemiologic and virologic studies, but also for patient management and public health policy.

A related problem is that when ELISA-positive specimens are sent for Western blot tests, ELISA-negative specimens are not routinely included for checking quality. As a result, people performing Western blot tests know that all the specimens they receive were positive by ELISA. We would not do this with any other test in a clinical laboratory, and it is time that ELISA-negative specimens be blindly included with ELISA-positive specimens that are sent for Western blot analysis.

COMPARISON OF EPIDEMIOLOGIC AND BENCH LABORATORY RESEARCH

Bench scientists and epidemiologists easily see the differences in the work they do, but understate the points they have in common. In general, in both animal research and epidemiologic research we study the relationship of an independent and a dependent variable. However, in the laboratory we can regulate the exposure dose very precisely, whereas in free-living human populations, precise measurement of exposure may be difficult. In the laboratory, we examine the effects of an independent variable and try to hold all other factors constant, thus controlling potential confounders in the design of the study. When studying free-

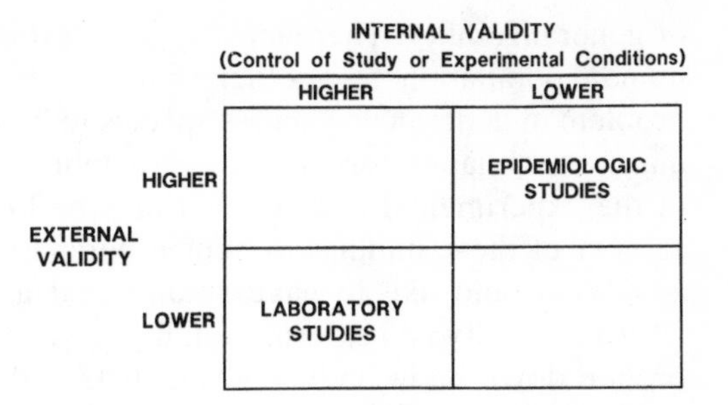

Figure 11. Comparison of internal validity and external validity of epidemiologic studies and laboratory studies of animals.

living human populations we cannot do this, because human beings are packages, in effect, of characteristics. We must therefore control for these other factors in the data analysis.

In animal research in a well-run laboratory, there are usually few if any losses to follow up. In epidemiology, losses to follow up and nonresponses unfortunately do occur. The problem is not only the percentage of nonresponders, but the degree to which those nonresponders are nonrepresentative of the total population. In the laboratory we can use inbred strains of animals to minimize genetic variability, whereas most epidemiologic studies are done with genetically heterogeneous populations. This could be viewed as an asset, since our ultimate objective is to generalize from our studies to heterogeneous populations living in the real world.

However, in animal studies, no matter how well done, we have the problem of extrapolating the findings across species. Are the results relevant to human disease? Although the human studies conducted by epidemiologists share the problems described, they are clearly relevant to human disease because they are carried out in free-living human populations.

Figure 11 compares internal validity and external validity of epidemiologic studies and laboratory studies in animals. Internal validity denotes the validity of the findings within the study, i.e., how well the study was done and the experimental conditions were controlled. In this figure, studies are categorized into higher internal validity and lower internal validity. The vertical axis shows external validity, which is the generalizability of the study and measures the extent to which the findings can be generalized. In this figure, studies are also categorized into higher generalizability and lower generalizability.

Many epidemiologic studies are lower on the scale of internal validity than many animal studies are because of the problems mentioned above. However, epidemiologic studies are high on the scale of external validity or generalizability. There may be some problem in generalizing from one human population to another, but we do not have the more difficult problem of generalizing across species to human populations. In contrast, many bench laboratory studies with animals are very high in the validity of the experimental findings, but may be lower in terms of the generalizability of these findings to human populations.

We would like to have studies that are high on the scale of both internal validity and external validity or generalizability. If barriers can be broken down between bench scientists and epidemiologists, as they are gradually being broken down today, and if we can enhance collaboration and incorporate more biomarkers and biologic measurers into epidemiologic studies, we will be able to achieve the goal of having both high internal and high external validity in studies of the etiology of chronic diseases due to infectious agents.

Literature Cited

1. **Burkitt, D.** 1962. A tumour safari in East and Central Africa. *Br. J. Cancer* **16:**379–386.
2. **Cowan, L. D., L. Gordis, J. A. Tonascia, and G. S. Jones.** 1981. Breast cancer incidence in women with a history of progesterone deficiency. *Am. J. Epidemiol.* **114:**209–217.
3. **Evans, A. S.** 1976. Causation and disease: the Henle-Koche postulates revisited. *Yale J. Biol. Med.* **49:**175–195.
4. **Evans, A. S., and G. W. Comstock.** 1981. Presence of elevated antibody titres to Epstein-Barr virus before Hodgkin's disease. *Lancet* **ii:**1183–1186.
5. **MacMahon, B., S. Yen, D. Trichopoulos, K. Warren, and G. Nardi.** 1981. Coffee and cancer of the pancreas. *N. Engl. J. Med.* **304:**630–633.
6. **U.S. Department of Health, Education and Welfare.** 1964. Criteria for judgment, p. 19–21. *In Smoking and Health. Report of the Advisory Committee to the Surgeon General of the Public Health Service.* U.S. Department of Health, Education and Welfare, Washington, D.C.

Part II.

ARENAVIRUS INFECTIONS

Chapter 6

Lymphocytic Choriomeningitis Virus Ambisense Coding: a Strategy for Persistent Infections?

David H. L. Bishop

INTRODUCTION

Viruses have been classified on the basis of the type of their genetic information (RNA or DNA), its form (single stranded or double stranded), and the strategy used for virus replication (19). After entry into a permissive cell, viruses use a variety of strategies to replicate their genomes. For the single-stranded RNA viruses (other than members of the *Retroviridae*, which synthesize a DNA intermediate during the replication cycle), the strategy involves either the immediate translation of proteins from the viral RNA (positive-stranded viruses) or the initial synthesis of virus-complementary mRNA species that are then translated into proteins (negative-stranded viruses). For the positive-stranded RNA viruses, proteins are made either by complete translation of the infecting viral RNA species (e.g., picornaviruses and caliciviruses) or by partial translation of the infecting viral RNA (e.g., coronaviruses and alphaviruses) and the subsequent synthesis of viral-sense subgenomic mRNA species that are made following the viral RNA replication process. Either way, viral-sense RNA species function as mRNA species. For all of the negative-stranded RNA viruses, the synthesis of mRNA species occurs at the onset of infection immediately after entry into a permissive cell and is catalyzed by virus-encoded enzymes present in the infecting virus particles. Double-stranded RNA viruses (e.g., members of the *Reoviridae* and *Birnaviridae*) similarly use a virion polymerase to copy into mRNA one

David H. L. Bishop • Natural Environment Research Council Institute of Virology, Oxford OX1 3SR, United Kingdom.

strand of each duplex of the genomic RNA. The infecting virion RNA of double-stranded RNA viruses is not translated by cellular ribosomes.

Examples of positive-stranded virus families include the *Picornaviridae*, *Caliciviridae*, *Coronaviridae*, *Togaviridae* (alphaviruses), *Flaviviridae*, and most of the plant virus groups that have either a single or segmented RNA genome (exceptions include the negative-stranded plant rhabdoviruses). Positive-stranded RNA viruses include both enveloped viruses (togaviruses, flaviviruses, and coronaviruses) and nonenveloped viruses (picornaviruses, caliciviruses, and most of the plant virus groups). By contrast, the negative-stranded RNA viruses are all enveloped (*Rhabdoviridae*, *Orthomyxoviridae*, *Paramyxoviridae*, *Bunyaviridae*, and *Arenaviridae*). Among these negative-stranded viruses are viruses with a single species of RNA (rhabdoviruses and paramyxoviruses) and viruses with seven or eight species of viral RNA (orthomyxoviruses), three species (bunyaviruses), or two species (arenaviruses).

The arenaviruses have a different coding strategy from that of most of the other virus groups, namely, an ambisense arrangement for the small (S) RNA species. The term ambisense is used to indicate that viral gene products are translated from virus-complementary as well as viral-sense mRNA sequences. In this regard arenaviruses differ from rhabdoviruses, paramyxoviruses, and orthomyxoviruses, although phleboviruses and probably uukuviruses (two of the *Bunyaviridae* genera) also exhibit an ambisense coding arrangement for their smallest RNA species, a characteristic not shared by other members of the *Bunyaviridae* (bunyaviruses, nairoviruses, and hantaviruses).

THE ARENAVIRUSES

Members of the *Arenaviridae* include lymphocytic choriomeningitis virus (LCMV, the prototype virus of the family) and Lassa (the etiologic agent of Lassa fever), Ippy, Mopeia, Mobala, Junin (the agent of Argentine hemorrhagic fever), Machupo (the agent of Bolivian hemorrhagic fever), Pichinde, Tacaribe, Tamiami, Latino, Amapari, Flexal, and Parana viruses (15, 19, 21). All are transmitted in nature by rodents (*Mus*, *Calomys*, *Praomys*, *Akodon*, *Mastomys*, *Oryzomys*, *Arvicanthis*, *Lemniscomys*, *Neacomys*, *Thomasomys*, and *Sigmodon* species), or, for Tacaribe virus, fruit-eating bats (*Artibeus* species) (13). LCMV and Lassa, Ippy, Mobala, and Mopeia viruses are often referred to as the Old World arenaviruses (or the lymphocytic choriomeningitis complex viruses); the others have been termed either the New World arenaviruses or the Tacaribe complex viruses.

In the normal nonfatal transuterine, transovarian, or neonatally

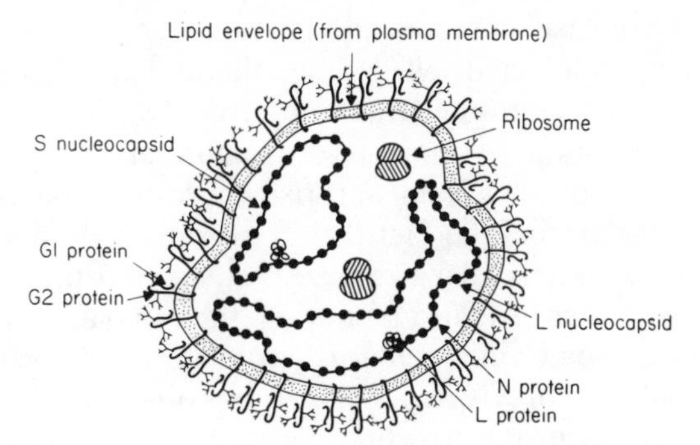

Figure 1. Schematic diagram of an arenavirus particle.

acquired infection of the host species, persistent lifelong infection by an arenavirus is common (13, 19, 21). Viremia and viruria contribute to the syndrome and to the transmission cycle. The acquisition of an arenavirus by an adult rodent species usually results in a temporary acute phase of infection with virus removal in the survivors. The virulence of the virus in the adult rodent species depends on the virus, the host species, and the route of infection. Infection within the rodent nest may involve transmission of virus to the offspring via milk, saliva, and/or urine (19). Although virus spread to other species is uncommon, human infections in nature (with LCMV and Lassa, Junin, and Machupo viruses) and in the laboratory (e.g., with LCMV and Lassa, Flexal, Junin, Machupo, Pichinde, and Tacaribe viruses) have been documented (15). The symptoms and severities of human arenavirus diseases vary with the virus (13).

As discussed by others, interest in arenaviruses has centered on the persistent nature of the normal rodent infection from both immunologic and molecular viewpoints, as well as on the unique attributes of the arenavirus replication processes (11, 21, 22). The effect of virus replication on the functions of infected cells in different organs of the host species is also a subject of study (20). Persistent infections are readily established in vitro (23).

Molecular analyses have shown that arenaviruses are enveloped and pleomorphic (Fig. 1) and 50 to 300 nm in diameter, with glycoprotein projections on the surface (either one or two [G1, G2] depending on the virus) and two internal nucleocapsid structures (11, 21, 22). As exemplified by Pichinde virus and LCMV, the nucleocapsids consist of viral RNA (large [L], approximately 2×10^6 daltons; S, 1.1×10^6 daltons), associated with many copies of the 62×10^3- to 63×10^3-dalton nucleocapsid protein (N) and small amounts of an estimated 180×10^3- to 200×10^3-dalton polypeptide (L protein) that may be a transcriptase-replicase component. There is as yet no evidence for synthesis in virus-infected cells of nonstructural proteins that are unrelated in primary sequence to the structural proteins. Ribosomes are usually incorporated within the envelope of the mature arenavirus particle (22). The presence of ribosomes is not surprising in view of the pleomorphic character of the virions and the fact that arenaviruses do not effectively inhibit host cell macromolecular syntheses. Also, ribosomes are frequently observed at the sites of virus maturation (the cell plasma membrane). The presence of ribosomes in arenavirus particles is a characteristic feature that sets this family of viruses apart from other virus families (19). Experiments involving the growth of Pichinde virus in cells in which the ribosomes are temperature sensitive have shown that the virion ribosomes are not essential for virus infectivity, since other cells can be infected at nonpermissive temperatures with viruses possessing temperature-sensitive ribosomes (18).

GENETIC STRUCTURE OF ARENAVIRUSES

Genetic experiments have shown that reassortant (recombinant) viruses can be generated between temperature-sensitive mutants or distinguishable wild-type strains of LCMV or Pichinde virus (16, 24, 25, 30, 31), although no intertypic reassortants have been recovered (25). For both viruses the genetic studies have identified two recombination groups of mutants (L and S mutants), in agreement with the existence of two species of viral RNA. In addition, genetic studies with mutants of LCMV have provided evidence for two groups of complementing mutants encoded by the viral S RNA, indicating that the S species contains the information for at least two proteins (25). These observations have been substantiated by nucleic acid sequencing and other molecular analyses (see below). Furthermore, viruses that contain multiple copies of the S RNA species have been identified among virus populations (showing that such viruses are genetically diploid [25]). This finding probably reflects the pleomorphic character of the arenavirus particle. Diploidy (polyploidy) may contribute to the biological stability of arenaviruses in nature and to the experimental difficulties encountered in recovering temperature-sensitive mutants (30).

The observation that there are at least three types of viral protein (N, L, and G1 or G2) but only two viral RNA species indicates that one of the viral RNA species must code for at least two proteins. Analyses of reassortant Pichinde viruses (derived from parent viruses with distinguishable RNA sequences and proteins) have shown that the L RNA codes for the L protein and the S RNA codes for both the viral glycoproteins and the N protein (12, 31). The fact that the S RNA codes for N and glycoprotein precursor (GPC) proteins accounts for the complementation observed between certain S RNA mutants of LCMV (25). Similar coding results have been obtained from analyses of reassortant LCMV generated from different virus strains (24).

AMBISENSE CODING STRATEGY OF ARENAVIRUS S RNA

The 5′ and 3′ ends of the S RNA species of Pichinde virus are complementary for approximately 20 residues (1–5), as are those of LCMV S RNA (1, 26, 27). The 3′ end sequences are comparable not only between the two viruses but also for the L and S RNA species of both viruses. Identification of the end sequences has allowed DNA clones of the viral RNA species to be obtained by using 3′ complementary oligodeoxynucleotide primers, viral RNA, and reverse transcriptase.

Sequence analyses of DNA copies of Pichinde virus and LCMV S RNA species have been reported (3, 4, 26–29). For Pichinde virus, the S RNA is 3,419 nucleotides in length, and for LCMV it is 3,375 nucleotides. The data indicated that the 1.1×10^6-dalton viral S RNA codes for a 62×10^3- to 63×10^3-dalton N protein in a virus-complementary sequence corresponding to the 3′ half of the S RNA. The 5′ half of the viral-sense RNA sequence codes for a glycoprotein precursor to the G1 and G2 proteins (unglycosylated primary gene product: GPC, 56×10^3 to 57×10^3 daltons). To characterize this coding arrangement, the term ambisense RNA has been coined (4, 6, 27). Northern (RNA) blot analyses undertaken to confirm the identities of the N and GPC mRNA species have shown that for both LCMV and Pichinde virus the N mRNA species is virus complementary in sequence and subgenomic in size. It corresponds to approximately half the length of the viral S RNA species (4, 27, 29). Thus, the N mRNA species of LCMV is approximately 1,770 nucleotides long, whereas that of Pichinde virus is some 1,800 nucleotides. The lack of binding of the viral N mRNA species to oligo(dT)-cellulose indicates that it does not possess 3′ polyadenosine tracts (3, 5). For both viruses, Northern analyses have confirmed that the GPC mRNA species have a viral-sense sequence and that they also correspond in length to approximately half the size of the S RNA species. The GPC mRNA species of LCMV consists of some 1,600 nucleotides; for Pichinde virus the GPC mRNA species is approximately 1,610 nucleotides. These results indicate, therefore, that the two S RNA gene products are encoded on separate subgenomic mRNA species and that they must be made from different template RNA species (Fig. 2). Whether the complete arenavirus

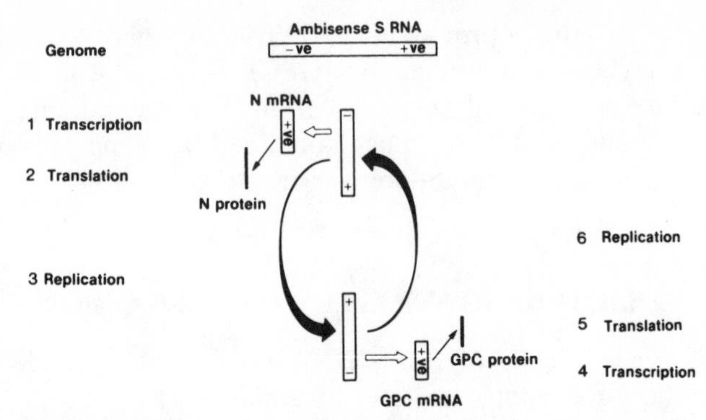

Figure 2. Coding, transcription, and replication strategies of arenavirus S RNA species. Whether the viral S RNA also functions as an mRNA is not known.

S RNA species can be translated directly by cellular ribosomes into GPC is not known. The existence of a subgenomic GPC mRNA species does not rule out this possibility, although the character of the 5′ end sequence of the viral RNA, or its association with the nucleoprotein, might preclude it.

Although only incomplete sequence data have been reported for the L RNA species of LCMV (26), the data indicate that the viral L protein is coded in the virus-complementary sequence. So far, only one L mRNA species has been identified by Northern analyses (D. H. L. Bishop and V. Romanowski, unpublished data). It appears to be similar in size to the viral L RNA. Whether the L RNA has a simple negative-strand strategy or whether it exhibits an ambisense arrangement must await complete L RNA sequence data and further molecular analyses.

Comparison of the sequences of the N proteins of Pichinde virus and LCMV indicates that, in a best-fit alignment, some 50% of their amino acids are homologous in type and relative position. This homology is a minimal estimate, since it does not take into consideration amino acid differences in the two proteins that represent substitutions of residues of similar charge or character. Taking these considerations into account, therefore, the data indicate that the N proteins of the two viruses probably have similar three-dimensional structures. The GPC primary gene products of Pichinde virus and LCMV exhibit approximately 40% sequence homology, which is somewhat lower than that observed for the N proteins. The DNA sequence of Lassa virus S RNA (5, 10) indicates that Lassa virus N protein is related by primary sequence to both LCMV and Pichinde virus N proteins. Some 62% of the N protein amino acids of

Lassa virus and LCMV are identical in type and relative position, as are some 50% of the N protein residues of Lassa and Pichinde viruses (10). Similarly, some 60% of the amino acids of Lassa virus GPC and that of LCMV are homologous, whereas only 45% of the Lassa virus and Pichinde virus GPC proteins are homologous. These data indicate that Lassa virus and LCMV are closer to each other, in an evolutionary sense, than either is to Pichinde virus.

In classical serological procedures such as the complement fixation test, which measures predominantly the antigenic epitopes on the viral nucleocapsid protein, different arenaviruses have been shown to be related to each other to different extents (9). However, by these procedures LCMV is reported to exhibit little relationship to the Tacaribe complex viruses, although in such tests LCMV has been shown to be distantly related to Lassa virus (9). The Pichinde virus-LCMV-Lassa virus protein relationships were not expected from this earlier serological information, nor from monoclonal antibody studies, which indicated that only a few N protein monoclonal antibodies are cross-reactive among these viruses (7, 8). The fact, though, that some are cross-reactive is in agreement with the reported sequence relationships. The observation that most of the monoclonal antibodies that have been analyzed are not cross-reactive suggests that the conserved regions of the viral N proteins may not be as antigenic as the nonconserved regions. The same may well hold true for the viral glycoproteins.

IMPLICATIONS OF AN AMBISENSE S RNA STRATEGY

The presence of RNA polymerase activities in arenavirus preparations (17) suggests that after attachment, penetration, and uncoating in permissive cells, arenaviruses synthesize virus-complementary N and (presumably) L mRNA species (primary transcription). It is assumed that these are translated to give the N and L proteins before RNA replication takes place (Fig. 2). The ambisense strategy of the arenavirus S RNA species precludes the synthesis of the subgenomic S-encoded GPC mRNA species until after viral RNA replication has been initiated. This is in complete contrast to the strategy used by members of the negative-stranded rhabdovirus, paramyxovirus, and orthomyxovirus groups. For viruses that have a single species of genomic RNA (i.e., rhabdoviruses and paramyxoviruses), all the virus-complementary, subgenomic mRNA species are made in a coordinated manner by transcription of the infecting viral RNA species (primary transcription) and, in larger quantities, after viral RNA replication has been initiated (secondary transcription). For the segmented-genome orthomyxoviruses, all of the viral mRNA species

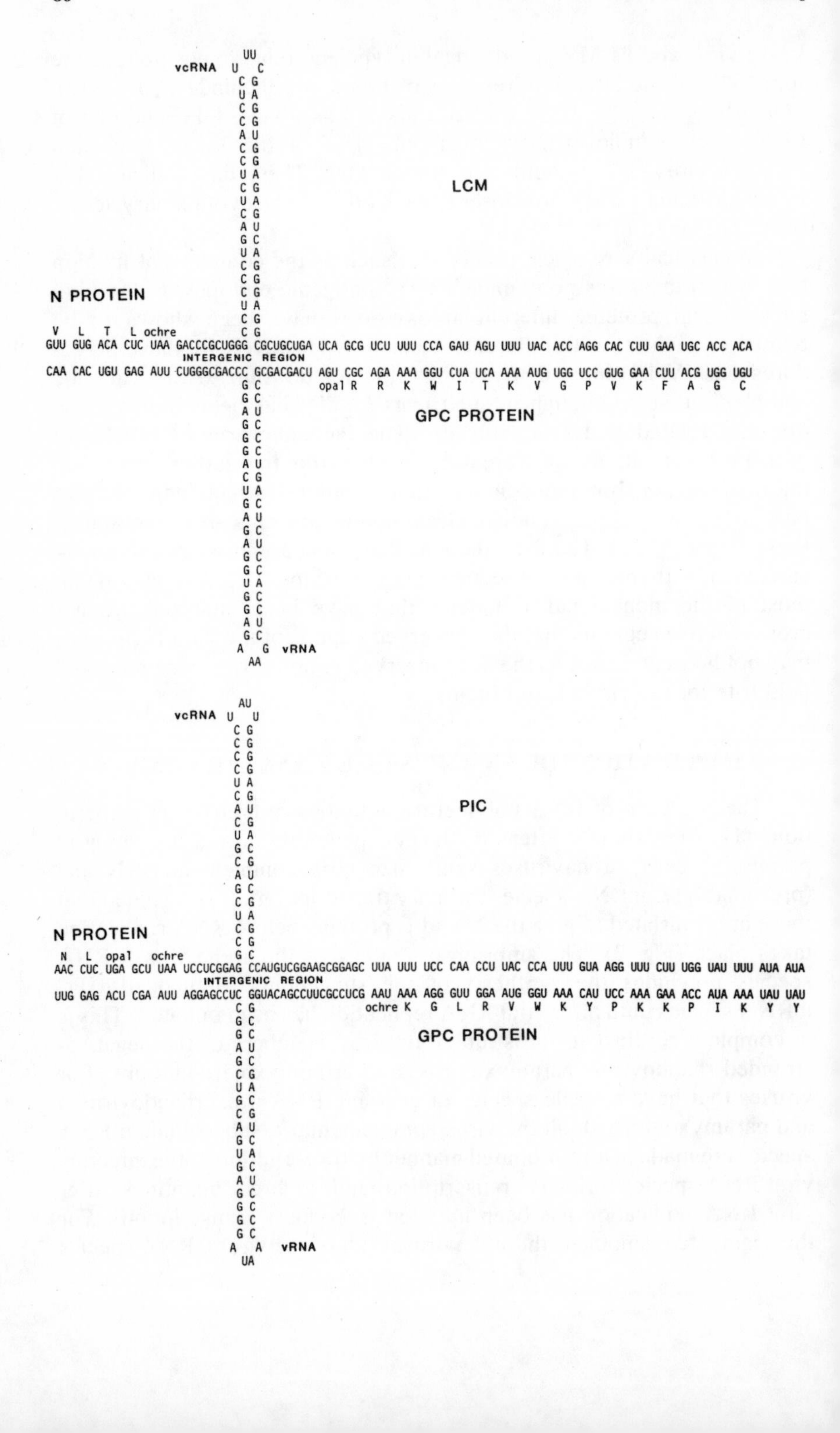

LCM
vcRNA
N PROTEIN
V L T L ochre
GUU GUG ACA CUC UAA GACCCGCUGGG CGCUGCUGA UCA GCG UCU UUU CCA GAU AGU UUU UAC ACC AGG CAC CUU GAA UGC ACC ACA
INTERGENIC REGION
CAA CAC UGU GAG AUU CUGGGCGACCC GCGACGACU AGU CGC AGA AAA GGU CUA UCA AAA AUG UGG UCC GUG GAA CUU ACG UGG UGU
opal R R K W I T K V G P V K F A G C
GPC PROTEIN
vRNA
PIC
vcRNA
N PROTEIN
N L opal ochre
AAC CUC UGA GCU UAA UCCUCGGAG CCAUGUCGGAAGCGGAGC UUA UUU UCC CAA CCU UAC CCA UUU GUA AGG UUU CUU UGG UAU UUU AUA AUA
INTERGENIC REGION
UUG GAG ACU CGA AUU AGGAGCCUC GGUACAGCCUUCGCCUCG AAU AAA AGG GUU GGA AUG GGU AAA CAU UCC AAA GAA ACC AUA AAA UAU UAU
ochre K G L R V W K Y P K K P I K Y Y
GPC PROTEIN
vRNA

(including those that are derived by splicing) have virus-complementary sequences.

The data obtained by analyses of the types and amounts of the intracellular S-encoded RNA species induced by arenaviruses indicate that the subgenomic N mRNA is made before the subgenomic GPC mRNA species. At later stages of an infection, when both S-encoded mRNA species are present, the subgenomic N mRNA species predominate (4; unpublished data). These observations support the hypothesis that the arenavirus N mRNA synthesis is regulated independently of the GPC mRNA synthesis. The obvious advantage of this strategy is that N and (presumably) L proteins are made at the beginning of the infection, when they are needed for viral RNA replication. Likewise, GPC mRNA and the viral glycoproteins are not made until they are needed, i.e., after the initiation of RNA replication. Presumably, the continued synthesis of N mRNA from the viral RNA template must compete at some stage with virus-complementary, replicative intermediate RNA synthesis. Nothing is known about the enzymology concerning the process of arenavirus RNA replication or its relationship to the mRNA transcription process. Conceivably, different forms of the same gene product (possibly the L protein) perform different functions (transcription and replication). The simplest hypothesis, however, is that mRNA transcription and RNA replication are catalyzed by the same enzyme, which is influenced in terms of its activity by the availability of newly translated N protein. Notwithstanding these points, it is evident that N mRNA transcription and RNA replication must compete for the same template (similarly for L mRNA transcription and L RNA replication). Likewise, the synthesis of GPC mRNA must compete with S viral RNA replication. However, the observations that much less GPC mRNA than S viral RNA is observed in infected cell extracts and that more N mRNA is present than S virus-complementary, replicative intermediate RNA indicate that the determinants of RNA synthesis are complex and are not simply regulated by just the form of the enzyme and the availability of N protein.

Figure 3. Intergenic sequences of the S RNA species of (top) the WE strain of LCMV and (bottom) Pichinde virus. The virus-complementary S RNA sequences are shown above the corresponding viral RNA sequences. The indicated amino acids, translation termination codons, and gene products of the two viral S RNA species are shown. The Pichinde virus S sequence (3, 4) has been corrected (D. A. Auperin, personal communication).

INTERGENIC REGION OF AMBISENSE S RNA SPECIES

The intergenic regions of Pichinde and Lassa virus and LCMV S RNA species have a unique feature, i.e., an inverted complementary sequence that may be arranged into a hairpin configuration. For Pichinde virus the intergenic region between N and GPC is 86 nucleotides in length, and 38 of these can be arranged into a hairpin configuration (Fig. 3) (4). For LCMV the intergenic region is shorter (64 nucleotides), and 42 nucleotides can be arranged into a hairpin structure (Fig. 3) (27). The corresponding sequence for Lassa virus is similar (5). Although no information has yet been published about the mRNA transcription initiation and termination processes for either the S or L RNA species of arenaviruses, we have found in experiments with oligonucleotides representing viral and virus-complementary sequences of the intergenic region of Pichinde virus S RNA that transcription of both N and GPC mRNA species terminates near the top of the intergenic hairpin (M. Galinski and D. H. L. Bishop, unpublished results) (Fig. 3). Thus, virus-sense oligonucleotides representing the 3′ half of the intergenic hairpin anneal to both N and GPC mRNA species, whereas those representing the 5′ half do not. Both types of oligonucleotide anneal to complete viral S RNA species (viral RNA and replicative intermediate S RNA). How transcription termination in the intergenic region occurs is not known.

ORIGINS OF VIRUSES WITH AMBISENSE RNA

Concerning the question of how arenaviruses with an ambisense coding strategy may have arisen, unless one invokes an origin from a DNA source in which the arrangement of proteins encoded on opposite strands of nucleic acid has been maintained, the simplest explanation is that a chimeric RNA was derived at some stage of arenavirus evolution. Such a chimeric RNA could have been formed during RNA replication and could represent a consolidation of genetic information (i.e., a virus with three species of RNA, each coding for a single gene product, gave rise to an arenavirus with a consolidated genome and, subsequently, only two RNAs by the exclusion of the redundant third RNA species). Such consolidation of genetic information could occur by a viral replicase copying the coding strand of one RNA species and, instead of terminating, continuing RNA synthesis on the noncoding strand of another RNA. This would result in a chimeric RNA molecule composed of two genes coded on opposite strands of the RNA. Together with the subsequent loss of the redundant RNA species, a virus would be generated in which all the original genetic information had been conserved. Of course, the reverse situation may have occurred, i.e., the formation of a virus with three RNA species from a virus having two, one of which originally had an ambisense coding arrangement. Whatever the origins, the ambisense coding arrangements that have been observed for arenaviruses (and phleboviruses [14]) open yet another dimension to the way in which viruses replicate in cells.

Literature Cited

1. **Auperin, D. D., R. W. Compans, and D. H. L. Bishop.** 1982. Nucleotide sequence conservation at the 3′ termini of the virion RNA species of New World and Old World arenaviruses. *Virology* **121:**200–203.
2. **Auperin, D. D., K. Dimock, P. Cash, W. E. Rawls, W.-C. Leung, and D. H. L. Bishop.** 1982. Analyses of the genomes of prototype Pichinde and a virulent derivative of Pichinde Munchique arenaviruses: evidence for sequence conservation at the 3′ termini of their viral RNA species. *Virology* **116:**363–367.
3. **Auperin, D. D., M. Galinski, and D. H. L. Bishop.** 1984. The sequences of the N protein gene and intergenic region of the S RNA of Pichinde arenaviruses. *Virology* **134:** 208–219.
4. **Auperin, D. D., V. Romanowski, M. Galinski, and D. H. L. Bishop.** 1984. Sequencing studies of Pichinde arenavirus S RNA indicate a novel coding strategy, an ambisense viral S RNA. *J. Virol.* **52:**897–904.
5. **Auperin, D. D., D. R. Sasso, and J. B. McCormack.** 1986. Nucleotide sequence of the glycoprotein gene and intergenic region of the Lassa virus S RNA. *Virology* **154:** 155–167.
6. **Bishop, D. H. L., and D. D. Auperin.** 1987. Arenavirus gene structure and organization. *Curr. Top. Microbiol. Immunol.* **133:**5–17.
7. **Buchmeier, M. J., H. A. Lewicki, O. Tomori, and K. M. Johnson.** 1980. Monoclonal antibodies to lymphocytic choriomeningitis virus react with pathogenic arenaviruses. *Nature* (London) **288:**486–487.
8. **Buchmeier, M. J., H. A. Lewicki, O. Tomori, and M. B. A. Oldstone.** 1981. Monoclonal antibodies to lymphocytic choriomeningitis and Pichinde viruses: generation, characterization and cross-reactivity with other arenaviruses. *Virology* **113:**73–85.
9. **Casals, J., S. M. Buckley, and R. Cedeno.** 1975. Antigenic properties of the arenaviruses. *Bull. W.H.O.* **52:**421–427.
10. **Clegg, J. C. S., and J. D. Oram.** 1985. Molecular cloning of Lassa virus RNA: nucleotide sequence and expression of the nucleocapsid protein gene. *Virology* **144:**363–372.
11. **Compans, R. W., and D. H. L. Bishop.** 1985. Biochemistry of arenaviruses. *Curr. Top. Microbiol. Immunol.* **114:**153–175.
12. **Harnish, D. G., K. Dimock, D. H. L. Bishop, and W. E. Rawls.** 1983. Gene mapping in Pichinde virus: assignment of viral polypeptides to genomic L and S RNAs. *J. Virol.* **46:**638–641.
13. **Howard, C. R.** 1986. *Arenaviruses. Perspectives in Medical Virology*, vol. 2. Elsevier Science Publishers BV, Amsterdam.
14. **Ihara, T., H. Akashi, and D. H. L. Bishop.** 1984. Novel coding strategy (ambisense genomic RNA) revealed by sequence analyses of Punta Toro phlebovirus S RNA. *Virology* **136:**293–306.
15. **Karabatsos, N.** 1985. *International Catalogue of Arboviruses*, 3rd ed. American Society of Tropical Medicine and Hygiene, San Antonio, Tex.

16. **Kirk, W. E., P. Cash, C. J. Peters, and D. H. L. Bishop.** 1980. Formation and characterization of an intertypic lymphocytic choriomeningitis recombinant virus. *J. Gen. Virol.* **51:**213–218.
17. **Leung, W.-C., M. K. F. L. Leung, and W. E. Rawls.** 1979. Distinctive RNA transcriptase, polyadenylic acid polymerase, and polyuridylic acid polymerase activities associated with Pichinde virus. *J. Virol.* **30:**98–107.
18. **Leung, W.-C., and W. E. Rawls.** 1977. Virion-associated ribosomes are not required for the replication of Pichinde virus. *Virology* **81:**174–176.
19. **Matthews, R. E. F.** 1982. Classification and nomenclature of viruses. *Intervirology* **17:**1–200.
20. **Oldstone, M. B. A., R. Ahmed, M. J. Buchmeier, P. Blount, and A. Tishon.** 1985. Perturbation of differentiated functions during viral infection *in vivo*. 1. Relationship of lymphocytic choriomeningitis virus and host strains to growth hormone deficiency. *Virology* **142:**158–174.
21. **Pedersen, I. R.** 1979. Structural components and replication of arenaviruses. *Adv. Virus Res.* **24:**277–330.
22. **Rawls, W. E., M. A. Chan, and S. R. Gee.** 1981. Mechanisms of persistence in arenaviruses infections: a brief review. *Can. J. Microbiol.* **27:**568–574.
23. **Rawls, W. E., and W.-C. Leung.** 1979. Arenaviruses, p. 157–192. *In* R. R. Wagner and H. Fraenkel-Conrat (ed.), *Comprehensive Virology*, vol. 14. Plenum Publishing Corp., New York.
24. **Riviere, Y., R. Ahmed, P. J. Southern, M. J. Buchmeier, F. J. Dutko, and M. B. A. Oldstone.** 1985. The S RNA segment of lymphocytic choriomeningitis virus codes for the nucleoprotein and glycoproteins 1 and 2. *J. Virol.* **53:**966–968.
25. **Romanowski, V., and D. H. L. Bishop.** 1983. The formation of arenaviruses that are genetically diploid. *Virology* **126:**87–95.
26. **Romanowski, V., and D. H. L. Bishop.** 1985. Conserved sequences and coding of two strains of lymphocytic choriomeningitis virus (WE and ARM) and Pichinde arenavirus. *Virus Res.* **2:**35–51.
27. **Romanowski, V., Y. Matsuura, and D. H. L. Bishop.** 1985. The complete sequence of the S RNA of lymphocytic choriomeningitis virus (WE strain) compared to that of Pichinde virus. *Virus Res.* **3:**101–114.
28. **Southern, P. J., and D. H. L. Bishop.** 1987. Sequence comparisons among arenaviruses. *Curr. Top. Microbiol. Immunol.* **133:**19–39.
29. **Southern, P. J., M. K. Singh, Y. Riviere, D. R. Jacoby, M. J. Buchmeier, and M. B. A. Oldstone.** 1987. Molecular characterization of the genomic S RNA from lymphocytic choriomeningitis virus. *J. Virol.* **157:**145–155.
30. **Vezza, A. C., and D. H. L. Bishop.** 1977. Recombination between temperature-sensitive mutants of the arenavirus Pichinde. *J. Virol.* **24:**712–715.
31. **Vezza, A. C., P. Cash, P. Jahrling, G. Eddy, and D. H. L. Bishop.** 1980. Arenavirus recombination: the formation of recombinants between prototype Pichinde and Pichinde Munchique viruses and evidence that arenavirus S RNA codes for N polypeptide. *Virology* **106:**250–260.

Chapter 7

Structure and Expression of Arenavirus Proteins

M. J. Buchmeier, K. E. Wright, E. L. Weber, and B. S. Parekh

INTRODUCTION

Lymphocytic choriomeningitis virus (LCMV), like all arenaviruses, has the capacity to establish persistent infection in vivo. Such persistent infection both guarantees survival of LCMV from one generation to the next and poses the challenging intellectual puzzle of unraveling the mechanism of virus persistence in vivo. Studies described below have focused on two primary aspects of arenavirus infections, i.e., their protein synthesis and expression and their antigenicity, and we have attempted to correlate in vitro biochemical and immunologic data with biological observations during acute and persistent infections. In this brief review, recent findings which highlight some of the important features of the protein structure of LCMV and other arenaviruses are described and related to biological and immunological phenomena observed during acute and persistent virus infections.

STRUCTURAL PROTEINS OF ARENAVIRUSES

The structural proteins of purified arenaviruses were first studied by Ramos et al. (22) with Pichinde virus and by Pedersen (20) with LCMV. Numerous other descriptive studies of the proteins of these agents followed, and to date, structural proteins of at least nine different arenaviruses have been examined (summarized in reference 9). Despite differences, a number of common features have emerged. Arenaviruses

M. J. Buchmeier, K. E. Wright, E. L. Weber, and B. S. Parekh • Department of Immunology, Research Institute of Scripps Clinic, La Jolla, California 92037.

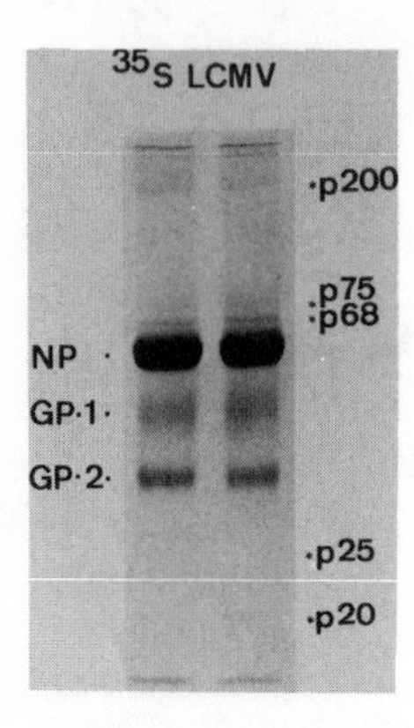

Figure 1. [^{35}S]methionine-labeled LCMV separated by sodium dodecyl sulfate-polyacrylamide gel electrophoresis. In addition to NP, GP1, and GP2, the positions of migration of several minor proteins are marked. One of these, p200, has been shown by us to correspond to the L gene-encoded 200-kilodalton protein (see text).

all contain a major dominating protein which is the viral nucleocapsid protein (NP; 60 to 68 kilodaltons). NP constitutes up to 58% of the protein in arenaviruses (28) and is easily detected in sodium dodecyl sulfate-polyacrylamide gel electrophoresis gels by protein staining or by radiolabeling with amino acid precursors such as [^{3}H]leucine or [^{35}S]methionine (Fig. 1). The viruses also contain either one, as reported for Tacaribe and Tamiami viruses (11), or two, as reported for LCMV and Pichinde virus (4, 28), glycoproteins of somewhat lower molecular weight than NP. Other minor proteins have also been detected, but their origin has until recently been largely a subject of conjecture. Predominant among these quantitatively minor proteins is a 180- to 200-kilodalton protein termed L in Pichinde virus (13) and LCMV (26a), which is likely to be the viral RNA polymerase.

Recent advances in our understanding of the molecular genetics of arenaviruses have cleared up some of the ambiguity surrounding the identity and derivation of the viral proteins. As recently detailed in reviews by Bishop and Auperin (2) and Southern and Bishop (27), molecular cloning approaches have definitively assigned NP and the glycoprotein precursor (GPC) of glycoproteins GP1 and GP2 to the S RNA segments of LCMV and Pichinde virus. These two open reading frames, which are arranged in an ambisense orientation (2), completely account for the coding capacity of S.

We sought to establish definitively the linear orientation of GP1 and GP2 within the GPC-coding region at the 5′ end of the viral S RNA. Using the methods described by Hunkapiller et al. (16), we first attempted to determine the N-terminal sequence for each of the glycoproteins isolated by sodium dodecyl sulfate-polyacrylamide gel electrophoresis separation and extraction from the gel matrix. Despite the presence of sufficient quantities of protein, we were unable to degrade the amino terminus, presumably because it was blocked. To resolve this problem, we developed an alternative strategy involving the use of synthetic peptides deduced from the nucleotide sequences of two strains of LCMV, strains ARM and WE.

Peptides were selected from predominantly hydrophilic stretches within the predicted LCMV ARM glycoprotein-coding sequence and were used to raise antisera in rabbits. By their reactivity with the viral glycoproteins GP1 and GP2, these antisera defined the orientation of the glycoproteins on the precursor and the probable site of proteolytic cleavage.

Previous studies (1, 4, 22) showed that the GP2 proteins of LCMV and Pichinde virus were probably integral membrane proteins. Dissociation of virions with mild detergents resulted in a complex of GP2 with viral nucleocapsid complexes, suggesting that the former spanned the membrane. GP1, in contrast, is highly exposed on the virion envelope, as indicated by its susceptibility to surface iodination and its targeting by neutralizing antibodies. Recent studies have delineated the sequences of several arenavirus GPC genes, and a number of common features among them have emerged (Fig. 2). All have prominent hydrophobic stretches of 20 to 30 amino acids near their carboxy termini; these stretches are thought to serve as membrane, and possibly transmembrane, anchor sequences. There are, however, two additional hydrophobic sequences within the amino terminal 50 amino acids. The first of these, spanning approximately the first 30 amino acids, is likely to be a signal sequence by analogy with other membrane glycoproteins. The second, spanning amino acids 30 to 50 of GPC, may provide an alternative membrane anchor sequence for the precursor or cleaved products or both.

Considerable structural homology is evident among arenaviruses in the carboxyl half of GPC, whereas much divergence is found in the amino-terminal domain (27). In particular, a motif appearing between LCMV GPC amino acids 225 and 285 was found to be repeated in Pichinde and Lassa viruses (Fig. 3). This motif consisted of the amino acid sequences $_{228}$LIIQNXTWEXHC$_{239}$ and $_{272}$ISDSXGXX XPGGYCL$_{286}$ bracketing a pair of basic amino acids, RR or RK, which appeared to be likely targets for membrane-bound or extracellular proteases. Antisera were prepared to each of the flanking conserved peptides and to a third peptide immediately to the left of the RR pair. The peptide closest to the carboxyl terminus elicited an antiserum reactive with GP2, whereas the two peptides to the left of the RR pair elicited GP1-specific antibodies (Fig. 4). Thus, it appears highly likely that the RR doublet at GPC amino acids 262 to 263 constitutes the site of proteolytic cleavage to

```
         ARM  MGQIVTMFEA LPHIIDEVIN IVIIVLIVIT GIKAVYNFAT CGIFALISFL
          WE  MGQIVTMFEA LPHIIDEVIN IVIIVLIIIT SIKAVYNFAT CGILALVSFL
          LA  MGQIVTFFQE VPHVIEEVMN IVLIALSVLA VLKGLYNFAT CGLVGLVTFL
          PV  MGQIVTLIQS IPEVLQEVFN VALIIVSVLC IVKGFVNLMR CGLFQLVTFL
   CONSERVED  MGQIVT      P    EV N    I         K   N    CG   L  FL

              LLAGRSCGMY GLKGPDIYKG VYQFKSVEFD MSHLNLTMPN ACSANNSHHY
              FLAGRSCGMY GLNGPDIYKG VYQFKSVEFD MSHLNLTMPN ACSVNNSHHY
              LLCGRSCT.. ....TSLYKG VYELQTLELN METLNMTMPL SCTKNNSHHY
              ILSGRSCDSM MIDRRHNLTH VEFNLTRMFD NL......PQ SCSKNNTHHY
               L GRSC               V                   P  C  NN HHY

              ISMGTS...G LELTFTNDSI ISHNFCNLTS AFNKKTFDHT LMSIVSSLHL
              ISMGSS...G LEPTFTNDSI LNHNFCNLTS ALNKKSFDHT LMSIVSSLHL
              IMVGNET..G LELTLTNTSI INHKFCNLSD AHKKNLYDHA LMSIISTFHL
              YKGPSNTTWG IELTLTNTSI ANETSGNFSN IGSLGYGNIS NCDRTREAGH
                       G  E T TN SI       N

              SIRGNSNYKA VSCDFNNG.. .......... .ITIQYNLTF SDAQSAQSQC
              SIRGNSNYKA VSCDFNNG.. .......... .ITIQYNLSS SDPQSAMSQC
              SIPNFNQYEA MSCDFNGG.. ....K..... .ISVQYNLSH SYAGDAANHC
              TLKWLLNELH FNVLHVTRHI GARCKTVEGA GVLIQYNLTV GDRGGEVGRH
                                                   QYNL

              RTFRGRVLDM F.RTAFGGKY MRSGWGWTGS DGKTTW.CSQ TSYQYLIIQN
              RTFRGRVLDM F.RTAFGGKY MRSGWGWTGS DGYTTW.CSQ TSYQYLIIQN
              GTVANGVLQT FMRMAWGGSY I......ALD SGRGNWDCIM TSYQYLIIQN
              LIASLAQIIG DPKIAWVGKC FNNCSGDTCR LTNCEGGTH. ..YNFLIIQN
                             A  G                           Y  LIIQN
                                               GP-1 ◄|► GP-2
              RTWENHCTYA ..GPFGMSRI LLSQEKTKFF TRRLAGTFTW TLSDSSGVEN
              RTWENHCRYA ..GPFGMSRI LFAQEKTKFL TRRLSGTFTW TLSDSSGVEN
              TTWEDHCQFS RPSPIGYLGL LSQRTRDIYI SRRLLGTFTW TLSDSEGKDT
              TTWENHCTYT ...PMATIRM ALQRTAYSSV SRKLLGFFTW DLSDSSGQHV
               TWE HC        P                   R L G FTW  LSDS G

              PGGYCLTKWM ILAAELKCFG NTAVAKCNVN HDAEFCDMLR LIDYNKAALS
              PGGYCLTKWM ILAAELKCFG NTAVAKCNVN HDEEFCDMLR LIDYNKAALS
              PGGYCLTRWM LIEAELKCFG NTAVAKCNEK HDEEFCDMLR LFDFNKQAIQ
              PGGYCLEQWA IIWAGIKCFD NTVMAKCNKD HNEEFCDTMR LFDFNQNAIK
              PGGYCL  W      A  KCF  NT  AKCN   H  EFCD  R L D N  A

              KFKEDVESAL HLFKTTVNSL ISDQLLMRNH LRDLMGVPYC NYSKFWYLEH
              KFKQDVESAL HVFKTTLNSL ISDQLLMRNH LRDLMGVPYC NYSKFWYLEH
              RLKAEAQMSI QLINKAVNAL INDQLIMKNH LRDIMGIPYC NYSKYWYLNH
              TLQLNVENSL NLFKKTINGL ISDSLVIRNS LKQLAKIPYC NYTKFWYIND
                                N L I D L   N  L      PYC NY K WY

              AKTGETSVPK CWLVTNGSYL NETHFSDQIE QEADNMITEM LRKDYIKRQG
              AKTGETSVPK CWLVTNGSYL NEIHFSDQIE QEADNMITEM LRKDYIKRQG
              TTTGRTSLPK CWLVSNGSYL NETHFSDDIE QQADNMITEM LQKEYMERQG
              TITGRHSLPQ CWLVHNGSYL NETHFKNDWL WESQNLYNEM LMKEYEERQG
                TG  S P  CWLV NGSYL NE HF          N   EM L K Y  RQG

              STPLALMDLL MFSTSAYLVS IFLHLVKIPT HRHIKGGSCP KPHRLTNKGI
              STPLALMDLL MFSTSAYLIS IFLHFVRIPT HRHIKGGSCP KPHRLTNKGI
              KTPLGLVDLF VFSTSFYLIS IFLHLVKIPT HRHIVGKSCP KPHRLNHMGI
              KTPLALTDIC FWSLVFYTIT VFLHIVGIPT HRHIIGDGCP KPHRITRNSL
               TPL L D     S   Y     FLH V IPT HRHI G  CP KPHR

              CSCGAFKVPG VKTVWKRR
              CSCGAFKVPG VKTIWKRR
              CSCGLYKQPG VPVKWKR
              CSCGYYKYQR NLTNG
              CSCG  K
```

ARM } LCMV } Old World
WE
LA = Lassa Virus
PV = Pichinde Virus = New World

Figure 2. Deduced polypeptide sequences of LCMV ARM and WE, as well as Lassa (LA) and Pichinde (PV) viruses. Conserved amino acids are more frequent in the carboxyl half of GPC, which corresponds to the GP2 glycoprotein after proteolytic cleavage. The peptide sequence identified as containing a group-specific antigen in LCMV is boxed.

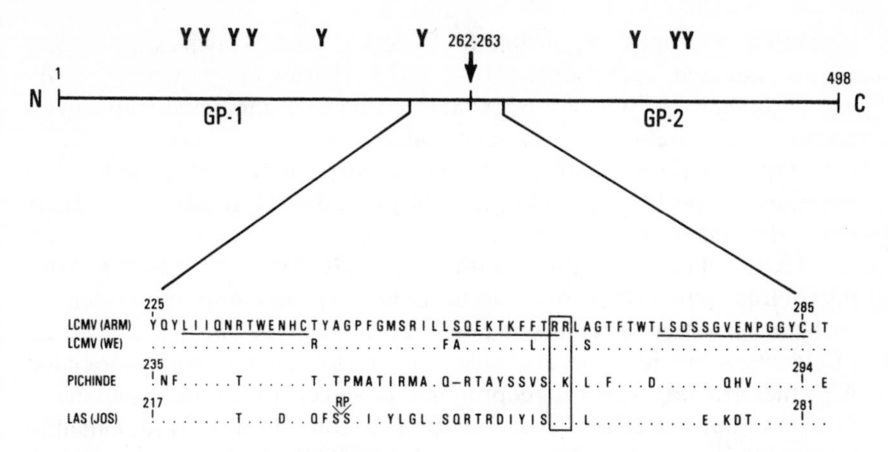

Figure 3. Structure around the GPC cleavage site of LCMV, Pichinde virus, and Lassa virus. Consensus N-linked glycosylation sites (N × S or N × T) are indicated as Y. Peptides synthesized to make antisera as described in the text and shown in Fig. 4 are underlined.

produce the mature glycoproteins from the GPC precursor. From other studies (4), we know that this cleavage occurs rapidly and is probably coincident with oligosaccharide processing of the mannose-rich GPC to the mannose-depleted viral proteins. We also have preliminary evidence that glycosylation is a prerequisite for this proteolytic processing. Such results are consistent with findings with a number of enveloped virus systems, including Sindbis and Semliki Forest virus E2 glycoproteins (12, 24) and for yellow fever virus NS and M proteins (23). Still, the

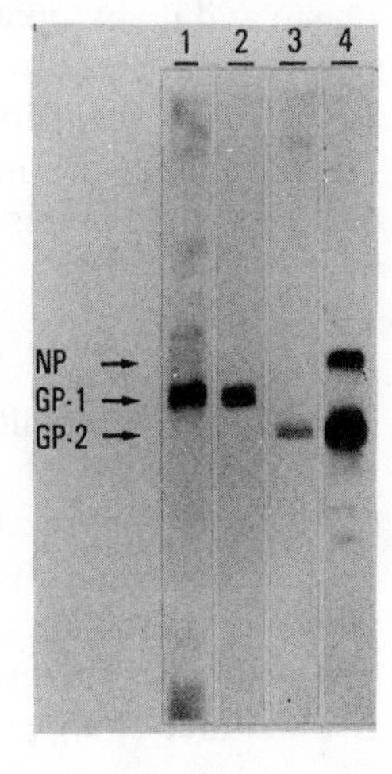

Figure 4. Western blot reactivity of peptide antisera showing the proteolytic cleavage site of LCMV GPC. Lanes 1 to 4 represent the reaction of antisera to LIIQNRTWENHC, SQEKTKFFTR, and LSDSSGVENPGGYC and polyclonal antiserum to LCMV, respectively.

proteolytic enzymes responsible for affecting these cleavages are largely uncharacterized. In the alphaviruses and flaviviruses in particular, cleavage is thought to involve intracellular action of a combined trypsin and carboxypeptidaselike activity functioning late in the secretory pathway (23, 24). Our results with LCMV are consistent with this model. We previously reported that only the fully processed glycoprotein products reach the cell surface (4).

Extracellular or postviral maturation cleavages of coronavirus, paramyxovirus, orthomyxovirus, and retrovirus glycoproteins are associated with the activation of cell fusion potential mediated by one of the products of cleavage. We have no evidence yet for biological activity associated with the cleaved LCMV glycoproteins; however, the amino terminus of GP2 liberated by cleavage at the RR site consists of a predominantly hydrophobic stretch of amino acids, LAGTPTWTL. Currently, we are studying this and other regions of the LCMV glycoproteins, taking advantage of synthetic peptide technology to define important functional regions. We have also used these techniques to definitively identify the gene product of the L RNA of LCMV as a 200-kilodalton polypeptide associated with the ribonucleoprotein complex (Singh et al., in press).

One might reasonably ask whether these studies of LCMV glycoprotein structure and processing are relevant to infection in vivo. During the transition from acute to persistent infection following neonatal inoculation of mice, we have observed a down-regulation or modulation of the expression of LCMV glycoproteins in infected cells (17). This down-regulation occurs predictably following either in vivo or in vitro infections and ultimately results in a cell or tissue that expresses little viral glycoprotein but that has abundant viral nucleoprotein within its cytoplasm and no cytopathology. This state provides potential advantages for persistence, since the load of viral antigen is low and cells are competent to perform normally enough for survival. Therefore, it is clear that understanding the mechanism of glycoprotein gene regulation in acute and persistent states is a significant piece of the puzzle of persistence.

ANTIGENICITY OF THE ARENAVIRUSES

Arenaviruses differ in their susceptibility to antibody-mediated neutralization. Neutralizing antibodies to LCMV have been shown to be directed against the GP1 glycoprotein. Similarly, monoclonal antibodies against the single glycoprotein of Tacaribe virus mediated highly efficient virus neutralization (15). Moreover, by using competitive binding assays and analysis of neutralization-resistant mutants, it was possible to map two distinct epitopes on Tacaribe virus type G. One epitope, character-

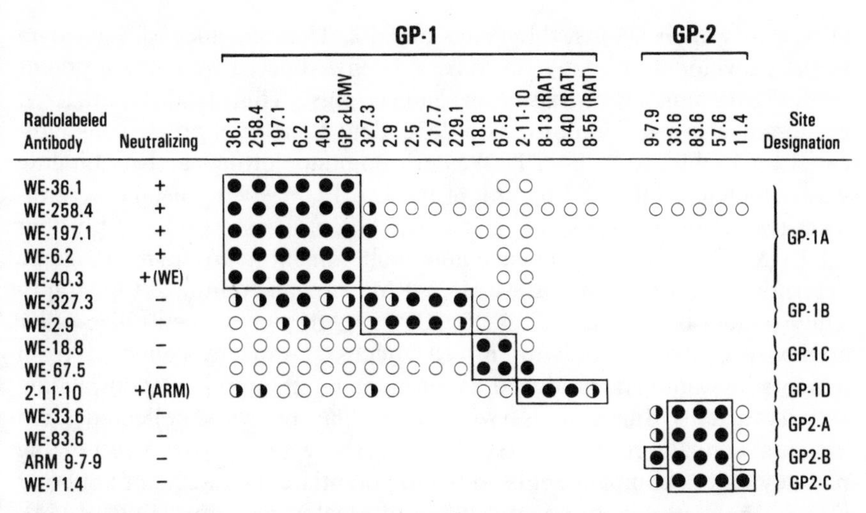

Figure 5. Summary of epitope mapping of LCMV monoclonal antibodies to GP1 and GP2 delineating four epitopes on GP1 and three on GP2. Symbols: ●, competitive binding greater than 90%; ○, competitive binding less than 20%; ◑, competitive binding of 40 to 80%. GP αLCMV indicates guinea pig immune serum (from reference 19).

ized by four monoclonal antibodies, was the target of highly efficient neutralization, whereas a single antibody to a second site was less so, leaving a large nonneutralizable persistent fraction. Failure to neutralize was not likely to be due to virus aggregation, since addition of a second antibody to the alternate site resulted in further reduction in virus titer. Analysis of neutralization kinetics for the highly efficient monoclonal antibody suggested that the reaction followed double-hit kinetics.

We have assessed the antigenic topography of the LCMV glycoproteins by using a large library of monoclonal antibodies against GP1 and GP2 to map the epitopes on these molecules (19). Elicitation of neutralizing monoclonal antibodies to LCMV in the BALB/c mouse was a relatively infrequent event. Only 6 of 46 antibodies to the LCMV glycoproteins neutralized virus infectivity in vitro. Five of these antibodies were raised against LCMV WE and were mapped by competition binding assay to a single conformation-dependent epitope (GP1a) shared by both ARM and WE strains and other LCMV strains (Fig. 5). The sixth neutralizing monoclonal antibody was uniquely specific for LCMV ARM, and its binding to that strain was only marginally affected by the other five antibodies, suggesting binding to a topographically related, but not identical, epitope (GP1d). Nonneutralizing monoclonal antibodies were found to be directed against two additional sites of GP1, (GP1b and

GP1c), as well as against three sites on GP2. The relevance of these data to the polyclonal antibody response was investigated by using a potent neutralizing antiserum raised in guinea pigs. This antibody reacted predominantly with conformation-dependent structures on GP1, as indicated by its failure to bind in Western immunoblotting, and its binding was completely inhibited by any of the five LCMV WE-specific neutralizing monoclonal antibodies against site GP1a. These results imply that the LCMV WE GP1 has a single immunodominant neutralizing antigenic determinant (GP1a) and that LCMV ARM bears an additional topographically related but not identical site (GP1d). Attempts to neutralize other arenaviruses have met with mixed success. Antisera collected from patients and antisera produced experimentally show potent neutralizing activity against Junin virus, whereas similar reagents collected from patients convalescent with Lassa fever show rather low neutralizing potency unless complement is added to potentiate the effects of antibody (21). Virus neutralization is discussed in greater depth by Howard (14), but from the brief treatment here it is evident that more information about the molecular nature of neutralizing antigenic determinants of arenaviruses is necessary before rational approaches to immunotherapy and immunization can be made. Obviously, one needs to define a structure which will elicit strong protective immune responses without the risk of triggering immunopathologic disease.

Toward this end, we have recently studied in detail a group-specific antigen conserved on the glycoproteins of all arenaviruses. In previous studies (6) we described a monoclonal antibody, 9-7.9, which cross-reacted between LCMV and the African arenavirus Mopeia (Mozambique) virus. Subsequently we isolated several more group-reactive monoclonal antibodies, and one of these, 33.6, is of particular interest. This monoclonal antibody reacts with GP2 glycoprotein of LCMV and cross-reacts with both New World (Pichinde, Junin, Tacaribe, Amapari, Parana, and Machupo) and Old World (LCMV and Lassa, Mopeia, and Mobala) viruses. Given this promiscuous cross-reactivity, we sought to define at the molecular level the binding site of 33.6. During this and other studies, we synthesized a series of nested peptides corresponding to over 90% of the GPC sequence. Monoclonal antibody 33.6 was screened for reactivity against this panel, and we found that it reacted only with one peptide, corresponding to GPC (GP2) residues 370 to 382 (Table 1). Further analysis demonstrated that both the 33.6 epitope and the previously described 9-7.9 epitope mapped within a nine-amino-acid segment spanning residues 370 to 378, which contains five amino acids conserved among LCMV, Lassa virus, and Pichinde virus (Fig. 6).

Assays with N-terminal deletions from this sequence suggested that

Table 1. Reactivity of Monoclonal Antibodies with Peptides 353 to 370, 370 to 382, and 378 to 391[a] in Enzyme-Linked Immunosorbent Assay

Monoclonal antibody	Reactivity with following peptide:		
	353–370	370–382	378–391
33.6[b]	<100	25,600	<100
83.6[b]	<100	6,400	<100
9-7.9[b]	<100	1,638,400	<100
2-11.10[c]	<100	<100	<100

[a] The peptide sequence is as follows:
353 370 382 391
GPC...DQLLMRNHLRDLMGVPYCNYSKFWYLEHAKTGETSVPKC...

[b] Monoclonal antibodies 33.6, 83.6, and 9-7.9 bind specifically to peptide 370 to 382.

[c] Negative control.

the minimal epitope recognized by the broadly cross-reactive monoclonal antibody 33.6 (epitope GP2a) consisted of five amino acids, GPC 374 to 378 (Lys-Phe-Trp-Tyr-Leu). Reactivity of a second monoclonal antibody, 9-7.9 (epitope GP2b), but not 33.6, was abolished when tyrosine was substituted for phenylalanine at position 375 in the antigenic sequence corresponding to a naturally occurring sequence difference between LCMV and Lassa virus.

The possibility of a single peptide antigen containing a universal arenavirus antigenic determinant was raised by these observations; therefore, we assayed convalescent-phase serum samples for evidence of reactivity with the peptide. Polyclonal serum samples from human patients and from animals experimentally infected with Junin virus,

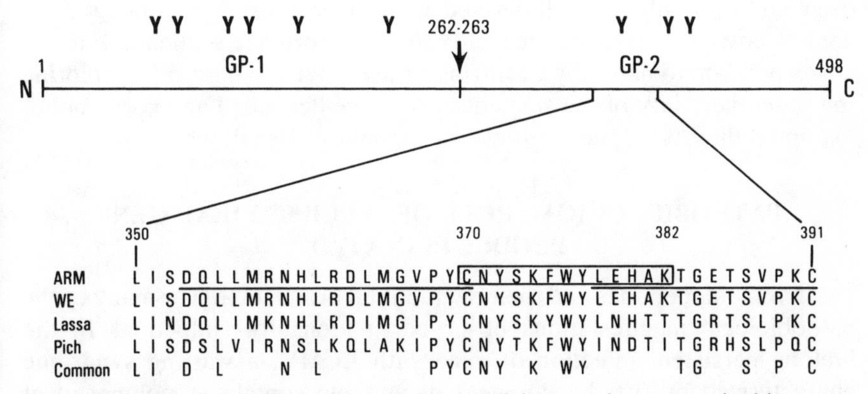

Figure 6. Schematic localization of a peptide containing an epitope recognized by monoclonal antibody 33.6 on all arenaviruses (box). Flanking peptides (underlined) were used as controls and were not reactive with 33.6 (see Table 1).

LCMV, and Lassa virus bound to the antigenic peptide GPC 370 to 382 but not to control peptides. As was the case with the monoclonal antibodies, this binding activity was abrogated by blocking with the antigenic peptide, but not with control peptides in solution.

The observation of conservation of this sequence across a broad spectrum of arenaviruses from the New World and Old World groups suggests that it represents an important functional or structural component of the virus. We have not as yet been able to assign functional significance to this site. We found that neither 33.6 nor 9-7.9 mediated complement-independent virus neutralization, although 9-7.9 showed modest neutralizing activity when guinea pig complement was added to the reaction mixture (7). In contrast, several monoclonal antibodies against GP1 have been shown to be strongly neutralizing in the absence of complement (3, 7, 19). The question of structural significance is a subject of current investigation.

In studies reported elsewhere in this volume, Whitton et al. have mapped at least one cytotoxic T-lymphocyte recognition epitope to LCMV GP2 amino acids 272 to 293. We are currently extending our studies of the sequence from 370 to 382 to determine whether it plays a significant role in either induction of or recognition by virus-primed helper or cytotoxic T lymphocytes.

The present studies may prove to be of practical importance. As described above, we have observed binding of polyclonal convalescent-phase serum samples directed against LCMV, Junin virus, and Lassa virus with the synthetic peptide, suggesting that it may be of use as a diagnostic antigen to detect antibody against arenaviruses. Such a peptide diagnostic reagent has several potential advantages over the currently used virus- or infected-cell-derived antigen preparations. Among these are low cost, stability, and the ability to rapidly provide sequence variants corresponding to naturally occurring viruses. We are currently exploring the potential utility of the sequences we have described and corresponding antibodies as diagnostic tools for arenavirus infections.

PATHOBIOLOGICAL ROLE OF SPECIFIC VIRAL GENE PRODUCTS IN VIVO

Viral polypeptides and their degradation products trigger many of the pathobiologic manifestations observed in arenavirus infection. In the lifelong persistent infection of mice with LCMV, a wasting syndrome characterized by the development of immune complexes composed of viral antigen and antiviral antibody has been well documented (8, 18). These complexes lodge in the renal glomeruli, where they trigger a

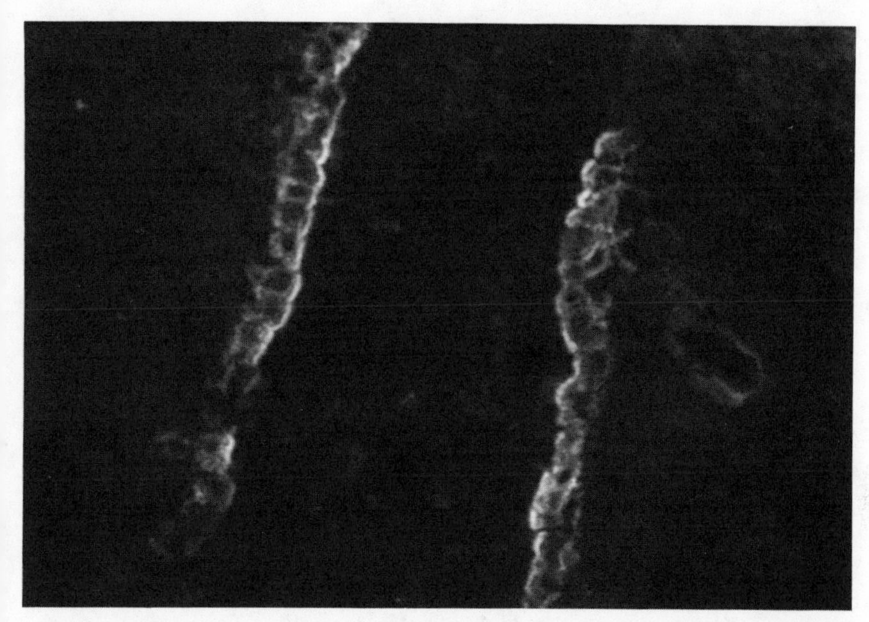

Figure 7. Immunofluorescent localization of LCMV GP1 along the ependymal cell layer of a BALB/c mouse infected 6 days earlier with virus. Monoclonal antibody 2-11.10 against GP1 was used.

chronic glomerulonephritis. At least one component of the virus has been identified in the glomeruli of diseased mice. Using a monospecific antibody to the NP of LCMV, we demonstrated colocalization of NP antigen and the host glomerular mesangium (8).

A role of NP in neuronal dysfunction during LCMV persistence has also been proposed. Rodriguez et al. (26) observed expression of NP in association with polyribosomes in the cytoplasm of neurons from widespread areas of the central nervous system. In contrast, no significant expression of viral glycoproteins was seen. Therefore, it was proposed that the presence of NP on the neuronal polyribosomes compromised their function.

Acute LCMV infection following intracerebral inoculation of the virus results in a fatal choriomeningitis (reviewed in reference 10). We have used monoclonal antibodies against individual viral structural proteins to study their expression in the central nervous system following acute infection (5). Viral GP1 is expressed on the apical surfaces of ependymal cells in the central nervous system (Fig. 7). At this site the glycoprotein (and perhaps also other virally encoded proteins) triggers the well-characterized immune response which results in choriomeningitis and death; this response appears to depend on the presence of virus-directed cytotoxic T lymphocytes.

Finally, the role of the viral L gene-encoded proteins in pathogenesis has recently been explored by Riviere et al. (25), who used genetic reassortants between strains of LCMV that differed in virulence for guinea pigs. They demonstrated that L RNA-encoded products were necessary for expression of the pathogenic potential of the virus. It is clear from such studies that understanding the molecular basis of viral persistence, regulation of viral gene products, and pathogenesis of arenavirus infections is an attainable goal.

Acknowledgments. We thank Rebecca Day, Gretchen Bloom, and Kaleo Wooddell for excellent technical assistance and Jim Johnston for manuscript preparation.

Portions of this work were supported by Public Health Service grant AI16102 from the National Institutes of Health and U.S. Army contract C6234. K. E. Wright was supported by a fellowship from the Arthritis Society of Canada.

Literature Cited

1. **Auperin, D. D., V. Romanowski, M. Galinski, and D. H. L. Bishop.** 1984. Sequencing studies of Pichinde arenavirus S RNA indicate a novel coding strategy, an ambisense viral S RNA. *J. Virol.* **52:**897–904.
2. **Bishop, D. H. L., and D. D. Auperin.** 1987. Arenavirus gene structure and organization. *Curr. Top. Microbiol. Immunol.* **133:**5–17.
3. **Bruns, M., J. Cihak, G. Muller, and F. Lehmann-Grube.** 1983. Lymphocytic choriomeningitis virus. VI. Isolation of a glycoprotein mediating neutralization. *Virology* **130:**247–251.
4. **Buchmeier, M. J., J. H. Elder, and M. B. A. Oldstone.** 1978. Protein structure of lymphocytic choriomeningitis virus: identification of the virus structural and cell associated polypeptides. *Virology* **89:**133–145.
5. **Buchmeier, M. J., and R. L. Knobler.** 1984. Experimental models for immune-mediated and immune-modulated diseases, p. 219–227. *In* P. Behan and F. Spreafico (ed.), *Neuroimmunology*. Raven Press, New York.
6. **Buchmeier, M. J., H. Lewicki, O. Tomori, and K. M. Johnson.** 1980. Monoclonal antibodies to lymphocytic choriomeningitis virus react with pathogenic arenaviruses. *Nature* (London) **288:**486–487.
7. **Buchmeier, M. J., H. A. Lewicki, O. Tomori, and M. B. A. Oldstone.** 1981. Monoclonal antibodies to lymphocytic choriomeningitis and Pichinde viruses: generation, characterization, and cross-reactivity with other arenaviruses. *Virology* **113:**73–85.
8. **Buchmeier, M. J., and M. B. A. Oldstone.** 1978. Virus-induced immune complex disease: identification of specific viral antigens and antibodies deposited in complexes during chronic lymphocytic choriomeningitis virus infection. *J. Immunol.* **120:**1297–1304.

9. **Buchmeier, M. J., and B. S. Parekh.** 1987. Protein structure and expression among arenaviruses. *Curr. Top. Microbiol. Immunol.* **133:**41–57.
10. **Buchmeier, M. J., R. M. Welsh, F. J. Dutko, and M. B. A. Oldstone.** 1980. The virology and immunobiology of lymphocytic choriomeningitis virus infection. *Adv. Immunol.* **30:**275–331.
11. **Gard, G. P., A. C. Vezza, D. H. L. Bishop, and R. W. Compans.** 1977. Structural proteins of Tacaribe and Tamiami virions. *Virology* **83:**84–95.
12. **Garoff, H., A. M. Frischauf, K. Simons, H. Lehrach, and H. Delius.** 1980. Nucleotide sequence of cDNA coding for Semliki Forest virus membrane glycoprotein. *Nature* (London) **288:**236–241.
13. **Harnish, D. G., K. Dimock, D. H. L. Bishop, and W. E. Rawls.** 1983. Gene mapping in Pichinde virus: assignment of viral polypeptides to genomic L and S RNAs. *J. Virol.* **46:**638–641.
14. **Howard, C. R.** 1987. Neutralization of arenaviruses by antibody. *Curr. Top. Microbiol. Immunol.* **134:**117–130.
15. **Howard, C. R., H. Lewicki, L. Allison, M. Salter, and M. J. Buchmeier.** 1985. Properties and characterization of monoclonal antibodies to Tacaribe virus. *J. Gen. Virol.* **66:**1383–1395.
16. **Hunkapiller, M. W., J. E. Strickler, and K. J. Wilson.** 1984. Contemporary methodology for protein structure determination. *Science* **226:**304–311.
17. **Oldstone, M. B. A., and M. J. Buchmeier.** 1982. Restricted expression of viral glycoprotein in cells of persistently infected mice. *Nature* (London) **300:**360–362.
18. **Oldstone, M. B. A., A. Tishon, and M. J. Buchmeier.** 1983. Virus-induced immune complex disease: genetic control of C1q binding complexes in the circulation of mice persistently infected with lymphocytic choriomeningitis virus. *J. Immunol.* **130:**912–918.
19. **Parekh, B. S., and M. J. Buchmeier.** 1986. Proteins of lymphocytic choriomeningitis virus: antigenic topography of the viral glycoproteins. *Virology* **153:**168–178.
20. **Pedersen, I. R.** 1973. LCM virus: its purification and its chemical and physical properties, p. 13–23. *In* F. Lehmann-Grube (ed.), *Lymphocytic Choriomeningitis Virus and other Arenaviruses*. Springer Verlag KG, Berlin.
21. **Peters, C. J.** 1984. Arenaviruses, p. 513–545. *In* R. G. Belshe (ed.), *Textbook of Human Virology*. PSG Publishing, Littleton, Mass.
22. **Ramos, B. A., R. J. Courtney, and W. E. Rawls.** 1972. Structural proteins of Pichinde virus. *J. Virol* **10:**661–667.
23. **Rice, C. M., E. M. Lenches, S. R. Eddy, S. J. Shin, R. L. Sheets, and J. H. Strauss.** 1985. Nucleotide sequence of yellow fever virus: implications for flavivirus gene expression and evolution. *Science* **229:**726–733.
24. **Rice, C. M., and J. H. Strauss.** 1981. Nucleotide sequence of the 26S mRNA of Sindbis virus and deduced sequence of the encoded virus structural proteins. *Proc. Natl. Acad. Sci. USA* **78:**2062–2066.
25. **Riviere, Y., R. Ahmed, P. J. Southern, M. J. Buchmeier, and M. B. A. Oldstone.** 1985. Genetic mapping of lymphocytic choriomeningitis virus pathogenicity: virulence in guinea pigs is associated with the L RNA segment. *J. Virol.* **55:**704–709.
26. **Rodriguez, M., M. J. Buchmeier, M. B. A. Oldstone, and P. W. Lampert.** 1983. Ultrastructural localization of viral antigens in the CNS of mice persistently infected with lymphocytic choriomeningitis virus (LCMV). *Am. J. Pathol.* **110:**95–100.

26a. **Singh, M. K., F. V. Fuller-Pace, M. J. Buchmeier, and P. J. Southern.** 1987. Analysis of the genomic L RNA segment from lymphocytic choriomeningitis virus. Virology **161:**448–456.

27. **Southern, P. J., and D. H. L. Bishop.** 1987. Sequence comparison among arenaviruses. *Curr. Top. Microbiol. Immunol.* **133:**20–39.

28. **Vezza, A. C., G. P. Gard, R. W. Compans, and D. H. L. Bishop.** 1977. Structural components of the arenavirus Pichinde. *J. Virol.* **23:**776–786.

Chapter 8

Role of Mononuclear Phagocytes in the Control of Lymphocytic Choriomeningitis Virus Infection of Mice

Fritz Lehmann-Grube

INTRODUCTION

If we were to look in standard textbooks of virology and immunology for information about how higher organisms protect themselves against viruses, we would probably read that they do so with a complex system of interacting cells and their products, in which cytotoxic T lymphocytes (CTL) and mononuclear phagocytes (MNP) play key roles. CTL are assumed to function by lysing infected cells in situ, just as they do in vitro, thereby preventing further virus replication (67), and MNP are assumed to function by forming the first line of defense, i.e., before activated elements of the immune system are available (38); MNP are also thought to assist T and B lymphocytes during the effector phase of virus elimination (5, 12, 46, 67).

There is plenty of evidence to support each of these possibilities, and the ability of MNP to diminish the spread of virus in experimentally infected animals before immunologically activated cells can intervene has been demonstrated in a number of examples, but otherwise, most of what is known about these cells comes from experiments performed in vitro. Nobody would deny that the use of animals should be avoided whenever possible. On the other hand, looking at the vast amount of information available and realizing that much of it was obtained under highly artificial conditions, it seems valid to ask whether and to what extent this knowledge is applicable to the intact organism. Experiments are de-

Fritz Lehmann-Grube • Heinrich-Pette-Institut für Experimentelle Virologie und Immunologie an der Universität Hamburg, 2000 Hamburg 20, Federal Republic of Germany.

scribed below that were designed to obtain at least a partial answer to this question. The findings seem to indicate that not all of the properties of these cells are expressed in like manner in vivo.

THE MODEL

The term infection denotes the entrance of an agent into a macroorganism or cell, followed by its multiplication, whereas (infectious) disease refers to the pathologic consequences. Although the course of infection and the events leading to illness may overlap to some extent, the two phenomena are fundamentally different and should be clearly distinguished. Studying the mechanism by which an organism controls a virus requires localization and quantitation of the latter in the tissues of the experimental animal. Illness and death, while conveniently registered, are unsuitable parameters for this purpose. In the ideal model, pathologic alterations would be totally absent and hence not able to confound the infectious process. I do not know whether such an ideal model exists, but the adult mouse infected intravenously (i.v.) with a moderate dose of strain WE lymphocytic choriomeningitis virus (LCMV) comes very close.

After i.v. inoculation of 10^3 mouse infectious units (IU), the virus multiplies in all major organs and reaches maximal titers within 4 to 6 days, after which it is cleared. The animals remain outwardly healthy, and histologic inspection reveals minor changes in most tissues, although hepatitis may be prominent. We have studied the spleen, because in this organ the virus rapidly attains concentrations as high as almost 10^9 IU/g of tissue and is subsequently as rapidly eliminated, the infectious titer being below detectable levels usually on day 10 (29). Furthermore, it had been shown that 8 days after infection of mice with LCMV the spleen MNP were activated, which seemed to suggest that they contributed to the ability of the spleen to cope with the virus (14).

EXPERIMENTAL RESULTS

The general applicability of the opinion that MNP, especially if activated, are important effector cells in the defense against virus infections was examined by monitoring the replication of LCMV in and its subsequent elimination from the spleens of mice whose MNP had been either activated or blocked by various treatments (30; F. Lehmann-Grube, I. Krenz, T. Krahnert, R. Schwachenwald, D. Moskophidis, J. Löhler, and C. J. Villeda Posada, *J. Immunol.*, in press). The functional state of the MNP was measured as the ability to control the replication of *Listeria monocytogenes* essentially as described by Blanden and Mims

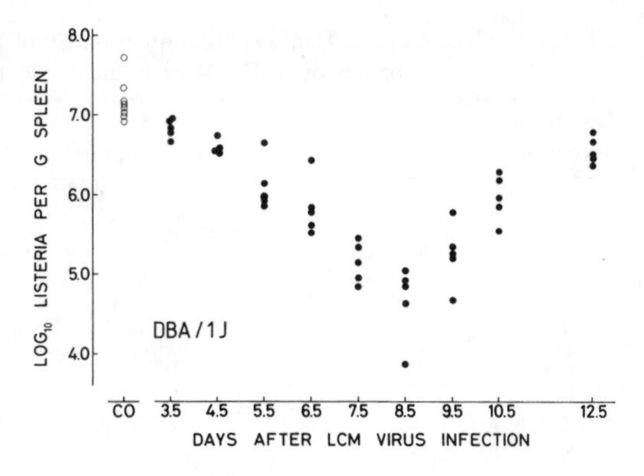

Figure 1. Activation of MNP in spleens of DBA/1J mice undergoing infection with LCMV. Beginning on day 3 after i.v. infection with 10^3 IU of LCMV, mice were injected i.v. at 1-day intervals with 10^5 CFU of *L. monocytogenes*. After an additional 24 h, individual spleens were homogenized, and viable listeria titers were determined by using tryptic soy agar. Each symbol denotes the number of bacteria per gram in the organ of one mouse.

(14); in a number of experiments it was also measured by determining the activity of cell-associated nonspecific esterase and acid phosphatase by standard methods (9, 25, 33). The findings to be reported were obtained with mice of different strains which vary considerably with respect to LCMV-specific cell-mediated immunity (31, 50). Every experiment was repeated at least once, often with mice of different strains; major dissimilarities of results were never recorded.

MNP Activation during LCMV Elimination from Murine Spleens

In the spleens of DBA/1J mice (high responders for LCMV-specific cell-mediated immunity) infected with LCMV, the replication of *L. monocytogenes* was greatly diminished, indicating marked activation of MNP (Fig. 1); the effect was maximal between days 7 and 10, i.e., the time during which virus was eliminated (29). Similar results were obtained with low-responder CBA/J mice (30) and high-responder NMRI mice (Lehmann-Grube et al., in press). Increase of MNP-specific enzymes (Table 1) was evident as early as 1 or 2 days after infection, which is noteworthy, because at this time the virus was still multiplying and virus-specific cell-mediated immunity could not yet be demonstrated.

Lack of MNP Activation during LCMV Elimination from Spleens of Immunized Mice

Transfer of spleen T lymphyocytes from a mouse just clearing LCMV into an infected recipient very effectively speeds up removal of the virus from the spleen of the recipient (29, 39, 68). Injection of much larger numbers than needed for efficient virus elimination did not influence the replication of *L. monocytogenes* in C57BL/6J mice (Fig. 2) or in CBA/J mice (Lehmann-Grube et al., in press), and levels of hydrolytic enzymes

Table 1. Cytochemical Demonstration of Increase of Intracellular Enzymes in MNP of Spleens of NMRI Mice Acutely Infected with LCMV[a]

Day after infection	Acid phosphatase[b]	Nonspecific esterase[b]
0[c]	(+)	(+)
1	+/++	+
2	++	+
3	++/+++	++
4	++/+++	++
5	+++	++/+++
6	+++	++/+++
7	+++	+++
8	+++	+++
9	+++	+++
10	++/+++	++/+++
12	++	++

[a] At the indicated days after i.v. infection with 10^3 IU, spleen cells were dispersed, deposited on slides by use of a cytocentrifuge, and stained for enzymes as described previously (9, 25, 33).
[b] Assessed on an arbitrary scale ranging from 0 (no expression) to +++ (marked enzyme-specific staining).
[c] No virus (control).

were not increased in the MNP of recipient spleens as revealed cytochemically in cytocentrifuge preparations.

Effect of MNP Activation on LCMV Replication

In conventionally maintained athymic nude mice, macrophages and related cells are permanently activated (16, 53, 66). We found that in the spleens of such animals, LCMV multiplied slightly less rapidly than in the spleens of corresponding *nu*/+ or +/+ mice. The differences were small, but were consistently found in several mouse strains (Lehmann-Grube et al., in press).

"*Corynebacterium parvum*" (now *Propionibacterium acnes*) is a potent activator of the mononuclear phagocyte system (24, 35). We observed maximal effects after i.v. inoculation of 1.4 mg of a suspension of Formalin-inactivated "*C. parvum*" per mouse; activation was highest on days 8 to 12 after injection (Fig. 3) and, again, was apparent when assessed on the basis of cytochemical demonstration of enzymes (Table 2). With this information, the replication of LCMV in the spleens of "*C. parvum*"-treated mice was monitored. In repeated experiments, considerably higher titers were found in treated than in untreated mice (Table 3), but the clearance rates were not affected by the immunomodulation (Lehmann-Grube et al., in press).

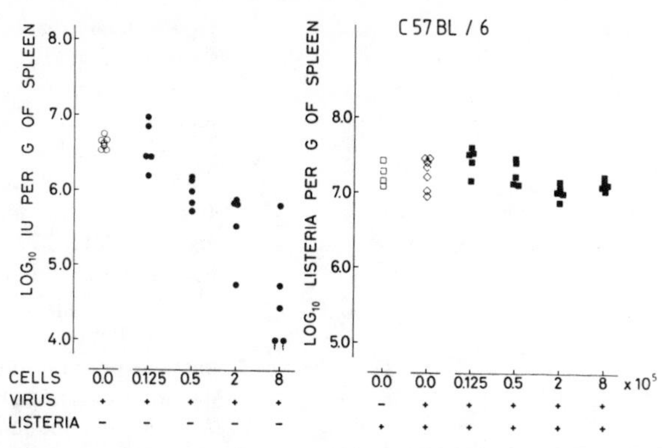

Figure 2. Absence of activation of MNP in spleens of LCMV-infected C57BL/6J mice during accelerated virus clearance due to adoptive immunization. Immune T lymphocytes were T cell-enriched spleen cells from mice 8 days after their i.v. infection with 10^3 IU. Syngeneic recipient mice were infected i.v. with 10^3 IU and injected i.v. 16 h later with immune T lymphocytes in the numbers indicated. Half of the mice were inoculated i.v. with 10^5 CFU of *L. monocytogenes* 8 h thereafter (24 h after infection). After an additional 24 h (48 h after infection with virus), spleens were homogenized and assayed, respectively, for LCMV and viable listerias. Each symbol denotes the number of IU of LCMV or bacteria per gram in the organ of one mouse. Note the control depicted on the right side of the figure, showing that the virus infection had not influenced the bacterial growth.

Effect of MNP Impairment on LCMV Replication

Of the substances assumed to selectively impair the function of cells of the MNP system, the best known are silica (7), carrageenan (10, 55), and dextran sulfate of molecular weight 500,000 (DS500) (19, 21). After preliminary experiments, we chose to use DS500, although it was found to be rather toxic: 100 mg/kg injected intraperitoneally (i.p.) killed a proportion of LCMV-infected mice, but inoculation of 50 mg/kg was well tolerated.

As mentioned above, MNP are activated in LCMV-infected mice. Inoculation of DS500 at 4, 5, or 6 days after virus inoculation reversed this effect for 3 days, 2 days, and 1 day, respectively (Fig. 4). Damage of MNP by DS500 contrasts with the effect of "*C. parvum*," which markedly increased the activity of these cells (see above). Despite this difference, treatment with DS500 resulted in enhanced multiplication of LCMV in the spleens (Table 4), although elimination was slightly retarded (Table 5).

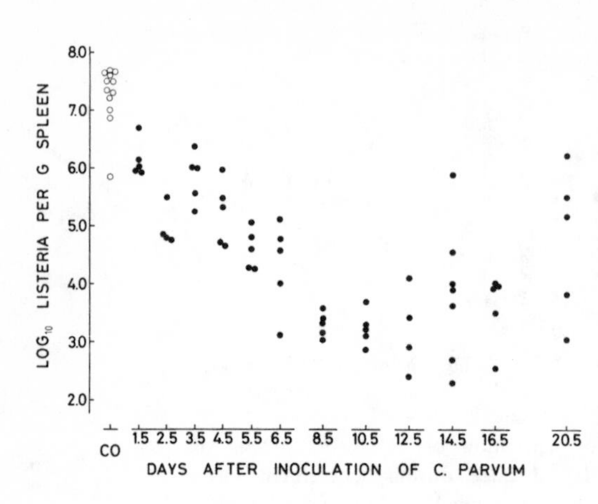

Figure 3. Activation of MNP in spleens of NMRI mice treated with "*C. parvum*." At 1-day intervals after i.v. inoculation of 1.4 mg of Formalin-inactivated "*C. parvum*," mice were injected with 10^5 viable *L. monocytogenes* organisms. After 24 h, spleens were homogenized and bacteria were counted on tryptic soy agar. Each symbol denotes the number of bacteria per gram in the organ of one mouse. Reproduced from the *Journal of Immunology* (Lehmann-Grube et al., in press) with permission from The American Association of Immunologists.

Hahn had demonstrated that the DS500-mediated block of elimination of *L. monocytogenes* from the murine spleen was vigorous but lasted only for a few days (19), and the data in Fig. 4 lead to the same conclusion. The stimulating effect of DS500 on virus replication was most marked when DS500 was injected 1 or 2 days after infection (Table 4), which was 3 to 4 days before virus titers were ascertained in the spleen and at the time during which the MNP were just recuperating from the damage caused by DS500 (Fig. 4). These findings made us suspect that the enhanced virus

Table 2. Cytochemical Demonstration of Increase of Intracellular Enzymes in MNP of Spleens of NMRI Mice after Injection of "*C. parvum*"[a]

Day after "*C. parvum*" injection	Acid phosphatase[b]	Nonspecific esterase[b]
0[c]	(+)	(+)
1	+	(+)
2	+++	++
4	+++	++ – +++
6	+++	++ – +++
8	+++	+++
10	+++	+++
14	+++	+++

[a] At the indicated days after i.v. inoculation of 1.4 mg of inactivated "*C. parvum*" per mouse, spleen cells were dispersed, deposited on slides by use of a cytocentrifuge, and stained for enzymes as described previously (9, 25, 33).
[b] Assessed on an arbitrary scale ranging from 0 (no expression) to +++ (marked enzyme-specific staining).
[c] No "*C. parvum*" (control).

Table 3. Effect of "*C. parvum*" Activation of MNP on Replication of LCMV in Spleens of NMRI Mice

"*C. parvum*" treatment[a]	Amt of LCMV (IU/g of spleen) at following day after infection[b]:	
	4[c]	5[d]
Yes	$(51.9 \pm 9.5) \times 10^7$	$(65.2 \pm 20.1) \times 10^7$
No (control)	$(9.4 \pm 4.2) \times 10^7$	$(5.4 \pm 2.1) \times 10^7$

[a] Six days before i.v. infection with 10^3 IU of LCMV, mice were each inoculated i.v. with 1.4 mg of "*C. parvum*." At the indicated intervals, virus concentrations in spleens were determined.
[b] Mean ± standard error for five mice.
[c] $P < 0.01$.
[d] $P < 0.05$.

replication following treatment with DS500 resulted from activation of MNP rather than their inactivation. In a further experiment, the effect of DS500 on the ability of spleen MNP to deal with *L. monocytogenes* (in uninfected mice) was determined. Impairment of function was evident only for 2 days (Fig. 5); 1 day later there was no significant difference between the two groups of mice, and during the following 4 days the numbers of colony-forming listeria organisms were lower in the spleens of treated animals, indicating that the MNP of these animals were activated. These findings suggest that multiplication of LCMV is increased in the

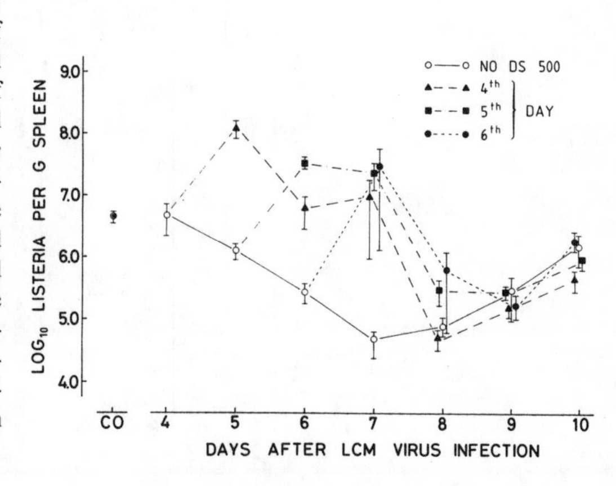

Figure 4. Effect of DS500 treatment of NMRI mice on LCMV-induced activation of MNP in spleens. Mice were infected i.v. with 10^3 IU and inoculated i.p. 4, 5, or 6 days later with 50 mg of DS500 per kg. At 12 h later and after additional time intervals of 24 h, the animals were inoculated with 10^5 viable *L. monocytogenes* organisms; 24 h thereafter, bacterial numbers in the spleens were determined. Datum points and vertical bars denote means and standard errors of results with five mice. Reproduced from the *Journal of Immunology* (Lehmann-Grube et al., in press) with permission from The American Association of Immunologists.

Table 4. Effect of DS500 Inoculated at Intervals before and after Virus on LCMV Replication in Spleens of NMRI Mice[a,b]

Dose of DS500 (mg/kg)[c]	Mean amt of LCMV ± SE (IU/g of spleen) at following day of DS500 inoculation[d] (no. of mice)						
	No DS500	−1	0	+1	+2	+3	+4
50	$(8.4 \pm 3.0) \times 10^7$ (5)	$(3.5 \pm 2.2) \times 10^8$ (4)	$(2.2 \pm 0.9) \times 10^8$ (5)	$(4.6 \pm 1.4) \times 10^8$ (5)	$(4.1 \pm 1.5) \times 10^8$ (5)	$(2.3 \pm 0.8) \times 10^8$ (5)	$(1.1 \pm 0.2) \times 10^8$ (3)
100	$(5.5 \pm 1.8) \times 10^7$ (5)	$(6.3 \pm 2.8) \times 10^8$ (5)	$(4.0 \pm 0.7) \times 10^8$ (3)	$(1.2 \pm 0.6) \times 10^9$ (4)	$(1.3 \pm 0.2) \times 10^9$ (5)	$(2.6 \pm 0.8) \times 10^8$ (5)	$(1.1 \pm 0.3) \times 10^8$ (5)

[a] Reproduced from the *Journal of Immunology* (Lehmann-Grube et al., in press) with permission from The American Association of Immunologists.
[b] Mice were each infected i.v. with 10^3 IU of LCMV, and 5 days later virus concentrations in the spleens were determined.
[c] Individual experiments for doses of 50 and 100 mg/kg.
[d] Day relative to virus infection at which DS500 was inoculated i.p.

Table 5. Effect of DS500 on Elimination of LCMV from Spleens of NMRI Mice[a]

Day after infection	Amt of LCMV (IU/g of spleen)[b] at following day of DS500 inoculation[c]			
	No DS500	4	5	6
4	$(1.3 \pm 0.7) \times 10^8$			
5	$(2.8 \pm 0.3) \times 10^8$	$(1.2 \pm 0.1) \times 10^8$		
6	$(1.9 \pm 1.1) \times 10^7$	$(6.5 \pm 3.3) \times 10^7$	$(2.0 \pm 1.9) \times 10^7$	
7	$(8.0 \pm 3.2) \times 10^5$	$(5.8 \pm 0.8) \times 10^7$	$(2.1 \pm 1.0) \times 10^7$	$(9.4 \pm 3.5) \times 10^5$
8	$(3.1 \pm 1.5) \times 10^5$	$(2.4 \pm 1.1) \times 10^5$	$(2.3 \pm 1.4) \times 10^5$	$(2.4 \pm 1.2) \times 10^5$

[a] Mice were infected i.v. with 10^3 IU of LCMV, and at the indicated intervals virus concentrations in the spleens were determined.
[b] Mean ± standard error for five mice.
[c] Day postinfection at which mice each received 50 mg of DS500 per kg of body weight i.p.

spleens of mice whose MNP are activated whether by treatment with "*C. parvum*" or by treatment (after some delay) with DS500.

Influence of Pretreatment with γ Rays or Cyclophosphamide on LCMV Elimination from Murine Spleens

It has been assumed that circulating monocytes may be more important in antiviral defense than tissue macrophages are (13), and rapidly multiplying cells in the bone marrow have been shown to be indispensable for the T-lymphocyte-mediated control of *L. monocytogenes* in the mouse (22); they are sensitive to irradiation and cyclophosphamide and probably represent the progenitors of circulating monocytes (20, 61). We found that neither irradiation nor treatment with cyclophosphamide prior to adoptive immunization diminished virus removal from the spleens of the recipients (Tables 6 and 7). Changes of mouse strain,

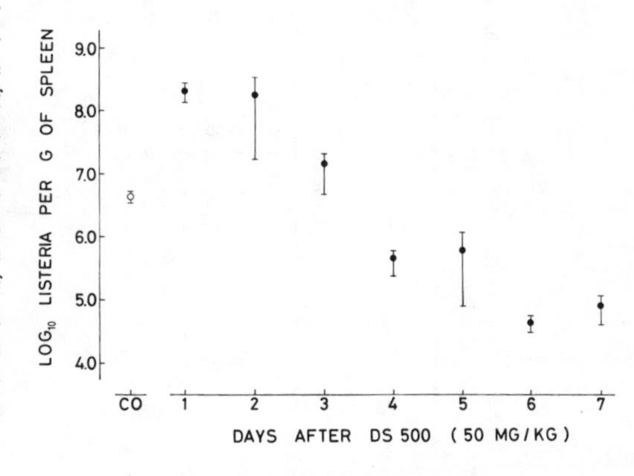

Figure 5. Effect of DS500 treatment of mice on the state of activity of spleen MNP. NMRI mice were injected i.p. with 50 mg of DS500 per kg. At 12 h later and after further intervals of 24 h, they were inoculated i.v. with 10^5 viable *L. monocytogenes* organisms; 24 h thereafter, the spleens were homogenized and numbers of listeriae were determined with tryptic soy agar. Datum points and vertical bars denote, respectively, means and standard errors from five mice.

Table 6. Effect of Whole-Body Irradiation prior to Adoptive Immunization of LCMV-Infected CBA/J Mice on Elimination of Virus from Spleens[a,b]

Day irradiated[c]	Mean amt of LCMV ± SE (IU/g of spleen) (no. of mice) after following dose of γ rays[d]					
	400 rad		600 rad		800 rad	
	Control[e]	Cells[f]	Control	Cells	Control	Cells
−10	$(5.6 \pm 1.1) \times 10^7$ (5)	$(3.7 \pm 2.7) \times 10^5$ (5)	ND[g]	ND	ND	ND
−8	$(14.4 \pm 6.0) \times 10^7$ (5)	$(0.8 \pm 0.2) \times 10^5$ (5)	$(8.9 \pm 3.2) \times 10^7$ (5)	$(50.7 \pm 18.8) \times 10^5$ (7)	ND	ND
−6	$(11.3 \pm 3.8) \times 10^7$ (5)	ca. 0.5×10^5 (5)	$(7.1 \pm 2.7) \times 10^7$ (5)	$(23.8 \pm 6.9) \times 10^5$ (6)	$(2.1 \pm 1.1) \times 10^7$ (5)	ca. 0.6×10^5 (6)
−4	$(21.3 \pm 2.6) \times 10^7$ (5)	$(5.8 \pm 3.4) \times 10^5$ (5)	$(11.3 \pm 3.4) \times 10^7$ (5)	$(52.8 \pm 24.7) \times 10^5$ (5)	$(6.3 \pm 1.3) \times 10^7$ (5)	ca. 0.6×10^5 (6)
−2	$(13.1 \pm 5.6) \times 10^7$ (5)	$(0.7 \pm 0.2) \times 10^5$ (5)	$(11.0 \pm 4.1) \times 10^7$ (5)	$(4.5 \pm 2.3) \times 10^5$ (5)	$(0.3 \pm 0.1) \times 10^7$ (5)	$<0.2 \times 10^5$ (5)

[a] Reproduced from the *Journal of Immunology* (Lehmann-Grube et al. in press) with permission from The American Association of Immunologists.
[b] At intervals after irradiation, mice were infected i.v. with 10^4 IU of LCMV and injected with immune cells 16 h later. At 40 h after cell transfer, spleens were homogenized and virus concentrations were determined.
[c] Time of irradiation relative to transfer of immune cells.
[d] Individual experiments for irradiation doses of 400, 600, and 800 rad.
[e] No cells.
[f] Inoculation i.v. of 10^6 day-8-immune T-lymphocyte-enriched spleen cells.
[g] ND, Not done.

Table 7. Effect of Cyclophosphamide on Elimination of LCMV from Spleens of CBA/J Mice Adoptively Immunized with Syngeneic Immune T Lymphocytes[a,b]

No. of cells transferred	Amt of LCMV (IU/g of spleen)[c] at following time (h) of cyclophosphamide administration[d]				
	No cyclophosphamide	−144	−120	−96	−72
0	$(5.7 \pm 1.4) \times 10^7$	$(1.0 \pm 0.6) \times 10^7$	$(1.9 \pm 1.2) \times 10^7$	$(10.2 \pm 5.0) \times 10^7$	$(4.2 \pm 1.9) \times 10^7$
5×10^5	$(4.3 \pm 2.1) \times 10^5$	$(1.2 \pm 0.9) \times 10^5$	$(1.8 \pm 0.8) \times 10^5$	$(10.3 \pm 5.8) \times 10^5$	$(2.5 \pm 0.9) \times 10^5$

[a] Reproduced from *Medical Microbiology and Immunology* (30) with permission from Springer-Verlag, Berlin.
[b] Donor mice were infected i.v. with 10^3 IU. Eight days later their spleen cells were enriched for T lymphocytes by passage through nylon wool columns and counted as "living" on the basis of trypan blue exclusion. Recipient mice were treated with cyclophosphamide and infected i.v. with 10^3 IU at intervals thereafter; 16 h after infection they were inoculated with day-8-immune T lymphocytes, and virus concentrations in the spleens were determined 40 h later (56 h after infection).
[c] Mean ± standard error for five mice.
[d] Time, relative to cell transfer, of i.p. inoculation of each recipient mouse with 200 mg of cyclophosphamide per kg of body weight.

time interval between treatment and cell transfer, and irradiation dose did not alter the results (Lehmann-Grube, in press). Thus, a fundamental difference seems to exist between the control by the mouse of *L. monocytogenes* on one hand and LCMV on the other.

Disparate results were reported by Thomsen and Volkert (59), who γ-irradiated C3H/Scs mice with 800 rad 24 h before infecting them i.p. with 10^3 50% lethal doses of LCMV Traub, injected them with 10^8 day-8-immune spleen lymphocytes 3 h later, and determined the virus titers in the livers of the recipients 5 days thereafter. In contrast to the situation with unirradiated controls, no adoptive immunization was achieved. In similar experiments, we found that preirradiation of the recipients with 600 rad did not reduce the ability of transfused immune spleen cells to hasten removal of the virus from the livers of i.v.-infected recipients (Table 8); on the contrary, we found that preirradiation greatly enhanced virus clearance from the liver as a result of adoptive immunization. I have no explanation for these differences, but it should be pointed out that neither the virus strain nor the protocol was identical.

Effect of DS500 Pretreatment on Adoptive Immunization

In contrast to monocyte progenitors, tissue macrophages remain functionally intact if exposed even to high doses of irradiation (17), and treatment with cyclophosphamide activates MNP in the spleen (61; F. Lehmann-Grube, unpublished observations). Thus, the experiments out-

Table 8. Effect of Adoptive Immunization of Irradiated C57BL/6J Mice with Day-8-Immune Spleen Cells on LCMV Concentrations in Spleens and Livers[a]

Day after transfer	Amt of LCMV (IU/g of tissue) in following organ[b]:					
	Spleen			Liver		
	10^8 Cells added	Control	% Residual[c]	10^8 Cells added	Control	% Residual[c]
1	$(3.4 \pm 1.1) \times 10^8$	$(2.6 \pm 0.5) \times 10^8$	132.0	ca. 4×10^5	$(4.7 \pm 0.7) \times 10^6$	ca. 9
2	$(5.5 \pm 2.2) \times 10^7$	$(2.5 \pm 0.4) \times 10^8$	22.4	ca. 9×10^4	$(1.0 \pm 0.2) \times 10^7$	ca. 1
3	$(7.8 \pm 4.4) \times 10^6$	$(1.7 \pm 0.4) \times 10^9$	0.5	$<7 \times 10^4$	$(1.6 \pm 1.1) \times 10^8$	<0.1
4	$(1.5 \pm 0.4) \times 10^6$	$(2.3 \pm 0.2) \times 10^9$	<0.1	$<7 \times 10^4$	$(1.3 \pm 0.4) \times 10^9$	<0.1

[a] Donor mice were infected by i.v. inoculation of 10^3 IU of virus. Eight days later, their spleen cells were counted as "living" on the basis of trypan blue exclusion and inoculated i.v. into syngeneic recipient mice that had been irradiated with 650 rads of γ rays and i.v. infected with 10^5 IU 5 days previously.
[b] Mean ± standard error for three mice.
[c] Residual infectious virus in recipients of immune cells.

lined above do not exclude a role for this subpopulation of MNP (even though they are not activated) in virus reduction due to adoptive immunization. Since DS500 is known to severely damage macrophages (21), the effect on infected recipients of DS500 pretreatment of immune cells to rapidly eliminate the virus was determined. Repeated experiments disclosed that DS500 diminished virus elimination from spleens of adoptively immunized mice (Table 9). The effect could be overcome by

Table 9. Effect of DS500 Treatment on Elimination of LCMV from Spleens of CBA/J Mice Adoptively Immunized with Syngeneic Immune Spleen Cells[a]

No. of cells transferred[b]	Amt of LCMV (IU/g of spleen)[c] (%)[d] after:	
	DS500 pretreatment	No pretreatment
0	$(1.8 \pm 0.7) \times 10^8$ (100)	$(9.9 \pm 1.1) \times 10^7$ (100)
1×10^8	$(4.5 \pm 1.9) \times 10^5$ (<1)	$(1.6 \pm 0.4) \times 10^5$ (<1)
2.5×10^7	$(3.0 \pm 1.7) \times 10^7$ (17)	$(2.6 \pm 0.8) \times 10^5$ (<1)
6.3×10^6	$(2.5 \pm 1.1) \times 10^7$ (14)	$(7.7 \pm 1.7) \times 10^5$ (<1)
1.6×10^6	$(2.1 \pm 1.1) \times 10^8$ (117)	$(1.3 \pm 0.4) \times 10^7$ (13)

[a] At 32 h before transfer of immune cells, recipient mice were infected i.v. with 10^3 IU of LCMV and inoculated i.p. with 50 mg of DS500 per kg of body weight 8 h later (24 h before transfer); 24 h after transfer, virus concentrations in the spleens were determined.
[b] Donor mice were immunized by i.v. infection with 10^3 IU of LCMV. Eight days later their spleen cells were counted as "living" on the basis of trypan blue exclusion and inoculated i.v.
[c] Mean ± standard error for five mice.
[d] Residual infectious virus in recipients of immune cells.

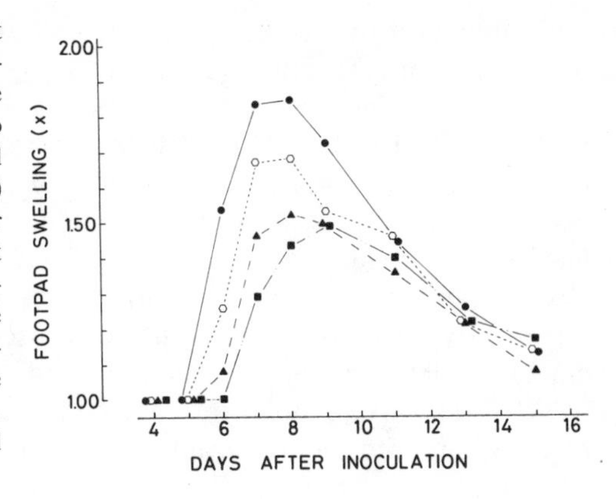

Figure 6. Effect of DS500 treatment of NMRI mice on LCMV-specific delayed-type hypersensitivity. Mice were infected subcutaneously into one hind foot with 10^5 IU (●) and inoculated i.p. 3 (○), 4 (▲), or 5 (■) days later with 50 mg of DS500 per kg. At intervals, thicknesses of feet were measured. Swelling is expressed as the factor with which the mean of the thicknesses of the inoculated feet exceeds the mean of the thicknesses of the uninoculated feet. Datum points signify values obtained with 10 mice.

increasing the numbers of transferred immune cells; however, this was probably not due to replenishment of damaged macrophages in the spleen by corresponding cells in the inoculum (Lehmann-Grube et al., in press).

For the control of LCMV in the acutely infected adult mouse, CTL and, possibly, T lymphocytes mediating delayed-type hypersensitivity are essential (8, 49, 59, 60, 68), and we considered the possibility that DS500 inhibited T-lymphocyte activity in vivo as it had been shown to do in vitro (63) and exerted its effect on adoptive immunization by damaging the injected immune T lymphocytes in the recipient. Repeated experiments revealed that DS500 decreased in LCMV-infected mice both virus-specific CTL activity (Lehmann-Grube et al., in press) and delayed-type hypersensitivity (measured as footpad swelling after local inoculation of the virus) (Fig. 6). Furthermore, CTL activities in spleens of mice that had received immune spleen cells 24 h previously were lower in DS500-treated than in untreated animals (Lehmann-Grube et al., in press), and we think that toxic injury of T lymphocytes is a plausible explanation for the deleterious effects of DS500 on LCMV cell-mediated immunity as well as on virus elimination after adoptive immunization.

DISCUSSION

The evidence for the widely held view that macrophages and other elements of the MNP system are intimately involved in the control of virus infections has been extensively summarized in a number of reviews (4–6, 12, 13, 18, 38, 41, 45, 46, 48; P. S. Morahan and D. M. Murasko, *in* D. S. Nelson, ed., *Natural Immunity*, in press). Besides presenting antigen to T lymphocytes (37, 62), MNP may be intrinsically resistant to virus infection by taking up and degrading virus internally, limiting its replication, or restricting its spread to other cells (46). As to our model, cultivated murine peritoneal macrophages are intrinsically resistant to LCMV, because each cell produces approximately 500 times less infectious virus than similarly cultivated and infected primary mouse fibroblasts do (56); furthermore, peritoneal macrophages taken from mice that had been treated with "*C. parvum*" did not produce any LCMV (I. Krenz, personal communication). Often intrinsic resistance as determined in vitro is associated with low susceptibility of the experimental animal, which sometimes is taken as evidence for the protective function of MNP. However, the experience with the *Mx* gene-mediated resistance of certain inbred mouse strains against several myxoviruses makes one wary, because in this case it has been conclusively shown that the same kind of correlation does not signify causality (23, 32). Macrophages also mediate extrinsic resistance (the ability to inhibit virus growth in neighboring cells) and furthermore have been shown to destroy virus-infected cells directly or through antibody-dependent cell-mediated cytotoxicity (46).

There can be little doubt that these antiviral properties reflect entirely different modes of action, and probably either type results from a number of quite dissimilar cellular functions. These have been thoroughly discussed by Morahan et al. in a number of reviews (46, 48; Morahan and Murasko, in press). Not all possible mechanisms are fully understood, and there may be more as yet unknown ways which enable MNP to interfere with virus replication, but one generalizing statement can safely be made: the expression of antiviral function, whether intrinsic or extrinsic, is usually dependent on the activation of these cells or is most clearly apparent when they are activated. If this is the rule, then our finding that LCMV multiplied extensively in the spleens of mice whose MNP were activated is a noteworthy exception. Indeed, much higher virus concentrations were attained in the spleens of "*C. parvum*"-treated animals than in the spleens of untreated animals. The results were similar with mice whose MNP had been impaired by DS500, which seemed a paradox until it was found that the damaging effect of DS500 was of short duration and was followed by activation. Thus, in this case also, it was probably functional activation rather than depletion of the MNP that led to enhanced replication of LCMV. There are other cases in which higher virus titers were found in organs of mice treated with compounds assumed to selectively damage cells of the MNP system, and these titers could be explained by such a rebound effect (47, 58, 69). Among these is the LCMV-infected mouse in whose liver the virus attained higher concentrations if the animal was pretreated with carrageenan (59).

Further findings reported here are at variance with the view that activated MNP aid in clearance. During the time the virus rapidly disappeared following adoptive immunization, MNP were not demonstrably activated, and the rapid decline of virus titers in the spleens of adoptively immunized mice was not affected by pretreatment of the recipients with γ rays or cyclophosphamide. The slightly slower diminution of virus concentrations in the spleens of adoptively immunized mice that had been pretreated with DS500 was probably due to direct effects of the drug on the injected CTL; Thomsen and Volkert (59) had not seen any effect on the accelerated clearance of LCMV from the livers as a result of transfusion of immune cells if the recipient mice had been pretreated with carrageenan, and we did not observe any increase in rate of virus elimination from the spleens when the MNP had been activated by treatment of the animals with "*C. parvum*" vaccine.

Although the evidence for assuming that MNP possess antiviral properties is impressive, most of it is based on experiments performed in vitro; in the work done in vivo, illness and death were usually made the basis of evaluation, which, as has been pointed out above, cannot be considered suitable for elucidating the course of an infection. The early work on the barrier function of macrophages against invading viruses has been summarized by Mims (38), and I found a few further communications in which MNP were shown to restrict in vivo dissemination of viruses (42–44). This list is not regarded as being complete, but reports of in vivo work directed at furthering our knowledge of the reduction of virus multiplication by MNP are indeed scarce. Only one account has come to my attention in which the effect of MNP on the effector phase of virus elimination was investigated. Blanden irradiated mice with 800 rad of γ rays and determined the effect on the elimination of ectromelia orthopoxvirus from livers and spleens after adoptive immunization. In preirradiated recipients of immune cells, the capability of clearing the virus was reduced, but was substantially higher than in irradiated mice that had not received immune cells (11). These findings do not clearly answer the question of whether radiosensitive elements of the MNP system, e.g., monocytes, had contributed to recovery, because in this model the virus itself is cytopathogenic, which makes it difficult to separate the effects on virus titers of scavenging by macrophages from more specific antiviral functions of MNP. Blanden also showed that rapid elimination of ectromelia orthopoxvirus from spleens of adoptively immunized mice was not accompanied by activation of spleen MNP (11).

As to extrapolating from in vitro findings, I have become very cautious. Cultivated peritoneal macrophages are readily infected and produce infectious LCMV (56, 57), and in the spleen of the infected but otherwise untreated mouse, MNP are susceptible to LCMV and probably contribute in large measure to the yield of infectious LCMV (33, 40); cultivated peritoneal macrophages taken from "*C. parvum*"-treated mice were found to be fully resistant to LCMV (Krenz, personal communication), yet in the spleens of mice whose MNP were activated by treatment with "*C. parvum*," much greater quantities of LCMV virus were produced than in the spleens of untreated controls (see above). MNP are heterogeneous in both cell type and function, which may differ between organs (3, 26, 36, 64). With regard to viral infections, an illustrative example was provided by Rodgers and Mims (54), who found marked differences of susceptibility to influenza virus between peritoneal and alveolar cultivated macrophages of the mouse. Furthermore, activation of MNP is a complex phenomenon involving many different capacities that differ qualitatively as well as quantitatively (1, 6, 52). It would not be surprising if maintenance of these cells in vitro leads to selected expression of certain functions or preferential survival of particular cells; furthermore, activation in vitro is not necessarily identical with activation in vivo.

Be that as it may, in the LCMV-infected mouse, activation of MNP in the spleen appears to accompany rather than promote virus elimination. MNP that are activated but do not contribute to the termination of the infection would be in respectable company. Plasma cells forming antiviral antibodies appear in large numbers during the time when virus is eliminated (51), yet antibodies have convincingly been shown not to participate directly in clearance of LCMV from the organs of mice (2, 27, 34); mutatis mutandis, the same is true with respect to natural killer cells, which are activated during infection of the mouse with LCMV (65) but do not appear to aid in controlling the infection (15).

If neither MNP, be they activated or not, nor antibodies or natural killer cells are instrumental, how is LCMV cleared from the murine spleen? All our findings are compatible with the assumption that LCMV-specific CTL alone block virus replication. They probably function by either releasing, or inducing other cells to release, antiviral soluble factors that diffuse into the surrounding tissue, where they block virus replication by some as yet unknown mechanism. Our findings do not favor the alternative explanation, namely, that virus replication in vivo is halted by lysis of virus-infected target cells (28; F. Lehmann-Grube, D. Moskophidis, and J. Löhler, *Ann. N.Y. Acad. Sci.*, in press).

Acknowledgments. Work done in my laboratory was aided by grants from the Deutsche Forschungsgemeinschaft. The Heinrich-Pette-Institut is financially supported by Freie und Hansestadt Hamburg and Bundesministerium für Jugend, Familie, Frauen und Gesundheit.

Literature Cited

1. **Adams, D. O., and T. A. Hamilton.** 1984. The cell biology of macrophage activation. *Annu. Rev. Immunol.* **2:**283–318.
2. **Ahmed, R., A. Salmi, L. D. Butler, J. M. Chiller, and M. B. A. Oldstone.** 1984. Selection of genetic variants of lymphocytic choriomeningitis virus in spleens of persistently infected mice. Role in suppression of cytotoxic T lymphocyte response and viral persistence. *J. Exp. Med.* **160:**521–540.
3. **Akagawa, K. S., and T. Tokunaga.** 1985. Lack of binding of bacterial lipopolysaccharide to mouse lung macrophages and restoration of binding by γ interferon. *J. Exp. Med.* **162:**1444–1459.
4. **Allison, A. C.** 1974. On the role of mononuclear phagocytes in immunity against viruses. *Prog. Med. Virol.* **18:**15–31.
5. **Allison, A. C.** 1974. Interactions of antibodies, complement components and various cell types in immunity against viruses and pyogenic bacteria. *Transplant. Rev.* **19:**3–55.
6. **Allison, A. C.** 1978. Macrophage activation and nonspecific immunity. *Int. Rev. Exp. Pathol.* **18:**303–346.
7. **Allison, A. C., J. S. Harington, and M. Birbeck.** 1966. An examination of the cytotoxic effects of silica on macrophages. *J. Exp. Med.* **124:**141–154.
8. **Assmann-Wischer, U., M. M. Simon, and F. Lehmann-Grube.** 1985. Mechanism of recovery from acute virus infection. III. Subclass of T lymphocytes mediating clearance of lymphocytic choriomeningitis virus from the spleens of mice. *Med. Microbiol. Immunol.* **174:**249–256.
9. **Barka, T., and P. J. Anderson.** 1962. Histochemical methods for acid phosphatase using hexazonium pararosanilin as coupler. *J. Histochem. Cytochem.* **10:**741–753.
10. **Bice, D. E., D. G. Gruwell, J. E. Salvaggio, and E. O. Hoffmann.** 1972. Suppression of primary immunization by carrageenan—a macrophage toxic agent. *Immunol. Commun.* **1:**615–625.
11. **Blanden, R. V.** 1971. Mechanisms of recovery from a generalized viral infection: mousepox. II. Passive transfer of recovery mechanisms with immune lymphoid cells. *J. Exp. Med.* **133:**1074–1089.
12. **Blanden, R. V.** 1974. T cell response to viral and bacterial infection. *Transplant. Rev.* **19:**56–88.
13. **Blanden, R. V., A. J. Hapel, P. C. Doherty, and R. M. Zinkernagel.** 1976. Lymphocyte-macrophage interactions and macrophage activation in the expression of antimicrobial immunity *in vivo*, p. 367–400. *In* D. S. Nelson (ed.), *Immunobiology of the Macrophage.* Academic Press, Inc., New York.
14. **Blanden, R. V., and C. A. Mims.** 1973. Macrophage activation in mice infected with ectromelia or lymphocytic choriomeningitis viruses. *Aust. J. Exp. Biol. Med. Sci.* **51:**393–398.
15. **Bukowski, J. F., J. F. Warner, G. Dennert, and R. M. Welsh.** 1985. Adoptive transfer studies demonstrating the antiviral effect of natural killer cells in vivo. *J. Exp. Med.* **161:**40–52.
16. **Cheers, C., and R. Waller.** 1975. Activated macrophages in congenitally athymic "nude" mice and in lethally irradiated mice. *J. Immunol.* **115:**844–847.
17. **Geiger, B., and R. Gallily.** 1974. Effect of X-irradiation on various functions of murine macrophages. *Clin. Exp. Immunol.* **16:**643–655.

18. **Gresser, I., and D. J. Lang.** 1966. Relationships between viruses and leucocytes. *Prog. Med. Virol.* **8:**62–130.
19. **Hahn, H.** 1974. Effects of dextran sulfate 500 on cell-mediated resistance to infection with *Listeria monocytogenes* in mice. *Infect. Immun.* **10:**1105–1109.
20. **Hahn, H.** 1975. Requirement for a bone marrow-derived component in the expression of cell-mediated antibacterial immunity. *Infect. Immun.* **11:**949–954.
21. **Hahn, H., and M. Bierther.** 1974. Morphological changes induced by dextran sulfate 500 in mononuclear phagocytes of *Listeria*-infected mice. *Infect. Immun.* **10:**1110–1119.
22. **Hahn, H., and S. H. E. Kaufmann.** 1981. The role of cell-mediated immunity in bacterial infections. *Rev. Infect. Dis.* **3:**1221–1250.
23. **Haller, O., H. Arnheiter, and J. Lindenmann.** 1979. Natural, genetically determined resistance toward influenza virus in hemopoietic mouse chimeras. Role of mononuclear phagocytes. *J. Exp. Med.* **150:**117–126.
24. **Halpern, B. N., A.-R. Prévot, G. Biozzi, C. Stiffel, D. Mouton, J. C. Morard, Y. Bouthillier, and C. Decreusefond.** 1963. Stimulation de l'activité phagocytaire du système réticuloendothélial provoquée par *Corynebacterium parvum. RES J. Reticuloendothel. Soc.* **1:**77–96.
25. **Leder, L.-D.** 1967. *Der Blutmonozyt.* Springer-Verlag KG, Berlin.
26. **Lee, K.-C., J. Kay, and M. Wong.** 1979. Separation of functionally distinct subpopulations of *Corynebacterium parvum*-activated macrophages with predominantly stimulatory or suppressive effect on the cell-mediated cytotoxic T cell response. *Cell. Immunol.* **42:**28–41.
27. **Lehmann-Grube, F.** 1971. Lymphocytic choriomeningitis virus. *Virol. Monogr.* **10:**1–173.
28. **Lehmann-Grube, F.** 1987. Mechanism of recovery from acute virus infection, p. 67–88. *In* H. Bauer, H.-D. Klenk, and C. Scholtissek (ed.), *Modern Trends in Virology.* Springer-Verlag, Berlin.
29. **Lehmann-Grube, F., U. Assmann, C. Löliger, D. Moskophidis, and J. Löhler.** 1985. Mechanism of recovery from acute virus infection. I. Role of T lymphocytes in the clearance of lymphocytic choriomeningitis virus from spleens of mice. *J. Immunol.* **134:**608–615.
30. **Lehmann-Grube, F., U. Assmann-Wischer, R. Schwachenwald, I. Krenz, T. Krahnert, and D. Moskophidis.** 1986. Questionable role of mononuclear phagocytes in the elimination of lymphocytic choriomeningitis virus from spleens of acutely infected mice. *Med. Microbiol. Immunol.* **175:**145–148.
31. **Lehmann-Grube, F., and J. Löhler.** 1981. Immunopathologic alterations of lymphatic tissues of mice infected with lymphocytic choriomeningitis virus. II. Pathogenetic mechanism. *Lab. Invest.* **44:**205–213.
32. **Lindenmann, J., E. Deuel, S. Fanconi, and O. Haller.** 1978. Inborn resistance of mice to myxoviruses: macrophages express phenotype in vitro. *J. Exp. Med.* **147:**531–540.
33. **Löhler, J., and F. Lehmann-Grube.** 1981. Immunopathologic alterations of lymphatic tissues of mice infected with lymphocytic choriomeningitis virus. I. Histopathologic findings. *Lab. Invest.* **44:**193–204.
34. **Marker, O., and M. Volkert.** 1973. Studies on cell-mediated immunity to lymphocytic choriomeningitis virus in mice. *J. Exp. Med.* **137:**1511–1525.
35. **Milas, L., and M. T. Scott.** 1978. Antitumor activity of *Corynebacterium parvum. Adv. Cancer Res.* **26:**257–306.
36. **Miller, G. A., M. W. Campbell, and J. L. Hudson.** 1980. Separation of rat peritoneal macrophages into functionally distinct subclasses by centrifugal elutriation. *RES J. Reticuloendothel. Soc.* **27:**167–174.

37. **Mills, K. H. G.** 1986. Processing of viral antigens and presentation to class II-restricted T cells. *Immunol. Today* **7:**260–263.
38. **Mims, C. A.** 1964. Aspects of the pathogenesis of virus diseases. *Bacteriol. Rev.* **28:**30–71.
39. **Mims, C. A., and R. V. Blanden.** 1972. Antiviral action of immune lymphocytes in mice infected with lymphocytic choriomeningitis virus. *Infect. Immun.* **6:**695–698.
40. **Mims, C. A., and F. A. Tosolini.** 1969. Pathogenesis of lesions in lymphoid tissue of mice infected with lymphocytic choriomeningitis (LCM) virus. *Br. J. Exp. Pathol.* **50:**584–592.
41. **Mims, C. A., and D. O. White.** 1984. *Viral Pathogenesis and Immunology*. Blackwell Scientific Publications, Ltd., Oxford.
42. **Mogensen, S.** 1977. Role of macrophages in hepatitis induced by herpes simplex virus types 1 and 2 in mice. *Infect. Immun.* **15:**686–691.
43. **Mogensen, S. C.** 1977. Genetics of macrophage-controlled resistance to hepatitis induced by herpes simplex virus type 2 in mice. *Infect. Immun.* **17:**268–273.
44. **Mogensen, S. C.** 1978. Macrophages and age-dependent resistance to hepatitis induced by herpes simplex virus type 2 in mice. *Infect. Immun.* **19:**46–50.
45. **Mogensen, S. C.** 1979. Role of macrophages in natural resistance to virus infections. *Microbiol. Rev.* **43:**1–26.
46. **Morahan, P. S., J. R. Connor, and K. R. Leary.** 1985. Viruses and the versatile macrophage. *Br. Med. Bull.* **41:**15–21.
47. **Morahan, P. S., L. A. Glasgow, J. L. Crane, and E. R. Kern.** 1977. Comparison of antiviral and antitumor activity of activated macrophages. *Cell. Immunol.* **28:**404–415.
48. **Morahan, P. S., and S. S. Morse.** 1979. Macrophage-virus interactions, p. 17–35. *In* M. R. Proffitt (ed.), *Virus-Lymphocyte Interactions: Implications for Disease*. Elsevier/North Holland Publishing Co., New York.
49. **Moskophidis, D., S. P. Cobbold, H. Waldmann, and F. Lehmann-Grube.** 1987. Mechanism of recovery from acute virus infection: treatment of lymphocytic choriomeningitis virus-infected mice with monoclonal antibodies reveals that Lyt-2^+ T lymphocytes mediate clearance of virus and regulate the antiviral antibody response. *J. Virol.* **61:**1867–1874.
50. **Moskophidis, D., and F. Lehmann-Grube.** 1983. The immune response of the mouse to lymphocytic choriomeningitis virus. III. Differences of numbers of cytotoxic T lymphocytes in spleens of mice of different strains. *Cell. Immunol.* **77:**279–289.
51. **Moskophidis, D., and F. Lehmann-Grube.** 1984. The immune response of the mouse to lymphocytic choriomeningitis virus. IV. Enumeration of antibody-producing cells in spleens during acute and persistent infection. *J. Immunol.* **133:**3366–3370.
52. **North, R. J.** 1978. The concept of the activated macrophage. *J. Immunol.* **121:**806–809.
53. **Rao, G. R., W. E. Rawls, D. Y. E. Perey, and W. A. F. Tompkins.** 1977. Macrophage activation in congenitally athymic mice raised under conventional or germ-free conditions. *RES J. Reticuloendothel. Soc.* **21:**13–20.
54. **Rodgers, B., and C. A. Mims.** 1981. Interaction of influenza virus with mouse macrophages. *Infect. Immun.* **31:**751–757.
55. **Sawicki, J. E., and P. J. Catanzaro.** 1975. Selective macrophage cytotoxicity of carrageenan *in vivo*. *Int. Arch. Allergy Appl. Immunol.* **49:**709–714.
56. **Schwartz, R., J. Löhler, and F. Lehmann-Grube.** 1978. Infection of cultivated mouse peritoneal macrophages with lymphocytic choriomeningitis virus. *J. Gen. Virol.* **39:**565–570.
57. **Seamer, J.** 1965. Mouse macrophages as host cells for murine viruses. *Arch. Gesamte Virusforsch.* **17:**654–663.

58. **Selgrade, M. K., and J. E. Osborn.** 1974. Role of macrophages in resistance to murine cytomegalovirus. *Infect. Immun.* **10:**1383–1390.
59. **Thomsen, A. R., and M. Volkert.** 1983. Studies on the role of mononuclear phagocytes in resistance to acute lymphocytic choriomeningitis virus infection. *Scand. J. Immunol.* **18:**271–277.
60. **Thomsen, A. R., M. Volkert, and K. Bro-Jørgensen.** 1983. Virus elimination in acute lymphocytic choriomeningitis virus infection. Correlation with virus-specific delayed-type hypersensitivity rather than cytotoxicity. *Scand. J. Immunol.* **17:**489–495.
61. **Tripathy, S. P., and G. B. Mackaness.** 1969. The effect of cytotoxic agents on the passive transfer of cell-mediated immunity. *J. Exp. Med.* **130:**17–30.
62. **Unanue, E. R.** 1984. Antigen-presenting function of the macrophage. *Annu. Rev. Immunol.* **2:**395–428.
63. **Vachek, H., and E. Kölsch.** 1975. Dextran sulfate stimulates the induction but inhibits the effector phase in T cell-mediated cytotoxicity. *Transplantation* **19:**183–185.
64. **Walker, W. S.** 1976. Functional heterogeneity of macrophages, p. 91–110. *In* D. S. Nelson (ed.), *Immunobiology of the Macrophage*. Academic Press, Inc., New York.
65. **Welsh, R. M.** 1978. Cytotoxic cells induced during lymphocytic choriomeningitis virus infection of mice. I. Characterization of natural killer cell induction. *J. Exp. Med.* **148:**163–181.
66. **Zinkernagel, R. M., and R. V. Blanden.** 1975. Macrophage activation in mice lacking thymus-derived (T) cells. *Experientia* **31:**591–593.
67. **Zinkernagel, R. M., and P. C. Doherty.** 1979. MHC-restricted cytotoxic T cells: studies on the biological role of polymorphic major transplantation antigens determining T-cell restriction-specificity, function, and responsiveness. *Adv. Immunol.* **27:**51–177.
68. **Zinkernagel, R. M., and R. M. Welsh.** 1976. H-2 compatibility requirement for virus-specific T cell-mediated effector functions in vivo. I. Specificity of T cells conferring antiviral protection against lymphocytic choriomeningitis virus is associated with H-2K and H-2D. *J. Immunol.* **117:**1495–1502.
69. **Zisman, B., M. S. Hirsch, and A. C. Allison.** 1970. Selective effects of anti-macrophage serum, silica and anti-lymphocyte serum on pathogenesis of herpes virus infection of young adult mice. *J. Immunol.* **104:**1155–1159.

Chapter 9

Function of Natural Killer Cells in Lymphocytic Choriomeningitis Virus Infection

Raymond M. Welsh, Hyekyung Yang, and Kim W. McIntyre

INTRODUCTION

Lymphocytic choriomeningitis virus (LCMV) is a relatively minor human pathogen, but the study of this virus infection in mice has led to a disproportionately large number of findings relating to immunology and viral pathogenesis. Important concepts on immunological tolerance, immune complex disease, virus-induced immunopathology, T-cell recognition, viral persistence, and virus-induced cellular dysfunction have been advanced by this model (10, 31, 49). LCMV infection has also been useful in studying natural killer (NK) cell phenotype, activation, accumulation, proliferation, and function and has provided one of the first demonstrations that NK cells become activated and respond to virus infection (48, 49).

NK cells are bone-marrow-derived, thymus-independent lymphocytes bearing a large granular lymphocyte morphology and lysing on contact a variety of target cells without specific immunological recognition or antigenic memory (48, 49). Little is known about either the putative NK receptor or the target structure to which it binds. The granules within NK cells contain cytolytic proteins, one of which, the perforin protein, bears some homology to the ninth component of complement (C9) (26, 27, 33, 37). The target cell range and cytotoxic capacity of NK cells are greatly expanded after exposure to interferon

---

Raymond M. Welsh, Hyekyung Yang, and Kim W. McIntyre • Department of Pathology, University of Massachusetts Medical School, Worcester, Massachusetts 01655.

(IFN) (47) or to interleukin-2 (25), and the activation of NK cells during LCMV infection has been shown to correlate directly with the levels of IFN in the spleen (46). The proliferation of NK cells, another IFN-dependent event, was first shown in the LCMV system (6–8). This proliferation parallels the IFN response, peaking at about days 2 to 5 postinfection, whereas the T-cell response and proliferation peak at days 7 to 9 postinfection (7, 8). The only analysis of T-cell receptor expression on in vivo-activated NK cells has been done with the LCMV system (7a). These stimulated NK cells lack detectable mRNA expression for T-cell-receptor variable α, β, and γ proteins and for the receptor-associated constant δ and ε proteins (Biron et al., submitted).

The LCMV system has been used to document the systemic activation of NK cells in the brain (3), bone marrow (44), peripheral blood (41), lungs (4), and liver (28), as well as in the more commonly examined spleen and peritoneal cavity (55). NK cells accumulate in LCMV-infected tissue, representing a very early cellular response to infection (28, 30). Peritoneal fluid from virus-infected mice is chemotactic to NK cells, partially explaining the accumulation of these cells at sites of infection (30).

Mice infected with LCMV at birth or in utero develop a lifelong persistent infection associated with the production of standard and defective-interfering virus in virtually all the major organs (10). The virus travels in the circulation in infectious immune complexes in association with antibody and complement (32), and this persistent infection has served as a model for virus-induced immune-complex disease (10). Associated with this persistent infection is a chronic low-level interferonemia (8 to 16 U/ml of plasma) (11, 37) and a slightly elevated spleen NK cell response (11). This model has been used to show that NK cell activity may remain elevated for prolonged periods under conditions of chronic inflammation.

The elevated NK cell response in acute and persistent LCMV infections would suggest a function for these cells in infection, but the available evidence indicates that they have little part in controlling LCMV synthesis. This is in contrast to results of work with murine cytomegalovirus and some other viruses, whose synthesis is probably regulated by NK cells (15, 48). LCMV Armstrong (ARM) synthesis during the first 3 days of infection is not elevated in C67BL/6 beige mice (53), which have an NK cell defect (36). Similarly, the synthesis of LCMV ARM or WE is not enhanced in C3H/St or C57BL/6 mice whose NK cells have been rather selectively depleted with antibody to asialo GM_1 (15, 50), a glycosphingolipid found on NK cells at very high concentrations, or whose NK cells have been nonselectively depleted by cyclophosphamide (50), a broad-spectrum immunosuppressive drug. When BALB/c mice persistently infected with LCMV are treated with antibody to asialo GM_1 over a 3-day period, the titer of virus in the blood is likewise not affected (15).

NK cells are also not required for the immunopathological lesions leading to death in mice acutely infected via the intracerebral route. NK cells are found in meningeal infiltrates and in cerebrospinal fluid, but do not appear to be required for lethal encephalitis (2, 3), which occurs in NK cell-depleted mice and which can be mediated by cloned cytotoxic T-cell (CTL) lines (17).

The reason why NK cells have little part in regulating LCMV infection may relate to two factors. The first is that LCMV-infected cells show little selective sensitivity to NK cell-mediated lysis in vitro (13, 51). The second, and perhaps more important, factor is that LCMV-infected cells become protected against NK cell-mediated lysis by IFN (13). An unusual paradox in the NK-cell system is that IFN can protect target cells from lysis mediated by IFN-activated NK cells (45). The reason for this is unknown, but IFN-treated target cells tend to bind to NK cells in a way which fails to trigger the release of cytotoxic factors (56). The NK cell-sensitive virus murine cytomegalovirus, perhaps by being cytopathic and altering cellular RNA and protein synthesis, renders target cells resistant to IFN-mediated protection (13). Thymocytes, some bone marrow cells, or tumor cells isolated from LCMV-infected or IFN-treated mice show remarkable resistance to lysis by NK cells (14, 22, 52). In contrast, these cells are considerably more sensitive to CTL-mediated lysis and express elevated levels of H-2 antigens, known to be targets of CTL and to be up-regulated by IFN (12, 14). Thus, during a virus infection, the induced IFN conditions target cells to become increasingly more resistant to NK cells as they become increasingly more sensitive to CTL. Since cells already infected with LCMV can be conditioned in this manner by IFN, it is easy to see how the infection in vivo may be highly sensitive to regulation by CTL but not by NK cells.

ROLE OF ADCC IN LCMV INFECTION

NK cells have Fc receptors and are able to mediate antibody-dependent cell-mediated cytotoxicity (ADCC) against virus-infected target cells coated with antiviral antibody (39). Antiviral ADCC in vitro has been widely demonstrated in many human and some murine virus systems (39). To date, there has been no definitive demonstration of a significant role for antiviral ADCC mediated by NK cells during virus infections, although some data are consistent with this hypothesis (23). Our attempts to demonstrate any type of ADCC in vitro against LCMV-

Table 1. Lack of Significant NK Cell-Mediated Antiviral ADCC during LCMV Infection[a]

Expt	Antiserum to LCMV	Treatment	% YAC-1 lysis	Log_{10} PFU/spleen
1	−	None	42	5.4 ± 0.5
	−	Cyclophosphamide	15	5.3 ± 0.3
	+	None	39	4.5 ± 0.4
	+	Cyclophosphamide	18	4.6 ± 0.1
2	−	None	26	5.3 ± 0.4
	−	Anti-asialo GM_1	0.7	5.1 ± 0.4
	+	None	38	4.1 ± 0.3
	+	Anti-asialo GM_1	2.5	4.1 ± 0.5
3	−	None	18	5.5 ± 0.3
	−	Cobra venom factor	28	5.6 ± 0.2
	+	None	22	3.5 ± 0.5
	+	Cobra venom factor	30	4.4 ± 0.5

[a] C57BL/6 mice (4 to 10 weeks old) were injected intravenously with 0.1 ml of anti-LCMV hyperimmune mouse serum and/or intraperitoneally with cyclophosphamide (200 mg/kg), antiserum to asialo GM_1 (Wako Laboratories, Dallas, Tex.), or cobra venom factor (15 U). Four hours later, mice were injected intraperitoneally with LCMV Armstrong. Spleens were harvested 3 days postinfection, and LCMV was titrated from spleen homogenates on Vero cell plaque assays.

infected L-929 cells have been unsuccessful when using a variety of mouse and guinea pig antisera, immunoglobulin fractions, and mouse monoclonal antibodies to LCMV cell surface glycoproteins, even though the same cells are sensitive to ADCC mediated by anti-H-2 antibodies. Nevertheless, some ADCC could still be occurring in vivo.

Mice were therefore injected with dilutions of mouse immune serum to LCMV and then challenged with virus. Titers of LCMV in the spleens were assessed 3 days later, and a dilution of antiserum which reduced the log_{10} titer by 1 to 2 was used in subsequent experiments. Treatment of mice with cyclophosphamide or with antibody to asialo GM_1 significantly depressed the NK cell response, but failed to elevate the titer of LCMV in the antiserum-treated mice (Table 1). This suggests that ADCC plays a minor role, if any, in controlling the LCMV infection. In contrast, mice treated with cobra venom factor, which is an analog of C3b and depletes complement activity (16), synthesized 10-fold more LCMV than did antiserum-treated mice with normal complement levels ($P < 0.5$). This suggests that the antibody may either neutralize or opsonize virus via a complement-dependent mechanism in vivo. Antibody-dependent neutralization of LCMV in vitro is greatly augmented by complement (54). Collectively, these data argue that complement, but not NK cells, is involved in antibody-dependent regulation of LCMV synthesis in vivo.

USE OF LCMV TO PROBE NK CELL-MEDIATED FUNCTION

Although it is a poor model to demonstrate the direct regulation of virus infections by NK cells, LCMV infection provides a unique system with which to clearly study other functions of NK cells in virus infections. Depletion of NK cell activity with antibody to asialo GM_1 results in elevated levels of virus synthesis with many other virus infections, including murine cytomegalovirus (15), murine hepatitis virus (15), vaccinia virus (15), influenza virus (40), Pichinde virus (R. J. Natuk and R. M. Welsh, unpublished data), and coxsackievirus (19). This elevated virus synthesis is usually associated with higher levels of virus-induced interferon, cytopathic effect, and inflammation (15, 19, 40). These side effects, resulting as a consequence of elevated virus synthesis, make it difficult to study other host functions that may have been influenced by the NK cell depletion. This is not a problem with the LCMV infection, which can be used to study a variety of parameters listed below.

REGULATION OF HEMATOPOIESIS

Bone marrow function is markedly depressed during LCMV infection, with its nadir correlating with the peak in NK cell activation and IFN synthesis (9, 44). NK cells lyse a subpopulation of bone marrow cells and inhibit the outgrowth of CFU in vitro (20, 21). NK cells may also function to inhibit outgrowth of bone marrow grafts in F1-anti-parent hybrids (18, 24). It has therefore been speculated that NK cells are important regulators of hematopoiesis. The activation and proliferation of NK cells during LCMV infection could, therefore, contribute to hematopoietic dysfunction. LCMV Traub grows well in the bone marrow and results in high levels of NK cell activation within the marrow during acute infection (44). Transfer of bone marrow cells from LCMV-infected mice into irradiated recipient mice results in poor growth of CFU in vivo, compared with results with bone marrow cells from uninfected mice. However, treatment of donor and recipient mice with antibody to asialo GM_1, which rather selectively depletes NK cells, blocks the LCMV-induced inhibition of bone marrow outgrowth (44). This suggests that during a virus infection the inhibition of bone marrow function, which contributes to the commonly observed leukopenia, may be regulated by NK cells activated during the infection.

REGULATION OF B- AND T-CELL FUNCTION

NK cells have been suggested to be major regulators of the immune system. Recent work suggests that NK cells may regulate antibody

Table 2. Enhanced CTL Response in NK Cell-Depleted Mice[a]

Expt	Source of CTL	% Virus-specific lysis for:	
		Normal mice	NK-depleted mice
1	C3H/St spleen	34	60
2	C3H/St spleen	47	65
3	C3H/St spleen	38	58
4	C3H/St spleen	38	68
5	C57BL/6 spleen	43	59
6	C57BL/6 spleen	53	63
7	C57BL/6 spleen	57	76
8	C57BL/6 spleen	72	87
	C57BL/6 liver	53	70

[a] Mice 4 to 10 weeks of age were either untreated and used as controls (normal mice) or injected intraperitoneally on day 0 and day 4 of LCMV Armstrong infection with antiserum to asialo GM_1 (NK-depleted mice). At 7 days postinfection, spleen or liver leukocytes were tested for cytotoxicity by the 6-h chromium release assay on LCMV-infected histocompatible targets. Lysis of uninfected targets was subtracted from that of infected targets to give the percent virus-specific lysis. Effector-to-target-cell ratios were 50 to 100:1 for the spleen leukocytes and 30:1 for the liver leukocytes.

production by lysing antigen-presenting dendritic cells (1, 38). It is therefore likely that the LCMV-specific antibody response during the LCMV infection is depressed by the highly activated NK cell population, although this has not yet been investigated. We have, however, examined the virus-specific CTL response in NK cell-depleted mice. A single injection of antibody to asialo GM_1 does not affect the proliferative response of T cells to the mitogen concanavalin A. However, during LCMV infection we have consistently noticed a 20 to 80% enhancement in LCMV-specific CTL activity in the spleen, as well as in the liver (Table 2). This probably represents an increase in the total number of CTL, since the number of large granular lymphocytes increases in the NK cell-depleted liver by a similar amount (normal liver, 2.5×10^6; NK cell-depleted liver, 3.7×10^6). Our interpretation of these results is that NK cells may inhibit CTL responses as they do B-cell responses, but more experiments are needed before definitive conclusions can be reached. Others have reported that the antiserum to asialo GM_1 inhibits the CTL response (42). We have never observed this in vivo and suggest either that the antiserum was used at too high a concentration (CTL do express some asialo GM_1) or that it was contaminated with another antibody. The CTL response to LCMV in NK cell-deficient beige mice is not elevated in terms of cytotoxicity (5, 53). However, beige mouse CTL, like the NK cells, have abnormal granules and reduced lytic activity on a per-cell basis (5). The number of beige mouse cells bearing the CTL marker *Lyt-2* and the number of Lyt-2^+ cells that bind to histocompatible LCMV-infected targets are significantly higher than in LCMV-infected beige/+ mice (5). This is consistent with the concept that NK cells may regulate CTL production, but it could also be the result of a compensatory immune response to overcome the beige CTL defect in lytic activity, as has been suggested previously (5).

IFN PRODUCTION BY NK CELLS

NK cells are reported to produce IFN-α, IFN-β, and IFN-γ, and it has been suggested that they may help regulate virus infection by secreting IFN (47). Depletion of NK cells in mice infected with murine cytomegalovirus or murine hepatitis virus actually leads to significant increases in IFN production (9), but this is probably only because much more virus is being made in the NK cell-depleted mice, resulting in much more IFN production by other cells in the body. In the LCMV system, in which more virus is not synthesized after NK cell depletion, there are no changes in IFN levels after this treatment (15). This suggests that the NK cells do not produce a sufficiently large amount of IFN to influence the LCMV infection. Antibody to IFN causes marked enhancements in LCMV titers in treated mice (35), so that if the NK cells were very important for IFN-α/β production, this should have been detected in this system.

INFLUENCE OF NK CELLS ON TUMOR GROWTH

It has long been known that infection of mice with LCMV is antagonistic to tumor growth (29). This could be due to any number of mechanisms, including direct IFN inhibition of tumor growth or cytotoxic mechanisms mediated by macrophages, CTL, or, of course, NK cells. Survival of radiolabeled tumor cells after intravenous injection into mice correlates with their ability to form tumors and is an indicator of in vivo cytotoxicity (34). Mice either acutely or persistently infected with LCMV display rapid in vivo lysis of injected [^{125}I]iododeoxyuridine-labeled NK-sensitive YAC-1 lymphoma cells, and this in vivo lysis is abrogated by treatment of mice with antiserum to asialo GM_1 (4, 11, 43). This argues that NK cells may mediate the LCMV-induced inhibition of tumor outgrowth.

CONCLUDING COMMENTS

Neither the acute nor the persistent LCMV infections of mice appear to be regulated by NK cells, either directly or via an ADCC mechanism. However, this model system has contributed much to our knowledge of virus-induced NK cell activation, proliferation, and regulation and, by virtue of the failure of NK cells to regulate LCMV synthesis, promises to continue to be a useful model with which to examine NK cell function under conditions of a very potent activating and proliferative stimulus.

Literature Cited

1. **Abruzzo, L. U., and D. A. Rowley.** 1983. Homeostasis of the antibody response: immunoregulation by NK cells. *Science* **222:**581–585.
2. **Allan, J. E., J. E. Dixon, and P. C. Doherty.** 1987. Nature of inflammatory process in the central nervous system of mice infected with lymphocytic choriomeningitis virus. *Curr. Top. Microbiol. Immunol.* **134:**131–143.
3. **Allan, J. E., and P. C. Doherty.** 1986. Natural killer cells contribute to inflammation but do not appear to be essential for the induction of clinical lymphocytic choriomeningitis virus. *Scand. J. Immunol.* **24:**153–162.
4. **Biron, C. A., S. Habu, K. Okumura, and R. M. Welsh.** 1984. Lysis of uninfected and virus-infected cells in vivo: a rejection mechanism in addition to that mediated by natural killer cells. *J. Virol.* **50:**698–707.
5. **Biron, C. A., K. F. Pedersen, and R. M. Welsh.** 1987. Aberrant T cells in beige mutant mice. *J. Immunol.* **138:**2050–2056.
6. **Biron, C. A., G. Sonnenfeld, and R. M. Welsh.** 1984. Interferon induces natural killer cell blastogenesis *in vivo*. *J. Leukocyte Biol.* **35:**31–37.
7. **Biron, C. A., L. R. Turgiss, and R. M. Welsh.** 1983. Increase in NK cell number and turnover rate during acute viral infection. *J. Immunol.* **131:**1539–1545.

7a. **Biron, C. A., P. van den Elsen, M. M. Tutt, P. Medvecsky, V. Kumar, and C. Terhorst.** 1987. Murine natural killer cells stimulated in vivo do not express the T-cell receptor α, β, γ, T3δ, or T3ε genes. *J. Immunol.* **139:**1704–1710.

8. **Biron, C. A., and R. M. Welsh.** 1982. Blastogenesis of natural killer cells during viral infection *in vivo*. *J. Immunol.* **129:**2788–2795.
9. **Bro-Jorgensen, K.** 1978. The interplay between lymphocytic choriomeningitis virus, immune function, and hemopioesis in mice. *Adv. Virus Res.* **22:**327–369.
10. **Buchmeier, M. J., R. M. Welsh, F. J. Dutko, and M. B. A. Oldstone.** 1980. The virology and immunobiology of lymphocytic choriomeningitis virus infection. *Adv. Immunol.* **30:**275–331.
11. **Bukowski, J. F., C. A. Biron, and R. M. Welsh.** 1983. Elevated natural killer cell-mediated cytotoxicity, plasma interferon, and tumor cell rejection in mice persistently infected with lymphocytic choriomeningitis virus. *J. Immunol.* **131:**991–996.
12. **Bukowski, J. F., and R. M. Welsh.** 1985. Interferon enhances the susceptibility of virus-infected fibroblasts to cytotoxic T cells. *J. Exp. Med.* **161:**257–262.
13. **Bukowski, J. F., and R. M. Welsh.** 1985. Inability of interferon to protect virus-infected cells against lysis by natural killer (NK) cells correlates with NK cell-mediated antiviral effects *in vivo*. *J. Immunol.* **135:**3537–3541.
14. **Bukowski, J. F., and R. M. Welsh.** 1986. Enhanced susceptibility to cytotoxic T lymphocytes of target cells isolated from virus-infected or interferon-treated mice. *J. Virol.* **59:**735–739.
15. **Bukowski, J. F., B. A. Woda, S. Habu, K. Okumura, and R. M. Welsh.** 1983. Natural killer cell depletion enhances virus synthesis and virus-induced hepatitis *in vivo*. *J. Immunol.* **131:**1531–1538.
16. **Cochrane, C. G., H. J. Muller-Eberhard, and B. S. Aikin.** 1970. Depletion of plasma complement *in vivo* by a protein of cobra venom: its effect on various immunological reactions. *J. Immunol.* **105:**55–69.

17. **Cole, G. A.** 1986. Production or prevention of neurological disease by continuous lines of arenavirus-specific cytotoxic T lymphocytes. *Med. Microbiol. Immunol.* **175:**197–199.
18. **Cudkowicz, G., and P. S. Hochman.** 1979. Do natural killer cells engage in regulated reactions against self to ensure homeostasis? *Immunol. Rev.* **44:**13–41.
19. **Godeny, E. K., and C. J. Gauntt.** 1986. Involvement of natural killer cells in Coxsackievirus B3-induced murine myocarditis. *J. Immunol.* **137:**1695–1702.
20. **Hansson, M., M. Beran, B. Andersson, and R. Kiessling.** 1982. Inhibition of *in vitro* granulopoiesis by autologous allogeneic human NK cells. *J. Immunol.* **129:**126–132.
21. **Hansson, M., R. Kiessling, and B. Andersson.** 1981. Human fetal thymus and bone marrow contain target cells for natural killer cells. *Eur. J. Immunol.* **11:**8–12.
22. **Hansson, M., R. Kiessling, B. Andersson, and R. M. Welsh.** 1980. Effect of interferon and interferon inducers on the NK sensitivity of normal mouse thymocytes. *J. Immunol.* **125:**2225–2231.
23. **Hirsch, R. L., D. E. Griffin, and R. T. Johnson.** 1979. Interactions between immune cells and antibody in protection from fatal Sindbis virus encephalitis. *Infect. Immun.* **23:**320–324.
24. **Kiessling, R., P. S. Hochman, O. Haller, G. M. Shearer, H. Wigzell, and G. Cudkowicz.** 1977. Evidence for a similar or common mechanism for natural killer cell activity and resistance to hemopoietic grafts. *Eur. J. Immunol.* **7:**655–663.
25. **Kuribayashi, K., S. Gillis, D. E. Kern, and C. S. Henney.** 1981. Murine NK cell cultures: effects of interleukin-2 and interferon on cell growth and cytotoxic reactivity. *J. Immunol.* **126:**2321–2327.
26. **Liu, C.-C., B. Perussia, Z. A. Cohn, and J. D.-E. Young.** 1986. Identification and characterization of a pore-forming protein of human peripheral blood natural killer cells. *J. Exp. Med.* **164:**2061–2076.
27. **Mallard, P. J., M. P. Henkart, C. W. Reynolds, and P. A. Henkart.** 1984. Purification and properties of cytoplasmic granules from cytotoxic rat LGL tumors. *J. Immunol.* **132:**3197–3204.
28. **McIntyre, K. W., and R. M. Welsh.** 1986. Accumulation of natural killer and cytotoxic T large granular lymphocytes in the livers of virus-infected mice. *J. Exp. Med.* **164:**1667–1681.
29. **Molomut, N., and M. Padnos.** 1965. Inhibition of transplantable and spontaneous murine tumors by the M-P virus. *Nature* (London) **208:**948–950.
30. **Natuk, R. J., and R. M. Welsh.** 1987. Accumulation and chemotaxis of large granular lymphocytes at sites of virus replication. *J. Immunol.* **138:**877–883.
31. **Oldstone, M. B. A.** 1987. The arenaviruses—an introduction. *Curr. Top. Microbiol. Immunol.* **133:**1–4.
32. **Oldstone, M. B. A., and F. J. Dixon.** 1971. Persistent lymphocytic choriomeningitis viral infection. III. Virus-anti-viral antibody complexes and associated chronic disease following transplacental infection. *J. Immunol.* **105:**829–837.
33. **Podack, E. R., and G. Dennert.** 1983. Assembly of two types of tubules with putative cytolytic function by cloned natural killer cells. *Nature* (London) **302:**442–445.
34. **Riccardi, C., P. Pucetti, A. Santoni, and R. B. Herberman.** 1979. Rapid *in vivo* assay of mouse natural killer cell activity. *J. Natl. Cancer Inst.* **63:**1041–1045.
35. **Riviere, Y., I. Gresser, J.-C. Guillon, and M. G. Tovey.** 1977. Inhibition by anti-interferon serum of lymphocytic choriomeningitis virus disease in suckling mice. *Proc. Natl. Acad. Sci. USA* **74:**2135–2139.
36. **Roder, J., and A. Duwe.** 1979. The beige mutation in the mouse selectively impairs natural killer cell function. *Nature* (London) **278:**451–453.
37. **Saron, M.-F., Y. Riviere, A. G. Hovanessian, and J.-C. Guillon.** 1982. Chronic

production of interferon in carrier mice congenitally infected with lymphocytic choriomeningitis virus. *Virology* **117:**253–256.
38. **Shaw, P. D., S. M. Gilbertson, and D. A. Rowley.** 1985. Dendritic cells that have interacted with antigen are targets for natural killer cells. *J. Exp. Med.* **162:**625–636.
39. **Sissons, J. G. P., and M. B. A. Oldstone.** 1980. Antibody-mediated destruction of virus-infected cells. *Adv. Immunol.* **29:**209–260.
40. **Stein-Streilein, J., and J. Guffee.** 1986. *In vivo* treatment of mice and hamsters with antibodies to asialo GM1 increases morbidity and mortality to pulmonary influenza infection. *J. Immunol.* **136:**1435–1441.
41. **Stitz, L., A. Althage, H. Hengartner, and R. Zinkernagel.** 1985. Natural killer cells vs. cytotoxic T cells in the peripheral blood of virus-infected mice. *J. Immunol.* **134:**598–602.
42. **Stitz, L., J. Baenziger, H. Pircher, H. Hengartner, and R. M. Zinkernagel.** 1986. Effect of rabbit anti-asialo GM1 treatment *in vivo* or with anti-asialo GM1 plus complement *in vitro* on cytotoxic T cell activities. *J. Immunol.* **136:**4674–4680.
43. **Talmadge, J. E., K. M. Meyers, D. J. Prieur, and J. R. Starkey.** 1980. Role of natural killer cells in tumor growth and metastasis: C57BL/6 normal and beige mice. *J. Natl. Cancer Inst.* **65:**929–935.
44. **Thomsen, A. R., P. Pisa, K. Bro-Jorgensen, and R. Kiessling.** 1986. Mechanisms of lymphocytic choriomeningitis virus-induced hemopoietic dysfunction. *J. Virol.* **59:**428–433.
45. **Trinchieri, G., and D. Santoli.** 1978. Anti-viral activity induced by culturing lymphocytes with tumor-derived or virus-transformed cells. Enhancement of natural killer activity by interferon and antagonistic inhibition of susceptibility of target cells to lysis. *J. Exp. Med.* **147:**1314–1333.
46. **Welsh, R. M.** 1978. Cytotoxic cells induced during lymphocytic choriomeningitis virus infection of mice. I. Characterization of natural killer cell induction. *J. Exp. Med.* **148:**164–181.
47. **Welsh, R. M.** 1984. Natural killer cells and interferon. *Crit. Rev. Immunol.* **5:**55–93.
48. **Welsh, R. M.** 1986. Regulation of virus infections by natural killer cells. *Nat. Immun. Cell Growth Regul.* **5:**169–199.
49. **Welsh, R. M.** 1987. Regulation and role of large granular lymphocytes in arenavirus infections. *Curr. Top. Microbiol. Immunol.* **134:**185–209.
50. **Welsh, R. M., C. A. Biron, J. F. Bukowski, K. W. McIntyre, and H. Yang.** 1984. Role of natural killer cells in virus infections in mice. *Surv. Synth. Pathol. Res.* **3:**409–431.
51. **Welsh, R. M., and L. A. Hallenbeck.** 1980. Effect of virus infections on target cell susceptibility to natural killer cell-mediated lysis. *J. Immunol.* **124:**2491–2497.
52. **Welsh, R. M., K. Karre, M. Hansson, L. A. Kunkel, and R. W. Kiessling.** 1981. Interferon-mediated protection of normal and tumor target cells against lysis by mouse natural killer cells. *J. Immunol.* **126:**219–225.
53. **Welsh, R. M., and R. W. Kiessling.** 1980. Natural killer cell response to lymphocytic choriomeningitis virus in beige mice. *Scand. J. Immunol.* **11:**363–367.
54. **Welsh, R. M., P. W. Lampert, P. A. Burner, and M. B. A. Oldstone.** 1976. Antibody-complement interactions with purified lymphocytic choriomeningitis virus. *Virology* **73:**59–71.
55. **Welsh, R. M., and R. M. Zinkernagel.** 1977. Hetero-specific cytotoxic cell activity induced during the first three days of acute lymphocytic choriomeningitis virus infection in mice. *Nature* (London) **268:**646–648.
56. **Wright, S. C., and B. Bonavida.** 1983. Studies on the mechanism of natural killer cell-mediated cytotoxicity. III. Interferon-induced inhibition of NK target cell susceptibility to lysis is due to a defect in their ability to stimulate release of natural killer cytotoxic factors (NKCF). *J. Immunol.* **130:**2960–2964.

Chapter 10

Specificity of Antiviral Cytotoxic T Lymphocytes

Kenneth L. Rosenthal

INTRODUCTION

In 1974, Zinkernagel and Doherty (54) demonstrated that virus-specific cytotoxic T lymphocytes (CTL) were restricted in their recognition of infected cells by self-surface glycoproteins encoded by the major histocompatibility complex (MHC). The discovery of MHC restriction provided a biologic role for MHC molecules and established a fundamental principle distinguishing antigen recognition mediated by immunoglobulin and T cells. Moreover, elucidation of this phenomenon led to the hypothesis of the single- versus dual-receptor model of T-cell recognition (17, 55), which served as a paradigm for research concerning the nature of the T-cell receptor. With the recent identification of the T-cell receptor and its associated accessory molecules (2), a clear picture of the recognition structure used by T cells is emerging.

Despite our knowledge of the structure of the T-cell receptor, we still do not have a clear understanding of what the T cell sees. Today, resolution of this issue is in sight. Advances in molecular biology, notably the use of DNA-mediated gene transfer, recombinant viral vectors, and synthetic peptides, are providing new insights into the nature of antigens recognized by T cells. Interestingly, two separate lines of investigation, one concerned with the identity of viral epitopes seen by class I MHC-restricted CTL and the other concerned with recognition of processed antigens by class II-restricted helper T cells, are resulting in an

---

Kenneth L. Rosenthal • Molecular Virology and Immunology Program, Department of Pathology, McMaster University Health Sciences Centre, Hamilton, Ontario, Canada L8N 3Z5.

exciting and unified concept concerning the form of antigen recognized by MHC-restricted T cells.

In this brief review I will highlight some of the research identifying viral antigenic determinants seen by CTL and discuss emerging concepts concerning MHC-restricted recognition of antigens by T cells. Clearly, determination of the antigenic epitopes of viruses seen by effector T cells will contribute to our basic understanding of immune recognition by the T-cell receptor and should lead to the development of more effective vaccines.

IN PURSUIT OF CROSS-REACTIVE CTL: INFLUENZA VIRUS

Although most CTL demonstrate specificity for the inducing virus, studies concerned with the fine specificity of CTL for closely related viruses demonstrated extensive cross-reactivity (56). This was shown for CTL generated by a number of viruses, including influenza A virus (10, 11, 16, 25, 30, 57, 58), alphavirus (31), and vesicular stomatitis virus (35). For example, the majority of CTL generated to a given strain of influenza A virus cross-reactively lysed syngeneic target cells infected with any influenza A virus but not influenza B viruses. This was shown to be true for both murine (11, 16, 18, 25, 57, 58) and human (10, 30) anti-influenza A virus CTL.

Observation of extensive cross-reactivity of antiviral CTL suggested that T cells and B cells recognized distinct antigenic determinants and indicated that antiviral CTL specificity could not be predicted from known serological relationships of particular viruses. More importantly, these observations raised questions concerning the precise identity of the shared antigenic determinant(s) recognized by CTL.

Notably, the work of Townsend et al. (39–46) in determining the exact specificity of cross-reactive anti-influenza virus CTL has provided major contributions and insights into CTL recognition. Identification of the cross-reactive determinants of influenza virus took on special importance, since it had been shown that CTL were important in recovery from infection. Thus, adoptive transfer of syngeneic influenza virus-specific CTL (50) or cloned CTL (28) reduced the virus titer in the lungs of mice infected with heterologous virus and protected against lethal infection (29).

A priori it seemed reasonable to assume that CTL would recognize native virus-induced cell surface molecules, such as glycoproteins, which are transported to and inserted in the target cell membrane. Indeed, although early studies (3, 12, 25) implicated the influenza virus hemagglutinin (HA) as a target for CTL, fully cross-reactive HA-specific CTL were not detected (41, 49).

Interestingly, anti-influenza virus CTL clones were isolated which, when tested on panels of recombinant influenza A virus-infected target cells, were shown to exclusively recognize a virus polymerase (PB2) (7) or the nucleoprotein (NP) (44, 46). It was unexpected that CTL could recognize nonglycoprotein molecules found predominantly in the nuclei and cytoplasm of infected cells. Since CTL clones may be highly selected and do not provide information on the fraction of CTL that display a given specificity, limiting-dilution assays were conducted. These experiments also demonstrated that the majority of cross-reactive influenza A virus-specific CTL did not recognize influenza glycoproteins, but rather reacted with internal virus determinants (24). Townsend and Skehel (45) then demonstrated that recognition of target cells mediated by polyclonal spleen cell cultures from influenza A virus-primed mice mapped to the NP gene.

The importance of the influenza A virus NP as a target for cross-reactive CTL was further demonstrated by using target cells transfected with the NP gene (42). Despite the lack of detectable surface expression of NP and low levels of expression of NP relative to infected cells, as assessed by antibody, targets transfected with the NP gene were efficiently lysed by the majority of cross-reactive CTL (42). In contrast, targets transfected with the influenza virus HA gene were lysed only by a minor subpopulation of strain-specific CTL (42). These results were confirmed by Yewdell et al. (52), who used recombinant vaccinia viruses capable of expressing individual influenza virus proteins in infected target cells. Similarly, recombinant vaccinia viruses were used to demonstrate that polyclonal murine (8) and human (22) CTL, which are fully cross-reactive for influenza A viruses, recognized internal viral proteins. They recognized the matrix M1, polymerase PB2, and NP of influenza virus in association with class I human lymphocyte antigen molecules (8, 22). There was no detectable recognition of influenza virus surface glycoproteins in target cells (8, 22).

These results raised questions concerning how nonglycosylated NP, which lacks an amino-terminal leader sequence, was transported to the cell surface. To examine this, a number of truncated NP genes were used to transfect target cells (40). These studies demonstrated that distinct fragments of NP were recognized by CTL from different strains of mice (40). Thus, CTL generated in $H\text{-}2^k$ mice predominantly recognized a determinant within the N-terminal 130 amino acids of NP, whereas $H\text{-}2^b$-restricted CTL recognized a determinant located between amino acids 328 and 386. These experiments showed that recognition and cell surface transport of NP were not dependent on a defined signal sequence and suggested that CTL recognize denatured peptide fragments of viral proteins associated with class I MHC molecules in a manner analogous to that of processed antigens presented to helper T cells in association with class II MHC molecules (1, 47).

Further work demonstrated that target cells could be sensitized for lysis by addition of short synthetic peptides containing as few as 14 amino acids of NP (6, 43). Thus, CTL epitopes can be defined with short sequential synthetic peptides in vitro. These results are also consistent with the idea that degraded viral molecules selectively associate with particular class I MHC alleles.

Indeed, even recognition of integral membrane glycoproteins does not appear to require surface expression of intact molecules. Recent studies (39) with recombinant vaccinia viruses demonstrated that although deletion of the signal sequence from influenza virus HA resulted in its inability to be detected at the surfaces of infected cells and its rapid degradation, it did not affect its recognition by HA-specific CTL. More recently, Braciale et al. (13) demonstrated that most HA-specific CTL clones recognized target cells that expressed a truncated HA gene that lacked the transmembrane and cytoplasmic domains of the glycoprotein. Furthermore, some clones appeared to recognize epitopes located within the transmembrane domain of HA (13). These data also support the contention that display of native proteins as an integral constituent of the target cell plasma membrane is not required for target cell recognition by class I MHC-restricted CTL.

SPECIFICITY OF ANTIVIRAL CTL: OTHER SYSTEMS

The molecular techniques described above and used so well to dissect the specificity of anti-influenza virus CTL are being increasingly applied to identify the target antigens for many viruses. One of the first applications of DNA-mediated gene transfer to the investigation of the antigenic determinants recognized by CTL was carried out by Tevethia et al. (38). They demonstrated that L cells transformed by full-sized and truncated simian virus 40 large tumor (T) antigen were recognized by CTL. Recognition by subpopulations of CTL was mapped by using target cells expressing the amino- or carboxy-terminal fragments of the simian virus 40 T antigen (21).

Interestingly, internal viral proteins in a number of other viruses appear to constitute the major CTL targets. For example, it has been demonstrated (33, 51) that the nucleoprotein of vesicular stomatitis virus (VSV) is a major target antigen for CTL. This may not be unexpected, since, as with influenza A viruses, VSV effectively induces CTL that cross-reactively lyse syngeneic target cells infected with distinct serotypes of VSV (35).

In addition, Bangham et al. (5) demonstrated that CTL from mice and humans recognized target cells infected with vaccinia virus recombinants expressing the NP of respiratory syncytial virus, but not those expressing the attachment glycoprotein (G). This work was recently extended by using a larger panel of vaccinia virus recombinants. Interestingly, it was found that BALB/c mice infected with recombinant vaccinia virus expressing the fusion (F) glycoprotein induced high levels of strain-specific anti-respiratory syncytial virus CTL (32). In contrast, only low levels of respiratory syncytial virus-specific CTL were induced by immunization with NP-expressing vaccinia virus recombinants (32). These results suggest that CTL recognition patterns vary in different strains and with different viruses.

Koszinowski and co-workers (26, 27, 48) demonstrated that a nonstructural immediate-early protein (pp89) of murine cytomegalovirus is recognized by CTL. Therefore, a herpesvirus phosphoprotein located in the nuclei of infected cells, where it is involved in viral transcriptional regulation, can also serve as a dominant target for CTL.

In contrast to the apparent importance of internal viral components as CTL targets for influenza virus, VSV, cytomegalovirus, and possibly respiratory syncytial virus, we have recently examined the specificity of herpes simplex virus (HSV)-specific CTL in mice. Our results (34) demonstrated that primary anti-HSV CTL did not lyse histocompatible target cells expressing glycoprotein gB or gE of HSV-1 or gD of HSV-2, but were able to markedly lyse target cells expressing low levels of HSV-1 gC. Recently, we demonstrated that gC is the major target antigen recognized by primary class I MHC-restricted CTL in the strain of mice examined (Rosenthal and Johnson, unpublished results).

RECOGNITION OF PEPTIDES BY CLASS II-RESTRICTED T CELLS

In 1959 Gell and Benacerraf (19) showed that denaturation greatly modified the antigenicity of protein for antibodies, but that denatured proteins were as efficient as native proteins in stimulating T-cell responses. Ziegler and Unanue (53) used *Listeria monocytogenes* to discover two features of antigen processing. First, they used fixation of antigen-presenting cells to demonstrate that antigen presentation to class II MHC-restricted T cells was an active time-dependent process. Second, they used lysosomotropic agents, such as chloroquine, to show that antigen processing involved an acidic intracellular compartment. From these results they concluded that antigen seen by T cells is not intact, but rather is degraded or denatured intracellularly before being exposed on the surfaces of antigen-presenting cells. Shimonkevitz et al. (37) provided direct support for this conclusion by demonstrating that protein enzymatically degraded in vitro could interact directly with fixed antigen-presenting cells and could subsequently trigger T cells.

Recent evidence (4, 14, 23) has provided insight into how antigenic peptides are presented by class II MHC molecules. Thus, Babbit et al. (4) showed that an antigenic peptide of hen egg lysozyme bound directly to a particular class II MHC molecule that served as its restriction element, but the peptide did not bind to a class II allele that was unable to present it. This work was extended by Guillet et al. (23) and Buus et al. (14), who showed that a number of peptides bind to various class II MHC molecules in patterns consistent with their ability to stimulate T cells. The binding was shown to be saturable and specific and had a surprisingly high binding affinity, averaging about 10^{-6} M. Furthermore, inhibition analysis indicated that all the peptides tested competed for binding to the same site on a given class II MHC molecule, suggesting that class II MHC molecules have a single binding site for antigen.

If many different antigenic peptides bind to a single site on a particular class II molecule, one might expect that they should share some structural similarities. It has been proposed by DeLisi and Berzofsky (15) that T-cell antigenic sites are likely to be amphipathic structures in an α-helical state. More recently, on the basis of sequence homology of three peptides and one class MHC II molecule, Guillet et al. (23) found that antigenic peptides for T cells share amino acid homology with the class II MHC molecules presenting them. These results suggest that the foreign peptide may displace an internal peptide sequence when it binds. This may represent the ultimate form of antigenic mimicry.

CONCLUSIONS

It is now clear that peptide antigens selectively bind to a single site on class II MHC molecules (4, 14, 23). Although direct binding of an antigenic epitope to class I MHC molecules has not yet been demonstrated, it seems likely to occur, since the epitopes of NP recognized by class I-restricted anti-influenza virus CTL can be defined with short peptides (6, 43). Thus, a major function of both class I and class II MHC molecules is to bind selected peptides to allow for possible interactions with the T-cell receptor. It is satisfying to think that immunoglobulin, T-cell receptor, and class I and class II MHC molecules, all members of the supergene family, appear to be involved in selective recognition and binding of antigens.

Although we already know that helper T cells recognize degraded forms of soluble protein antigens in the context of class II MHC products, we can now be certain that class I-restricted CTL also recognize degraded forms of viral proteins rather than intact molecules expressed at the cell surface. This is supported by the finding that influenza virus NP was recognized by CTL despite their inability to detect surface expression with antibodies (42). Indeed, this explains why antiviral antibodies are unable to block CTL-mediated lysis of virus-infected cells (56). Furthermore, target cells expressing fragments (40) or short synthetic peptides (6, 43) of influenza virus NP were recognized by class I-restricted CTL. Finally, leader sequence (39)- or transmembrane sequence (13)-deleted influenza virus HA was not expressed at the target cell surface, yet it was still recognized by HA-specific CTL.

The demonstration that CTL recognize processed fragments of proteins that may normally never reach the plasma membrane in an intact state suggests that most, if not all, cells have some ability to process antigens. In light of this, one can suggest new ways by which viruses may avoid immune destruction and attain a state of persistent infection. For example, we have shown that interleukin-2-dependent natural killer clones maintained in vitro can be persistently infected with VSV (36). Persistently infected natural killer cells were found to be resistant to lysis by VSV-specific CTL, even though they expressed both VSV G glycoprotein and NP (36). One explanation for these findings could be that natural killer cells lack appropriate processing enzymes or, owing to an excess of enzymes, are unable to properly process viral molecules for association with class I molecules.

Finally, it remains to be determined where and how viral proteins bind to class I and class II MHC molecules. It has been suggested that endogenously synthesized antigens, such as viral antigens, are class I restricted, whereas exogenous soluble antigens are class II restricted. It was recently argued (9, 20) that these selective associations may be due to different pathways of degradation in cells. Alternatively, a clear resolution of the selective association of antigenic epitopes with class I or class II MHC molecules may be determined by the nature of the binding site on a given MHC molecule. This information should be provided by X-ray crystallographic studies of the MHC molecules.

Literature Cited

1. **Allen, P. M.** 1987. Antigen processing at the molecular level. *Immunol. Today* **8:**270–273.

2. **Allison, J. P., and L. L. Lanier.** 1987. Structure, function, and serology of the T-cell antigen receptor complex. *Annu. Rev. Immunol.* **5:**503–540.
3. **Askonas, B. A., and R. G. Webster.** 1980. Monoclonal antibodies to the hemagglutinin and to H-2 inhibit the cross-reactive T-cell populations induced by influenza. *Eur. J. Immunol.* **10:**151–156.
4. **Babbit, B. P., P. M. Allen, G. Matsueda, E. Haber, and E. R. Unanue.** 1985. Binding of immunogenic peptides to Ia histocompatibility molecules. *Nature* (London) **317:** 359–361.
5. **Bangham, C. R. M., P. J. M. Openshaw, L. A. Ball, A. M. Q. King, G. W. Wertz, and B. A. Askonas.** 1986. Human and murine cytotoxic T cells specific to respiratory syncytial virus recognize the viral nucleocapsid (N), but not the major glycoprotein (G), expressed by vaccinia virus recombinants. *J. Immunol.* **137:**3973–3977.
6. **Bastin, J., J. Rothbard, J. Davey, I. Jones, and A. Townsend.** 1987. Use of synthetic peptides of influenza nucleoprotein to define epitopes recognized by class I-restricted cytotoxic T lymphocytes. *J. Exp. Med.* **165:**1508–1523.
7. **Bennick, J. R., J. W. Yewdell, and W. Gerhard.** 1982. A viral polymerase involved in recognition of influenza virus-infected cells by a cytotoxic T-cell clone. *Nature* (London) **296:**75–76.
8. **Bennick, J. R., J. W. Yewdell, G. L. Smith, and B. Moss.** 1987. Anti-influenza virus cytotoxic T lymphocytes recognize the three viral polymerases and a nonstructural protein: responsiveness to individual viral antigens is major histocompatibility complex controlled. *J. Virol.* **61:**1098–1102.
9. **Bevan, M. J.** 1987. Class discrimination in the world of immunology. *Nature* (London) **325:**192–194.
10. **Biddison, W. E., S. Shaw, and D. L. Nelson.** 1981. Virus specificity of human influenza virus immune cytotoxic T cells. *J. Immunol.* **122:**660–664.
11. **Braciale, T. J.** 1977. Immunologic recognition of influenza virus-infected cells. I. Generation of a virus strain-specific and a cross-reactive subpopulation of cytotoxic T cells in the response to type A influenza viruses of different serotypes. *Cell. Immunol.* **33:**423–426.
12. **Braciale, T. J., V. L. Braciale, T. J. Henkel, J. Sambrook, and M.-J. Gething.** 1984. Cytotoxic T lymphocyte recognition of the influenza hemagglutinin gene product expressed by DNA mediated gene transfer. *J. Exp. Med.* **159:**341–354.
13. **Braciale, T. J., V. L. Braciale, M. Winkler, I. Stroynowski, L. Hood, J. Sambrook, and M.-J. Gething.** 1987. On the role of the transmembrane anchor sequence of influenza hemagglutinin in target cell recognition by class I MHC-restricted, hemagglutinin-specific cytolytic T lymphocytes. *J. Exp. Med.* **166:**678–701.
14. **Buus, S., A. Sette, S. M. Colon, G. Miles, and H. M. Grey.** 1987. The relation between major histocompatibility complex (MHC) restriction and the capacity of Ia to bind immunogenic peptides. *Science* **235:**1353–1358.
15. **DeLisi, C., and J. A. Berzofsky.** 1985. T-cell antigenic sites tend to be amphipathic structures. *Proc. Natl. Acad. Sci. USA* **82:**7048–7052.
16. **Doherty, P. C., R. B. Effros, and J. Bennink.** 1977. Heterogeneity of the cytotoxic response of thymus-derived lymphocytes after immunization with influenza viruses. *Proc. Natl. Acad. Sci. USA* **74:**1209–1213.
17. **Doherty, P. C., and R. M. Zinkernagel.** 1974. T cell-mediated immunopathology in viral infection. *Transplant. Rev.* **19:**89–120.
18. **Effros, R. B., P. C. Doherty, W. Gerhard, and J. Bennink.** 1977. Generation of both cross-reactive and virus-specific T cell populations after immunization with serologically distinct influenza A virus. *J. Exp. Med.* **145:**557–568.

19. **Gell, P. G. H., and B. Benacerraf.** 1959. Studies on hypersensitivity. II. Delayed hypersensitivity to denatured proteins in guinea pigs. *Immunology* **2:**64–70.
20. **Germain, R. N.** 1986. The ins and outs of antigen processing and presentation. *Nature* (London) **322:**687–689.
21. **Gooding, L. R., and K. A. O'Connell.** 1983. Recognition by cytotoxic T lymphocytes of cells expressing fragments of SV40 tumour antigen. *J. Immunol.* **131:**2580–2586.
22. **Gotch, F. M., A. J. McMichael, G. L. Smith, and B. Moss.** 1987. Identification of the viral molecules recognized by influenza-specific human cytotoxic T lymphocytes. *J. Exp. Med.* **165:**408–416.
23. **Guillet, J.-G., M.-Z. Lai, T. J. Briner, S. Buus, A. Sette, H. M. Grey, J. A. Smith, and M. L. Gefter.** 1987. Immunological self, nonself discrimination. *Science* **235:**865–870.
24. **Kees, U., and P. H. Krammer.** 1984. Most influenza A virus-specific memory cytotoxic T lymphocytes react with antigenic epitopes associated with internal virus determinants. *J. Exp. Med.* **159:**365–377.
25. **Koszinowski, U. H., H. Allen, M.-J. Gething, M. D. Waterfield, and H.-D. Klenk.** 1980. Recognition of viral glycoproteins by influenza A specific cross-reactive cytotoxic T lymphocytes. *J. Exp. Med.* **151:**945–958.
26. **Koszinowski, U. H., G. M. Keil, H. Schwarz, J. Schickedanz, and M. J. Reddehase.** 1987. A nonstructural polypeptide encoded by immediate-early transcription unit 1 of murine cytomegalovirus is recognized by cytolytic T lymphocytes. *J. Exp. Med.* **166:**289–294.
27. **Koszinowski, U. H., M. J. Reddehasse, G. M. Keil, and J. Schickedanz.** 1987. Host immune response to cytomegalovirus: products of transfected virus immediate-early genes are recognized by cloned cytotoxic T lymphocytes. *J. Virol.* **61:**2054–2058.
28. **Lin, Y. L., and B. A. Askonas.** 1981. Biological properties of an influenza A virus specific killer T cell clone. *J. Exp. Med.* **154:**225–234.
29. **Lukacher, A. E., V. L. Braciale, and T. J. Braciale.** 1984. In vivo effector function of influenza virus-specific cytotoxic T lymphocyte clones is highly specific. *J. Exp. Med.* **160:**814–826.
30. **McMichael, A. J., and B. A. Askonas.** 1978. Influenza virus-specific cytotoxic T cells in man; induction and properties of the cytotoxic cell. *Eur. J. Immunol.* **8:**705–711.
31. **Mullbacher, A., I. D. Marshall, and R. V. Blanden.** 1979. Cross reactive cytotoxic T cells to alphavirus infection. *Scand. J. Immunol.* **10:**291–296.
32. **Pemberton, R. M., M. J. Cannon, P. J. M. Openshaw, L. A. Ball, G. W. Wertz, and B. A. Askonas.** 1987. Cytotoxic T cell specificity for respiratory syncytial virus proteins: fusion protein is an important target antigen. *J. Gen. Virol.* **68:**2177–2182.
33. **Puddington, L., M. J. Bevan, J. K. Rose, and L. Lefrancois.** 1986. N protein is the predominant antigen recognized by vesicular stomatitis virus-specific cytotoxic T cells. *J. Virol.* **60:**708–717.
34. **Rosenthal, K. L., J. R. Smiley, S. South, and D. C. Johnson.** 1987. Cells expressing herpes simplex virus glycoprotein gC but not gB, gD, or gE are recognized by murine virus-specific cytotoxic T lymphocytes. *J. Virol.* **61:**2438–2447.
35. **Rosenthal, K. L., and R. M. Zinkernagel.** 1980. Cross reactive cytotoxic T cells to serologically distinct vesicular stomatitis virus. *J. Immunol.* **124:**2301–2308.
36. **Rosenthal, K. L., R. M. Zinkernagel, H. Hengartner, P. Groscurth, G. Dennert, D. Takayesu, and L. Prevec.** 1986. Persistence of vesicular stomatitis virus in cloned interleukin-2-dependent natural killer cell lines. *J. Virol.* **60:**539–547.
37. **Shimonkevitz, R., J. Kappler, P. Marrack, and H. Grey.** 1983. Antigen recognition by H-2-restricted T cells. I. Cell-free antigen processing. *J. Exp. Med.* **158:**303–316.
38. **Tevethia, S. S., M. J. Tevethia, A. J. Lewis, V. B. Reddy, and S. M. Weissman.** 1983.

Biology of simian virus 40 (SV40) transplantation antigen (TrAg). IX. Analysis of TrAg in mouse cells synthesizing truncated SV40 large T antigen. *Virology* **128:**319–330.

39. **Townsend, A. R. M., J. Bastin, K. Gould, and G. G. Brownlee.** 1986. Cytotoxic T lymphocytes recognize influenza hemagglutinin that lacks a signal sequence. *Nature* (London) **324:**575–577.
40. **Townsend, A. R. M., F. M. Gotch, and J. Davey.** 1985. Cytotoxic T cells recognize fragments of the influenza nucleoprotein. *Cell* **42:**457–467.
41. **Townsend, A. R. M., and A. J. McMichael.** 1985. Specificity of cytotoxic T lymphocytes stimulated with influenza virus. *Prog. Allergy* **36:**10–43.
42. **Townsend, A. R. M., A. J. McMichael, N. P. Carter, J. A. Huddleston, and G. G. Brownlee.** 1984. Cytotoxic T cell recognition of the influenza nucleoprotein and hemagglutinin expressed in transfected mouse L cells. *Cell* **39:**13–25.
43. **Townsend, A. R. M., J. Rothbard, F. M. Gotch, G. Bahadur, D. Wraith, and A. J. McMichael.** 1986. The epitopes of influenza nucleoprotein recognized by cytotoxic T lymphocytes can be defined with short synthetic peptides. *Cell* **44:**959–968.
44. **Townsend, A. R. M., and J. J. Skehel.** 1982. Influenza A specific cytotoxic T cell clones that do not recognize viral glycoproteins. *Nature* (London) **300:**655–657.
45. **Townsend, A. R. M., and J. J. Skehel.** 1984. The influenza A virus nucleoprotein gene controls the induction of both subtype specific and cross-reactive cytotoxic T cells. *J. Exp. Med.* **160:**552–563.
46. **Townsend, A. R. M., J. J. Skehel, P. M. Taylor, and P. Palese.** 1984. Recognition of influenza A virus nucleoprotein by an H-2 restricted cytotoxic T cell clone. *Virology* **133:**456–459.
47. **Unanue, E. R., D. I. Beller, C. Y. Lu, and P. M. Allen.** 1984. Antigen presentation: comments on its regulation and mechanisms. *J. Immunol.* **132:**1–5.
48. **Volkmer, H., C. Bertholet, S. Jonjic, R. Wittek, and U. H. Koszinowski.** 1987. Cytolytic T lymphocyte recognition of the murine cytomegalovirus nonstructural immediate-early protein pp89 expressed by recombinant vaccinia virus. *J. Exp. Med.* **166:**668–677.
49. **Wraith, D. C.** 1987. The recognition of influenza A virus-infected cells by cytotoxic T lymphocytes. *Immunol. Today* **8:**239–246.
50. **Yap, K. L., G. L. Ada, and I. F. C. McKenzie.** 1978. Transfer of specific cytotoxic T lymphocytes protect mice inoculated with influenza virus. *Nature* (London) **273:**238–239.
51. **Yewdell, J. W., J. R. Bennick, M. Mackett, L. Lefrancois, D. S. Lyles, and B. Moss.** 1986. Recognition of cloned vesicular stomatitis virus internal and external gene products by cytotoxic T lymphocytes. *J. Exp. Med.* **163:**1529–1538.
52. **Yewdell, J. W., J. R. Bennick, G. L. Smith, and B. Moss.** 1985. Influenza A virus nucleoprotein is a major target antigen for cross-reactive anti-influenza A virus cytotoxic T lymphocytes. *Proc. Natl. Acad. Sci. USA* **82:**1785–1789.
53. **Ziegler, H. K., and E. Unanue.** 1982. Decrease in macrophage antigen catabolism caused by ammonia and chloroquine is associated with inhibition of antigen presentation to T cells. *Proc. Natl. Acad. Sci. USA* **79:**175–178.
54. **Zinkernagel, R. M., and P. C. Doherty.** 1974. Restriction of *in vitro* T cell mediated cytotoxicity in lymphocytic choriomeningitis within a syngeneic or semiallogeneic system. *Nature* (London) **248:**701–702.
55. **Zinkernagel, R. M., and P. C. Doherty.** 1974. Immunological surveillance against altered self components by sensitized T lymphocytes in lymphocytic choriomeningitis. *Nature* (London) **251:**547–548.
56. **Zinkernagel, R. M., and K. L. Rosenthal.** 1981. Experiments and speculation on antiviral specificity of T and B cells. *Immunol. Rev.* **58:**131–155.

57. **Zweerink, H. J., B. A. Askonas, D. Millican, S. A. Courtneidge, and J. J. Skehel.** 1977. Cytotoxic T-cells to type A influenza virus: viral haemagglutinin induces A-strain specificity while infected cells confer cross-reactive cytotoxicity. *Eur. J. Immunol.* **7:**630–635.
58. **Zweerink, H. J., S. A. Courtneidge, J. J. Skehel, M. Crumpton, and B. A. Askonas.** 1977. Cytotoxic T cells kill influenza virus infected cells, but do not distinguish between serologically distinct type viruses. *Nature* (London) **267:**354–356.

Chapter 11

Molecular Analyses of Cytotoxic T-Lymphocyte Responses to Lymphocytic Choriomeningitis Virus Infection

J. Lindsay Whitton, Hanna Lewicki, John R. Gebhard, Antoinette Tishon, Peter J. Southern, and Michael B. A. Oldstone

INTRODUCTION

Lymphocytic choriomeningitis virus (LCMV) provides us with an excellent prototype for the study of persistent virus infections. It is an ideal model system for several reasons. First, it can be studied in its natural host, the mouse, thus avoiding the need to "mouse-adapt" the virus; second, considerable advances have recently been made in its molecular characterization; and third, the required conditions for the establishment of virus persistence have been studied in some detail. Mice infected in utero or neonatally will carry the virus for life, but can be cured of infection by adoptive transfer of LCMV-specific immune memory cells of the $Lyt2^{+}$ $L3T4^{-}$ phenotype, i.e., cytotoxic T lymphocytes (CTL). The critical role played by these cells is underlined by the finding that adult mice, which normally mount a brisk major histocompatibility complex (MHC) class I-restricted CTL response to primary LCMV infection, become persistently infected if they are depleted of CTL or if virus variants are used which appear to abrogate the host CTL response. Given the importance of CTL in determining the outcome of LCMV infection,

---

J. Lindsay Whitton, Hanna Lewicki, John R. Gebhard, Antoinette Tishon, Peter J. Southern, and Michael B. A. Oldstone • Department of Immunology, Scripps Clinic and Research Foundation, La Jolla, California 92037.

we have set out to study the nature of the host CTL response. Specifically, what virus components are involved, and how are they restricted by the host MHC (*H-2* in the mouse)?

LCMV, a member of the *Arenaviridae* family, has a bisegmented RNA genome encoding at least three polypeptides: a polymerase, a glycoprotein (GP), and a nucleoprotein (NP) (2). The short (S) segment of LCMV Armstrong (ARM) has been cloned and sequenced in its entirety (20) and encodes two proteins, a 558-amino-acid NP and a 498-amino-acid precursor GP, GPC (3, 15), which is posttranslationally cleaved to generate the two mature structural GPs GP1 (residues 1 to 262) and GP2 (residues 263 to 498) (4). Interstrain amino acid sequence comparisons of each of these molecules (19) have shown that the greatest degree of variability between different strains lies in GP1, in which a major antibody neutralization site has been identified (14). By using segmental reassortants between two different strains of LCMV, both induction of and target cell recognition and lysis by MHC class I-restricted CTL have been mapped to the virus S segment (16). To estimate the relative CTL responses to GP and NP, we have expressed these proteins in vaccinia virus (VV); this vector was chosen primarily because, like LCMV, it has a cytoplasmic replication cycle with no known nuclear phase and we wished to avoid any potential problems generated by exposing the LCMV RNA sequences to a nuclear environment, where splicing and/or other posttranscriptional processing might occur.

We describe here the expression of LCMV NP and a family of truncated GP molecules in VV. We have analyzed the primary CTL responses mounted to LCMV, using these recombinants as target antigens, and we show that the responses vary markedly depending on the strain of mouse used to induce CTL. Furthermore, we have analyzed in detail the CTL response to LCMV GP. Using synthetic peptides and the GP deletions expressed in VV, we have precisely located a major CTL epitope at the N terminus of GP2.

CONSTRUCTION AND EXPRESSION OF GP DELETIONS AND NP IN VV

The cloning of a truncated LCMV GP gene, encoding amino acid residues 1 to 363, has been described previously (26). The resultant plasmid contains the GP translation initiation codon and coding region up to residue 363, followed by six plasmid-encoded amino acid residues preceding a translation termination codon. This vector was used as the parental plasmid from which a family of serial GP truncations were made. In all cases the GP coding regions are linked to the same plasmid

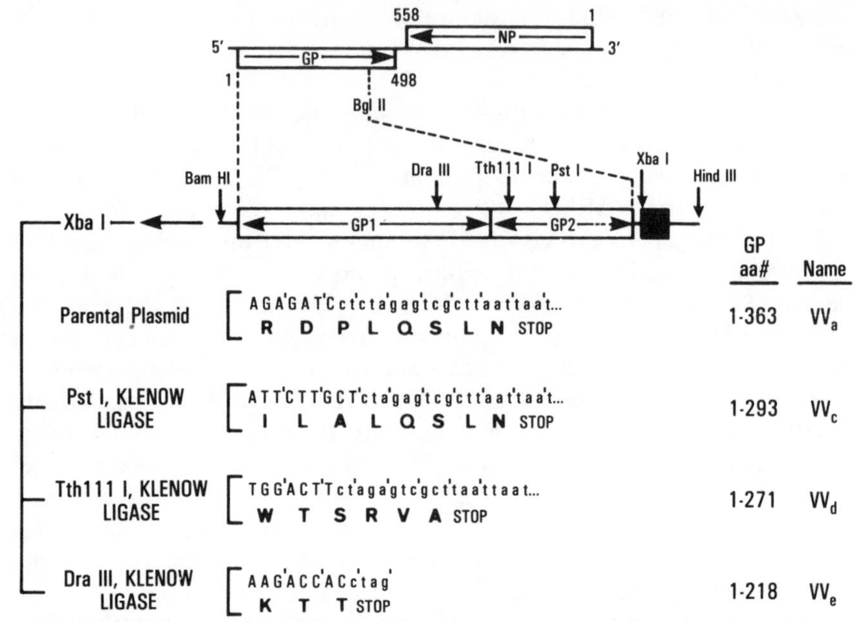

Figure 1. Construction of the family of C-terminal truncations of LCMV GP. The LCMV S segment is shown, encoding GP and NP; – – –, location in the S segment of the *Bam*HI-*Bgl*II GP fragment used to make the parental plasmid. The construction of this plasmid is presented in expanded form; ■, universal translation terminator sequence which contains stop codons in all three reading frames (J. L. Whitton, unpublished results). Restriction enzyme sites used to make the deletions are shown above the expanded area. DNA sequences at the cloning junctions were confirmed by sequencing and are shown as follows: capital letters represent GP-derived nucleotides, and lower-case letters represent plasmid-derived bases. The vertical dashes above the DNA sequences indicate the open reading frame coming in from the GP, and the encoded amino acids are shown below the DNA sequences in single-letter code. The name of the resultant VV recombinant and the number of LCMV GP residues encoded therein are shown.

sequences and termination codons. Figure 1 shows the nucleic acid and consequent amino acid sequences at the cloning junction of all four LCMV GP truncations. The nucleic acid sequences shown were established by DNA sequencing of the plasmids. In all four plasmids, the DNA fragment containing the truncated GP sequences was excised along with the translation termination codons by digestion with restriction enzymes *Bam*HI and *Hin*dIII. These fragments were then cloned into the *Sma*I site of plasmid pSC11 (6) (Fig. 2). Full-length LCMV NP was recombined into VV in a similar manner (6,7). One further recombinant was made by recombining pSC11 (without any LCMV sequences) into VV to yield

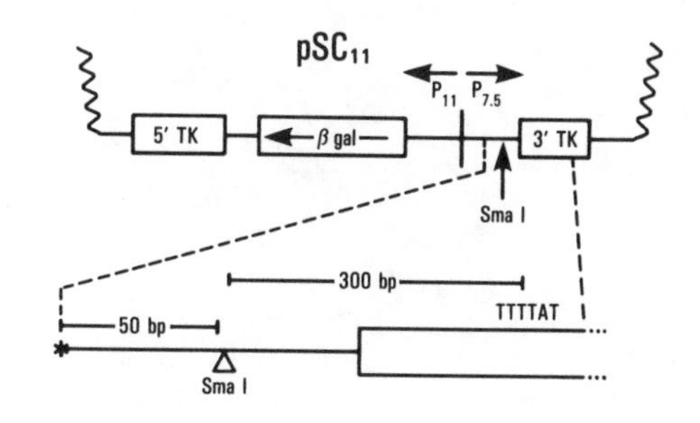

Recombinant Vaccinia	LCMV Protein Encoded	Length of DNA Insert (bp)	Expected VV mRNA Size
VV_a	GP 1-363	1200	1550
VV_c	GP 1-293	1000	1350
VV_d	GP 1-271	930	1280
VV_e	GP 1-218	770	1120
VV_{NP}	NP 1-558	1750	2100

Figure 2. Transcriptional strategy of plasmid pSC11, used to predict the approximate lengths of VV-LCMV recombinant mRNA molecules in the recombinant VV. The layout of pSC11 is shown. Abbreviations: 5′ TK and 3′ TK, 5′ and 3′ portions, respectively, of the VV thymidine kinase gene; β gal, gene encoding β-galactosidase; P_{11} and $P_{7.5}$, two divergently oriented VV promoters. The insertion site for LCMV DNA is arrowed. The asterisk indicates the 5′ end of the mRNA initiated on the $P_{7.5}$ promoter, and TTTTAT is the transcription terminator in the 3′ TK region. The *Sma*I site of pSC11 lies some 50 base pairs downstream of the transcription start site and approximately 300 base pairs upstream of the 3′ end, which is defined by transcription termination signals in the thymidine kinase gene sequences. Note that the expected VV mRNA size (bottom, column 4) does not include the poly(A) tail, which may be some 200 to 400 bases in length.

$VV_{SC_{11}}$; this was a TK^- β-Gal^+ recombinant for use as a control in cytotoxicity assays. We next tested the recombinant viruses for correct transcription of the LCMV sequences. Four components make up the length of the mRNA (Fig. 2); three are invariant [50 bases at the 5′ end, 300 bases at the 3′ end, and the poly(A) tail, which is added by a VV-encoded enzyme and may be several hundred residues long (11,17)], and the fourth is variable (the size of the LCMV cDNA insert). The expected sizes [before addition of poly(A)] of the five GP mRNAs expressed from VV are shown in Fig. 2. LCMV GP-specific and NP-specific DNA probes were hybridized to a Northern (RNA) blot of cytoplasmic RNA from recombinant-infected cells to identify the VV-encoded LCMV-specific RNA molecules. The bands detected (Fig. 3) closely approximate the expected sizes. Analysis of protein expression by immunofluorescence techniques indicated that each virus recombinant was expressing the

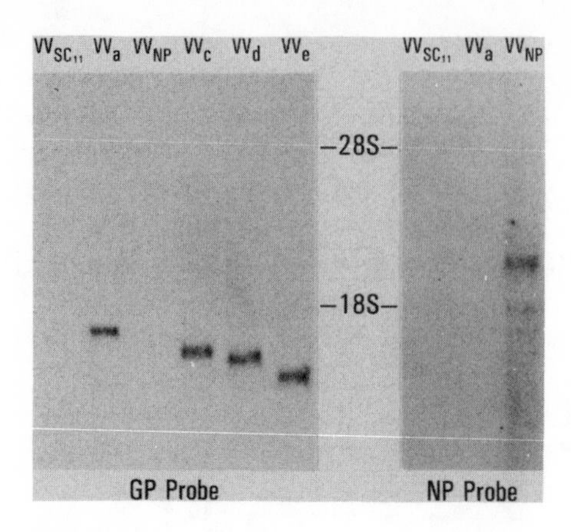

Figure 3. Northern blot analysis of six VV recombinants. Samples of total cytoplasmic RNA from cells infected with $VV_{SC_{11}}$, VV_a, VV_c, VV_d, VV_e, and VV_{NP} were electrophoresed and transferred to nitrocellulose as described in the text. The blot was hybridized with a probe specific for LCMV GP (left panel) or NP (right panel), washed, and exposed at −70°C to Kodak XAR5 film with a Cronex Lightning-Plus intensifying screen (Du Pont Co.). The positions of the 28S rRNA (ca. 5,200 bases) and the 18S rRNA (ca. 2,000 bases) are shown.

encoded LCMV polypeptide; in all cases abundant cytoplasmic expression was detected, although cell surface expression was weak.

HOST MHC DETERMINES WHICH LCMV PROTEIN WILL BE USED TO INDUCE A CTL RESPONSE

First we wished to establish that VV recombinants expressing NP or GP (residues 1 to 363; VV_a) could function as targets for anti-LCMV CTL. This was indeed found to be the case (Fig. 4), since on the $H\text{-}2^{bb}$ background, target cells expressing both GP and NP were efficiently lysed; thus, on this haplotype, LCMV infection induces CTL against both of these moieties.

Ahmed et al. have previously reported that CTL recognition of several LCMV strains varies in a host-dependent manner and that the variability maps to the MHC locus (1). Therefore, we wished to see whether, by switching between MHC haplotypes, we could identify changes in the pattern of CTL recognition of LCMV NP and GP. A change of MHC background does indeed result in a change in the pattern of recognition of LCMV proteins (Fig. 4). Cells infected with the VV_{NP} recombinant were efficiently lysed on all three haplotypes; this internal protein therefore induces CTL on all MHC haplotypes we have tested, and its interaction with CTL presumably relies upon the surface exposure of an as yet unidentified epitope. In contrast, VV_{GP}-infected targets, although productively infected on all three haplotypes (they are lysed by syngeneic anti-VV CTL) are lysed by anti-LCMV CTL only on $H\text{-}2^{bb}$; they exhibit little or no lysis when challenged with $H\text{-}2^{dd}$ or $H\text{-}2^{qq}$ anti-LCMV CTL, suggesting that anti-GP CTL are at best a minor component of the $H\text{-}2^{dd}$ and $H\text{-}2^{qq}$ splenocyte populations.

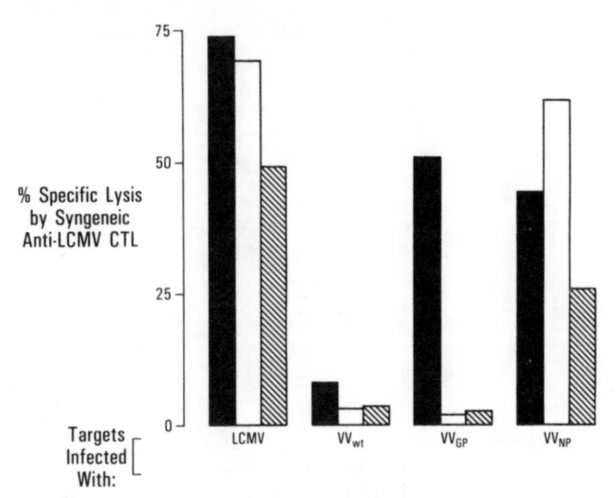

Figure 4. Effect of the MHC in determining which LCMV protein(s) will induce CTL. Cytotoxicity assays were carried out as described previously (5). In all cases, controls were included to confirm *H-2* restriction (not shown). The percent lysis by anti-LCMV CTL from $H\text{-}2^{bb}$ (solid bars), $H\text{-}2^{dd}$ (open bars), or $H\text{-}2^{qq}$ (hatched bars) mice is shown, measured on syngeneic targets infected as indicated. Note that LCMV- and VV_{NP}-infected cells are lysed on all three haplotypes, whereas cells infected with VV_{GP} (VV_a; GP residues 1 to 363) are lysed only on $H\text{-}2^{bb}$.

We have further analyzed the anti-GP response on $H\text{-}2^{bb}$ and have found that it makes up around one-third of the overall anti-LCMV response and appears to be largely cross-reactive (i.e., anti-ARM GP CTL can lyse targets infected with a serologically distinct strain of LCMV).

USE OF CLONED CTL AND GP DELETIONS TO MAP LCMV GP EPITOPES

Having established that LCMV GP constituted a major CTL target on $H\text{-}2^{bb}$, we continued our analysis by making a series of C-terminal truncations of GP expressed in VV (Fig. 1 to 3) and using these to determine the specificities of a battery of 18 independently isolated CTL clones. The percent specific lysis of the four GP deletions by bulk CTL and by three representative CTL clones is shown in Fig. 5. This figure shows that the bulk splenocytes from a C57BL/6 mouse infected 7 days previously with LCMV ARM were able to lyse syngeneic targets infected with each of the four recombinant viruses which express LCMV GP,

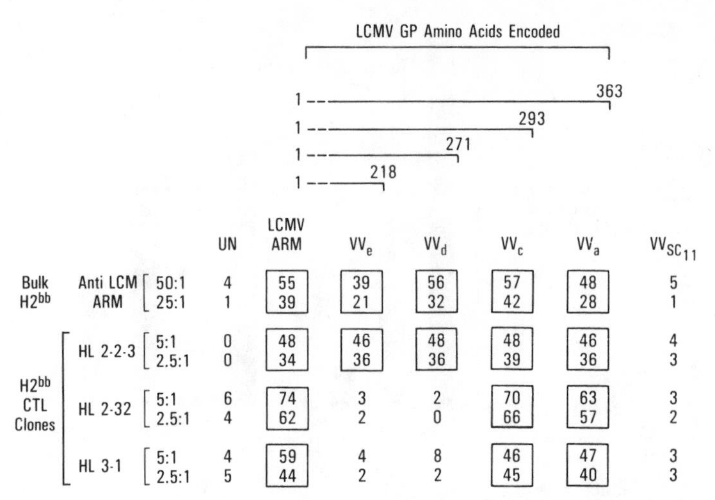

Figure 5. Cytotoxicity assay showing lysis of *H-2^{bb}* targets infected with LCMV, with one of the four VV_{GP} recombinants, or with $VV_{SC_{11}}$. Effectors are bulk splenocytes or CTL clones representing the two anti-GP families (see text). HL 2-2-3 is one of the two clones which recognize all four VV_{GP} constructs, whereas HL 2-32 and HL 3-1 are representative of the larger family which recognizes VV_a and VV_c, but neither VV_d nor VV_e. The percent specific lysis considered significant is boxed.

indicating that even the shortest truncation (VV_e residues 1 to 218) contains a CTL epitope. This lysis was LCMV specific, since cells infected with $VV_{SC_{11}}$ were not lysed (although these cells were productively infected and were efficiently lysed by anti-VV CTL), and was *H-2* restricted (since *H-2^{bb}* bulk CTL did not lyse allogeneic targets infected with LCMV ARM, and allogeneic effector cells did not lyse the *H-2^{bb}* targets infected with recombinant VV). Also (Fig. 6), the great majority (17 of 18) of the anti-LCMV CTL clones were GP specific, the single remaining clone being NP specific. These results are striking in the preponderance of GP recognition. In, for example, the influenza virus system, the great majority of CTL recognition is directed toward internal proteins (NP, NS1, and the polymerase complex); however, in the LCMV natural infection, none of 18 CTL clones are directed toward polymerase and only one recognizes NP. Although at this stage we cannot exclude the possibility that some degree of selection in favor of certain clones has occurred during the cloning procedure, these results are consistent with our earlier observations (Fig. 4) indicating that a significant part of the *H-2^{bb}* primary CTL response was against the GP moiety.

Further analysis (Fig. 6, right panel) of the 17 CTL clones which

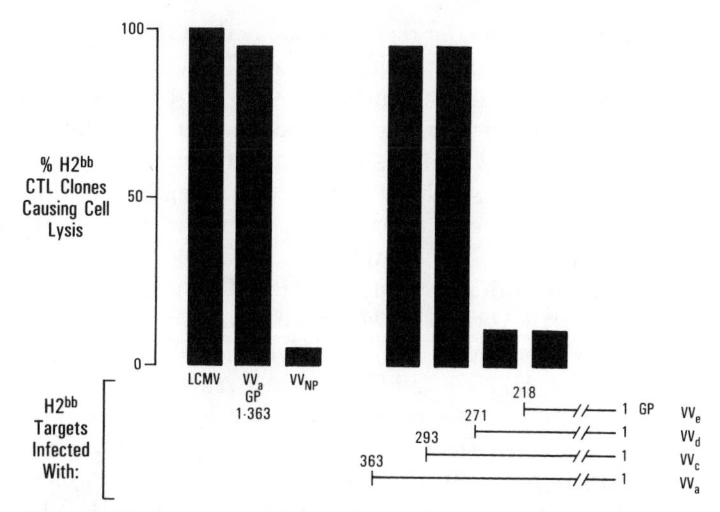

Figure 6. Diagrammatic representation of the pattern of killing by the 18 CTL clones, measured by cytotoxicity assays as described previously (5). Left panel: the great majority of clones recognize VV_a (GP residues 1 to 363); only one recognizes VV_{NP}. Right panel: all clones which recognize VV_a also recognize VV_c. When VV_d is used as a target, most of the clones lose their lytic ability; only two clones can kill VV_d and VV_e.

recognize the LCMV GP moiety shows that they fall cleanly into two families. All clones recognize VV_a (GP residues 1 to 363)- and VV_c (GP residues 1 to 293)-infected cells; however, when cells infected with VV_d (GP residues 1 to 271) or VV_e are used as targets, there is a dramatic reduction in the number of clones capable of lytic activity. Only two of the clones (exemplified by HL 2-2-3; Fig. 5) lyse syngeneic target cells infected with all four GP recombinants. These clones therefore recognize the epitope (or epitopes) inferred above, which are contained in GP residues 1 to 218. The remaining clones fall into a large group which recognizes VV_a and VV_c, but fails to recognize VV_d or VV_e (Fig. 6; data for representative clones HL 2-32 and HL 3-1 are given in Fig. 5). Thus, the majority of the CTL clones are in a family which recognizes VV_c (GP residues 1 to 293) but fails to lyse VV_d (GP residues 1 to 271), and hence amino acids 272 to 293 are required to allow CTL recognition by these clones. This region is near the N terminus of GP2.

CONFIRMATION BY SYNTHETIC PEPTIDE RECOGNITION THAT GP RESIDUES 272 TO 293 CONTAIN A CTL EPITOPE

The demonstration that GP residues 272 to 293 are required to allow target cell recognition and lysis could be interpreted in three ways. The simplest explanation is that the region contains the CTL epitope. Alternatively, the epitope may span residues 271 and 272 and be interrupted in VV_d. Finally, the epitope may lie outside this region, but this region may be needed to allow, for example, correct processing or presentation of the epitope. The most direct way of distinguishing among these hypotheses was to examine whether a synthetic peptide corresponding to this region could render syngeneic (uninfected) target cells susceptible to recognition and lysis by CTL (Fig. 7). The peptide was tested by coincubation with syngeneic targets and with either bulk LCMV CTL or one of three CTL clones, HL 2-2-8, HL 2-32 (both of which recognize VV_c but not VV_e), or HL 2-1 (which is NP specific). Cells coated with the peptide are efficiently lysed by CTL clones HL 2-2-8 and HL 2-32, as well as by bulk CTL. The specificity of the effect of this peptide is shown by the failure of these effector cells to lyse cells treated with an irrelevant peptide (a peptide from human MHC, HLA B27). The specificity of the two CTL clones HL 2-2-8 and HL 2-32 is reconfirmed by the inability of clone HL 2-1 to lyse targets treated with GP residues 272 to 293. These findings show conclusively that residues 272 to 293 contain a CTL epitope (or epitopes) and confirm the validity of using serially truncated molecules expressed from VV to map CTL epitopes.

CONCLUSIONS

The data presented here allow several points to be made, both about the primary CTL response to LCMV infection and about the general nature of class I MHC-restricted CTL recognition.

First, we have shown that, on the $H\text{-}2^{bb}$ background, both GP and NP moieties of LCMV can efficiently induce primary CTL responses. Thus, the GP, which contains the major antibody neutralization site of the virus, also contains at least one major CTL epitope on the $H\text{-}2^{bb}$ background. During acute LCMV infection, both GP1 and GP2 moieties are expressed on the cell membrane and are detectable by fluorescent-antibody analyses. As infection continues, however, expression of GP decreases dramatically both in vitro and in vivo (12,25). It is not known whether this decreased GP expression has any part to play in the establishment and/or maintenance of persistence, but the mapping to GP of sites critical to both humoral and cellular immune response may be of interest in this regard.

Second, we find that the host MHC exerts a profound influence in selecting the virus determinants used to induce CTL. Previous work with congenic mouse strains differing only at the *H-2* locus has shown that patterns of CTL cross-reactivity among different LCMV strains are dependent on the class I restricting molecules (1), and the observations

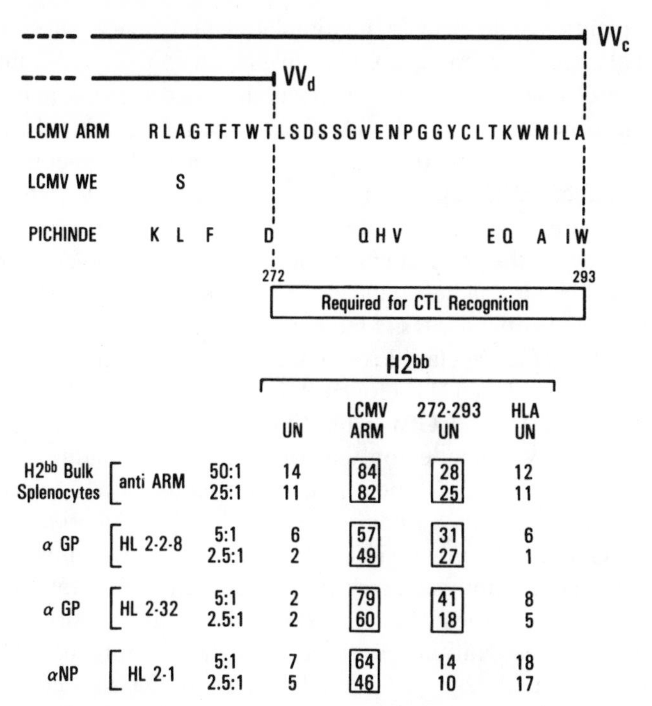

			H2bb			
			UN	LCMV ARM	272-293 UN	HLA UN
H2bb Bulk Splenocytes	anti ARM	50:1	14	84	28	12
		25:1	11	82	25	11
α GP	HL 2-2-8	5:1	6	57	31	6
		2.5:1	2	49	27	1
α GP	HL 2-32	5:1	2	79	41	8
		2.5:1	2	60	18	5
αNP	HL 2-1	5:1	7	64	14	18
		2.5:1	5	46	10	17

Figure 7. A synthetic peptide (9) representing GP residues 272 to 293 appropriately directs the killing of uninfected syngeneic target cells. The amino acid sequences of LCMV ARM and WE and Pichinde virus are illustrated in single-letter code. The GP coding boundaries of VV_c and VV_d are indicated, as is the 22-amino-acid region required for CTL killing (see text). This region was synthesized as a peptide and used in a cytotoxicity assay, the results of which are presented in the lower half of the figure. The effector cells used are shown: both of the anti-GP clones used were VV_c^+ VV_e^-. The target cells were uninfected (UN), infected with LCMV ARM (LCMV ARM), and uninfected and treated with 40 μg of either GP residues 272 to 293 (272–293 UN) or an irrelevant peptide (HLA UN). The percent ^{51}Cr release considered significantly above background is boxed.

we present here confirm that CTL responses to a particular virus protein (in this case LCMV GP) vary greatly in mice with different MHC backgrounds. These results complement similar recent findings in the influenza virus system (8,13) and have important implications for vaccine development. We would suggest that vaccination with the VV_{GP} recombinant may induce a CTL response only in $H\text{-}2^{bb}$ mice, whereas $H\text{-}2^{dd}$ and

H-2^{qq} recipients would remain unstimulated and hence susceptible to LCMV challenge. We are currently investigating this possibility. Our results counsel caution in the use of subunit vaccines, which may be effective in generating a CTL response only on certain MHC backgrounds, and suggest that to maximize the chance of a vaccine's being universally effective in a population, a variety of antigenic determinants may have to be provided.

With regard to the general nature of viral antigens presented by class I MHC molecules, it has been suggested that although class II molecules restrict processed antigen, class I molecules instead restrict intact (native) antigen. LCMV CTL are class I restricted, and the recognition of VV_e by bulk splenocytes and by CTL clones RG-1 and HL 2-2-3 clearly demonstrates that native LCMV GP1 cannot be required for CTL recognition and lysis, since VV_e encodes only a truncated GP1 (residues 1 to 218). Furthermore, the epitope(s) near the N terminus of GP2 is seen by LCMV-specific CTL clones when presented either as a peptide (Fig. 7) or by VV_c, which encodes only residues 263 to 293 of the GP2 moiety. Thus, native GP2 also is not required for successful presentation of this epitope(s). Together, these findings show that native protein is not a prerequisite for recognition of class I-restricted antigens. The VV_c recombinant that expresses GP2 residues 263 to 293 also contains five non-LCMV amino acids attached to residue 293 (Fig. 1). Despite these foreign components, VV_c is efficiently recognized by the CTL clones. It is not known whether this region is processed during VV_c infection to remove the non-LCMV amino acids prior to presentation by the MHC or whether the T-cell receptor of the CTL clone is able to recognize the presented moiety despite these five accessory residues. Whichever is the case, the recognition of this CTL epitope in VV_c attests to the flexibility of the antigen recognition system, which remains able to present and recognize a specific epitope despite the nearby covalent attachment of several foreign amino acids. These observations support the concept that class I MHC restriction may be similar to class II MHC restriction (18, 24) in the presentation of small, processed regions of virus polypeptide. This finding suggests a testable molecular explanation for the pattern of CTL induction by LCMV GP on the three MHC haplotypes studied (Fig. 4). We expect that the class I MHC molecules of *H-2*^{bb} will be capable of binding GP residues 272 to 293, whereas *H-2*^{dd} and *H-2*^{qq} MHC molecules will be unable to do so; failure of the MHC to select and bind GP epitopes would result in the absence of CTL induction. Our studies of the CTL interaction with LCMV GP complement and extend the initial observations of Townsend et al., who studied transfected deletion mutants of influenza virus NP (22) and more recently studied an influenza virus hemagglutinin apparently lacking cell membrane expression (21).

It has also been suggested that class I MHC alleles restrict only endogenously synthesized molecules, whereas class II alleles restrict only exogenously applied molecules (10). Our observation that a peptide along with a syngeneic uninfected target cell can direct CTL killing confirms studies of Townsend et al., who used NP in the influenza virus system (23). These observations, as well as further refuting any requirement for native protein in CTL recognition, indicate that endogenous synthesis is not required to sensitize target cells to CTL recognition and lysis in vitro. However, several important questions about the mechanism of sensitization by the peptide in vitro remain unanswered. For example, does the peptide bind directly to the MHC molecule on the cell surface, or is it first internalized by the presenting cell? If it is internalized, is it subjected to processing prior to presentation?

In summary, these studies of natural infection have begun to unravel the complexities of the interrelationship between host and virus. In the future we hope to determine the variety of T-cell epitopes present in virus proteins and the precise mechanism of their interaction with host MHC and T-cell receptor. As a result of these studies, we anticipate a better understanding both of the means by which the host responds to the virus and of the mechanisms which allow the virus to evade these responses.

Literature Cited

1. **Ahmed, R., J. A. Byrne, and M. B. A. Oldstone.** 1984. Virus specificity of cytotoxic T lymphocytes generated during acute lymphocytic choriomeningitis virus infection. Role of the H-2 region in determining cross-reactivity for different lymphocytic choriomeningitis virus strains. *J. Virol.* **51:**34–41.
2. **Bishop, D. H., and D. D. Auperin.** 1987. Arenavirus gene structure and organization. *Curr. Top. Microbiol. Immunol.* **133:**5–17.
3. **Buchmeier, M. J., and M. B. A. Oldstone.** 1979. Protein structure of lymphocytic choriomeningitis virus: evidence for a cell associated precursor of the virion glycopeptides. *Virology* **99:**111–120.
4. **Buchmeier, M. J., P. J. Southern, B. S. Parekh, M. K. Wooddell, and M. B. A. Oldstone.** 1987. Site-specific antibodies define a cleavage site conserved among arenavirus GP-C glycoproteins. *J. Virol.* **61:**982–985.
5. **Byrne, J. A., R. Ahmed, and M. B. A. Oldstone.** 1984. Biology of cloned cytotoxic T lymphocytes specific for lymphocytic choriomeningitis virus. I. Generation and recognition of virus strains and H-2^b mutants. *J. Immunol.* **133:**433–439.
6. **Chakrabarti, S., K. Brechling, and B. Moss.** 1985. Vaccinia virus expression vector: coexpression of β-galactosidase provides visual screening of recombinant virus plaques. *Mol. Cell. Biol.* **5:**3403–3409.
7. **Mackett, M., G. L. Smith, and B. Moss.** 1984. General method for production and selection of infectious vaccinia virus recombinants expressing foreign genes. *J. Virol.* **49:**857–864.

8. **McMichael, A. J., F. M. Gotch, and J. Rothbard.** 1986. HLA B37 determines an influenza A virus nucleoprotein epitope recognized by cytotoxic T lymphocytes. *J. Exp. Med.* **164:**1397–1406.
9. **Merrifield, R. B.** 1963. Solid phase peptide synthesis. I. The synthesis of a tetrapeptide. *J. Am. Chem. Soc.* **85:**2149–2154.
10. **Morrison, L. A., A. E. Lukacher, V. L. Braciale, D. P. Fan, and T. J. Braciale.** 1986. Difference in antigen presentation to MHC class I and class II restricted influenza virus-specific cytolytic T lymphocyte clones. *J. Exp. Med.* **163:**903–921.
11. **Moss, B., and E. N. Rosenblum.** 1974. Vaccinia virus polyriboadenylate polymerase: covalent linkage of the product with polyribonucleotide and polydeoxyribonucleotide primers. *J. Virol.* **14:**86–98.
12. **Oldstone, M. B. A., and M. J. Buchmeier.** 1982. Restricted expression of viral glycoprotein in cells of persistently infected mice. *Nature* (London) **300:**360–362.
13. **Pala, P., and B. A. Askonas.** 1986. Low responder MHC alleles for Tc recognition of influenza nucleoprotein. *Immunogenetics* **23:**379–384.
14. **Parekh, B. S., and M. J. Buchmeier.** 1986. Proteins of lymphocytic choriomeningitis virus antigenic topography of the viral glycoproteins. *Virology* **153:**168–178.
15. **Riviere, Y., R. Ahmed, P. J. Southern, M. J. Buchmeier, F. J. Dutko, and M. B. A. Oldstone.** 1985. The S RNA segment of lymphocytic choriomeningitis virus codes for the nucleoprotein and glycoproteins 1 and 2. *J. Virol.* **53:**966–968.
16. **Riviere, Y., P. J. Southern, R. Ahmed, and M. B. A. Oldstone.** 1986. Biology of cloned cytotoxic T lymphocytes specific for lymphocytic choriomeningitis virus. V. Recognition is restricted to gene products encoded by the viral S RNA segment. *J. Immunol.* **136:**304–307.
17. **Rohrmann, G., L. Yuen, and B. Moss.** 1986. Transcription of vaccinia virus early genes by enzymes isolated from vaccinia virions terminates downstream of a regulatory sequence. *Cell* **46:**1029–1035.
18. **Schwartz, R. H.** 1987. Antigenic presentation: fugue in T lymphocyte recognition. *Nature* (London) **237:**738–739.
19. **Southern, P. J., and D. H. Bishop.** 1987. Sequence comparison among arenaviruses. *Curr. Top. Microbiol. Immunol.* **133:**19–39.
20. **Southern, P. J., M. K. Singh, Y. Riviere, D. R. Jacoby, M. J. Buchmeier, and M. B. A. Oldstone.** 1987. Molecular characterization of the genomic S RNA segment from lymphocytic choriomeningitis virus. *Virology* **157:**145–155.
21. **Townsend, A. R. M., J. Bastin, K. Gould, and G. G. Brownlee.** 1986. Cytotoxic T lymphocytes recognize influenza hemagglutinin that lacks a signal sequence. *Nature* (London) **324:**575–577.
22. **Townsend, A. R. M., F. M. Gotch, and J. Davey.** 1985. Cytotoxic T cells recognize fragments of the influenza nucleoprotein. *Cell* **42:**457–467.
23. **Townsend, A. R. M., J. Rothbard, F. Gotch, G. Bahadur, D. C. Wraith, and A. J. McMichael.** 1986. The epitopes of influenza nucleoprotein recognized by cytotoxic T lymphocytes can be defined with short synthetic peptides. *Cell* **44:**959–968.
24. **Unanue, E. R., and P. M. Allen.** 1987. The basis for the immunoregulatory role of macrophages and other accessory cells. *Science* **236:**551–557.
25. **Welsh, R. M., and M. J. Buchmeier.** 1979. Protein analysis of defective interfering lymphocytic choriomeningitis virus and persistently infected cells. *Virology* **96:**503–515.
26. **Whitton, J. L., P. J. Southern, and M. B. A. Oldstone.** 1988. Analyses of the cytotoxic T lymphocyte responses to glycoprotein and nucleoprotein components of lymphocytic choriomeningitis virus. *Virology* **162:**321–327.

Chapter 12

Curing of a Congenitally Acquired Chronic Virus Infection: Acquisition of T-Cell Competence by a Previously "Tolerant" Host

Rafi Ahmed

INTRODUCTION

Chronic virus infections now constitute a major health problem. Although the reported incidence of several acute virus infections has declined in recent years, the incidence of persistent virus infections has been increasing (10). This is most dramatically illustrated by the spread of the acquired immunodeficiency syndrome (8). The incidence of chronic hepatitis B virus infections has also steadily increased, with current estimates of over 200 million virus carriers (7, 14). In addition, persistent virus infections have been associated with certain neoplasms and immunopathological diseases (10). To better understand the problem of chronic infections and to design the most effective treatment strategies, it is essential to examine the immunological mechanisms that are effective in clearing persistent and disseminated virus infections.

Infection of mice with lymphocytic choriomeningitis virus (LCMV) is a classic model of viral persistence and immunological tolerance (3, 13, 17, 19, 20). This model system allows one to test the potential of specific immunotherapy to clear virus from a chronically infected host, to study the effector mechanisms responsible for clearing such infections, and to determine whether a previously unresponsive (tolerant) host can acquire immunocompetence.

Rafi Ahmed • Department of Microbiology and Immunology, University of California-Los Angeles School of Medicine, Los Angeles, California 90024.

Adult mice infected intravenously or intraperitoneally with LCMV mount a vigorous primary cellular and humoral response against the virus and clear the infection within 2 weeks. After the primary immune response has subsided, these mice develop immunity and contain memory B and T cells. In contrast to the acute infection seen in adult mice, mice infected with LCMV at birth or in utero become persistently infected with lifelong viremia, and most of their major organs contain high levels of infectious virus and viral antigen. The persistence of LCMV in these carrier mice is accompanied by T-cell unresponsiveness to the virus. Such persistently infected mice show limited or no detectable cytotoxic T-lymphocyte (CTL) or delayed-type hypersensitivity responses against LCMV, and they also have either a paucity or a lack of T cells that will proliferate when stimulated with the virus (2, 3, 13, 17; R. Ahmed and L. Butler, unpublished data).

Volkert (20) was the first to show that the adoptive transfer of spleen cells from LCMV-challenged adult mice (30 to 60 days postinfection; i.e., immune mice) results in the reduction of infectious virus in carrier mice. In the study described below, we identified the effector cells mediating the clearance of this persistent and disseminated virus infection. More importantly, the presence of host-derived LCMV-specific CTLs in the cured carrier mice was documented, and it has been shown that these mice were then able to resist a second challenge with LCMV.

IMMUNOTHERAPY OF PERSISTENTLY INFECTED MICE

The lifelong infection established in neonatally or congenitally infected mice can be eliminated by adoptive transfer of lymphocytes from LCMV-immune mice (1, 9, 18, 20). In our experiments, the immunotherapy was effective in the majority (68 of 105) of the treated carrier mice. Both infectious virus and viral antigen were eliminated from most tissues (Fig. 1). However, distinct patterns of viral clearance and histopathology in various organs were observed. For example, clearance from the liver occurred within 30 days and was accompanied by extensive mononuclear cell infiltrates and necrosis of hepatocytes. Infectious virus and viral antigen were eliminated concurrently. This pattern of viral clearance was also seen in most other tissues (i.e., lungs, spleens, lymph nodes, pancreases, etc.). In contrast, a different pattern of clearance was observed in brain tissue. Infectious virus was eliminated within 30 days, but viral antigen persisted in the central nervous systems of treated carrier mice for up to 90 days. The urinary system was the most resistant to immunotherapy. Elimination of infectious virus and viral antigen from the kidneys took >200 days.

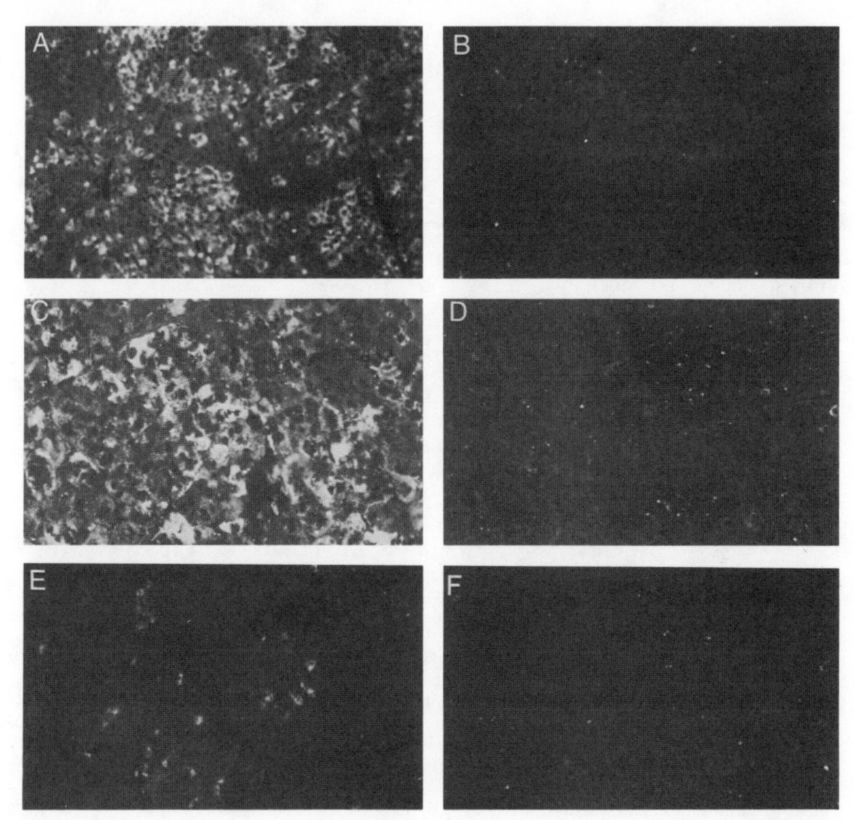

Figure 1. Clearance of viral antigen from tissues of carrier mice after immune therapy. Tissue sections were stained with guinea pig anti-LCMV serum followed by fluorescein isothiocyanate-labeled rabbit anti-guinea pig immunoglobulin G. (A) Liver from untreated carrier mouse. (B) Liver from carrier mouse 30 days after transfer of immune cells. (C) Salivary gland from untreated carrier mouse. (D) Salivary gland from carrier mouse 30 days after cell transfer. (E) Brain from untreated carrier mouse. (F) Brain from carrier mouse 183 days after cell transfer. Magnifications: A, B, C, E, and F, ×200; D, ×100.

IDENTIFICATION OF EFFECTOR CELLS MEDIATING VIRAL CLEARANCE

By using mice that are recombinant in the *H-2* region and by selectively depleting lymphocyte subpopulations, we found that viral clearance was mediated by LCMV-specific $Lyt2^+$ T cells that are restricted to the class I genes of the major histocompatibility complex (MHC). The results of experiments showing virus specificity, MHC restriction, and surface phenotype of the effector cells that clear virus

Table 1. Requirement of Virus-Specific MHC Class I-Restricted Lyt2+ T Cells for Clearance of Persistent LCMV Infection

Characteristic	Immune donor					Treatment of immune cells	LCMV carrier recipient				Viral clearance[a]
	Virus	*H-2*					*H-2*				
		K	*I*	*D*	*L*		*K*	*I*	*D*	*L*	
Virus specficity[b]	Mock	*d*	*d*	*d*	*d*	None	*d*	*d*	*d*	*d*	No
	LCMV	*d*	*d*	*d*	*d*	None	*d*	*d*	*d*	*d*	Yes
	Pichinde virus	*d*	*d*	*d*	*d*	None	*d*	*d*	*d*	*d*	No
MHC restriction[c]	LCMV	*d*	*d*	*d*	*d*	None	*b*	*b*	*b*	*b*	No
	LCMV	*d*	*d*	*d*	*d*	None	*b*	*b*	*d*	*d*	Yes
	LCMV	*b*	*b*	*d*	*d*	None	*b*	*b*	*d*	*d*	Yes
Surface phenotype[d]	LCMV	*b*	*b*	*b*	*b*	C′	*b*	*b*	*b*	*b*	Yes
	LCMV	*b*	*b*	*b*	*b*	αLyt2 + C′	*b*	*b*	*b*	*b*	No
	LCMV	*b*	*b*	*b*	*b*	αL3T4 + C′	*b*	*b*	*b*	*b*	Yes
	LCMV	*b*	*b*	*b*	*b*	αLyt1 + C′	*b*	*b*	*b*	*b*	No

[a] Mice were checked for the presence of infectious LCMV and viral antigen in serum and major organs for up to 6 months posttransfer. The titer of infectious LCMV was determined by plaque assay on Vero cell monolayers. Viral antigen was checked by staining tissue sections with guinea pig anti-LCMV serum followed by fluorescein isothiocyanate-labeled rabbit anti-guinea pig immunoglobulin G.

[b] Four- to six-week-old BALB/c (H-2^d) carrier mice infected at birth with LCMV were treated with lymphocytes (5×10^7 cells per mouse given intravenously) from control (mock-infected) BALB/c mice or from mice that were previously immunized with LCMV or Pichinde virus (another member of the arenavirus group). Viral clearance was observed only in carrier mice that received cells from LCMV-immune mice.

[c] Adoptive transfer of immune lymphocytes from B.10.D2 ($K^dI^dD^d$) mice had no effect on virus levels in C57BL/6 ($K^bI^bD^b$) carrier mice, but was effective in reducing virus titers in B.10.D2 (R107) ($K^bI^bD^d$) carrier mice. These results show that homology at the *D* end of the *H-2* locus is sufficient for viral clearance and that the effector cells mediating clearance are restricted to the class I MHC genes. (In these experiments, the LCMV carrier mice were irradiated [600 rads] 1 day prior to adoptive transfer to prevent rejection of the transferred cells.)

[d] For depletion of Lyt2+ cells, prior to transfer the immune lymphocytes were treated in vitro with monoclonal anti-Lyt2.2 antibody plus low-toxicity rabbit complement (C′). For depletion of L3T4+ cells, a protocol involving both in vivo and in vitro treatments was followed. LCMV-immune mice were injected intraperitoneally with a total of 1.5 mg of purified GK 1.5 antibody (three doses of 500 μg each on days −7, −4, and −2). On day 0 the immune mice treated in vivo with GK 1.5 antibody were sacrificed, and spleens and lymph nodes were removed. Single-cell suspensions were prepared and treated in vitro with RL172.4 antibody plus complement. After this combined in vivo and in vitro treatment, the immune lymphocytes contained <5% L3T4+ cells.

from LCMV carrier mice are summarized in Table 1. Immune cells treated with Lyt2 plus complement or Lyt1 plus complement were unable to clear virus from persistently infected mice. The Lyt1 antigen is present on the surfaces of most T cells, whereas the Lyt2 molecule is a marker for CTLs (6, 12). Although Lyt2+ T cells constituted only 8.5 to 11.0% of the total spleen cells, removal of this subpopulation abrogated the ability of

Table 2. Lyt2⁺ T-Cell-Mediated Clearance of Infectious Virus from Tissues of Persistently Infected Mice

Group[a]	Treatment of immune cells before transfer[b]	LCMV titer (log_{10} PFU/g or ml) at 54 days post-cell transfer in[c]:								
		Serum	Liver	Lung	Lymph node	Spleen	Thymus	Brain	Kidney	Salivary gland
I → C	Complement only	<1.6	<2.4	3.4	3.3	<3.0	<3.3	3.1	5.4	<3.0
I → C	αLyt2.2 + complement	4.4	5.5	5.7	5.2	4.8	4.6	5.3	6.3	7.3
U → C	None	4.2	5.4	6.0	5.1	5.4	4.4	4.5	5.9	6.8

[a] I → C, C57BL/6 LCMV carrier mice that received 2×10^7 spleen and lymph node cells intraperitoneally from syngeneic LCMV-immune mice; U → C, untreated carrier mice.
[b] The procedures for the antibody-plus-complement-mediated cell depletions have been described previously (11).
[c] Mice were sacrificed, the various organs were harvested and homogenized, and titers were determined on Vero cell monolayers. The data shown are the average of results with five to seven mice per group.

immune spleen cells to clear virus from chronically infected mice (Table 2). In contrast to the results seen upon removal of Lyt2⁺ T cells, depletion of L3T4⁺ cells (T helper cells) or B cells had no effect on the ability of immune cells to eliminate virus. Taken together, these results show that LCMV-specific Lyt2⁺ T cells were essential for effective immunotherapy of persistently infected mice.

ACQUISITION OF T-CELL COMPETENCE BY CURED CARRIER MICE

The results presented so far show that infectious virus as well as viral antigen is cleared from most tissues of carrier mice following immune therapy. It was of interest to determine whether these cured carriers were now able to generate a host-derived LCMV-specific CTL response. To address this question, we used congenic mice (Thy-1.1 and Thy-1.2) in the adoptive transfer experiments, making it possible to distinguish between activity due to immune donor T cells (Thy-1.1) and activity due to carrier host T cells (Thy-1.2).

Lymphocytes from LCMV-immune B6.PL-Thy-1ᵃ (*H-2ᵇ*; Thy-1.1) mice were transferred into C57BL/6 (*H-2ᵇ*; Thy-1.2) carrier mice. Three to four months after the cell transfer, when virus had been eliminated from most tissues, the treated carrier mice were injected with a monoclonal antibody (19E1.2) against the Thy-1.1 antigen to deplete the donor T cells. These cured carrier mice, which no longer contained any donor T

Table 3. Resistance of Cured Carrier Mice to a Second Challenge with LCMV[a]

Group	LCMV titer 8 days postinfection (log_{10} PFU/g or ml) in:			
	Serum	Liver	Spleen	Lung
Cured carrier mice	<1.6	<1.6	<1.6	<1.6
Normal mice	<1.6	<1.6	<1.6	<1.6
Normal mice (given 600 rads)	3.0	5.4	5.3	4.2
Normal mice ($Lyt2^+$ T-cell depleted)	3.2	5.9	5.1	4.0

[a] Mice were injected intraperitoneally with 10^5 PFU of LCMV ARM, and virus titers were determined by a plaque assay on Vero cells. The data shown are mean values for two to six mice per group. The groups are as follows. (i) Cured carrier mice (C57BL/6; Thy-1.2) depleted of immune donor T cells (Thy1.1). At 98 days after immune cell transfer, the cured mice were treated in vivo with a monoclonal antibody against Thy-1.1 (19E1.2) to deplete the immune donor T cells. After five injections of anti-Thy-1.1 antibody, the cured carriers were challenged with LCMV. These mice contained no detectable $Thy1.1^+$ cells (<0.1%) as determined by staining of spleen and lymph node cells with fluorescein isothiocyanate-conjugated anti-Thy-1.1 antibody. (ii) Normal adult C57BL/6 mice (8 to 12 weeks old). (iii) Adult C57BL/6 mice irradiated with 600 rads 1 day prior to LCMV infection. (iv) Adult C57BL/6 mice treated with a monoclonal antibody to deplete $Lyt2^+$ T cells.

cells, were then challenged with LCMV. The amount of virus present in various tissues was determined 8 days postinfection. The cured carriers were able to control the virus infection similarly to normal adult mice. Both groups (normal mice and cured carriers) contained no virus in the tissues tested (Table 3). In contrast, adult mice made immunodeficient by whole-body irradiation (600 rads) or by depletion of $Lyt2^+$ T cells were unable to control LCMV infection and contained high levels of virus.

Cured carrier mice can generate a host-derived LCMV-specific CTL response. As in the above experiment, donor immune T cells (Thy-1.1) were depleted from cured carriers (Thy-1.2), and the lymphocytes were tested for their ability to generate an LCMV-specific CTL response. Spleen and lymph node cells from cured carriers were stimulated in vivo with LCMV, and the CTL response was checked 8 days postinfection by a ^{51}Cr release assay. The killing was mediated by $Lyt2^+$ cells of carrier host (Thy-1.2) origin (Fig. 2). The cytotoxicity was also specific for LCMV-infected cells (no killing of uninfected cells or Pichinde virus-infected cells) and MHC-restricted cells (no killing of LCMV-infected *H-2^d^* cells). Congenitally infected carrier mice were used in the experiment shown in Fig. 2. Similar results were obtained with neonatally infected carriers (data not shown).

CONCLUSIONS

Our results demonstrating the acquisition of T-cell competence by a previously unresponsive and persistently infected host have implications

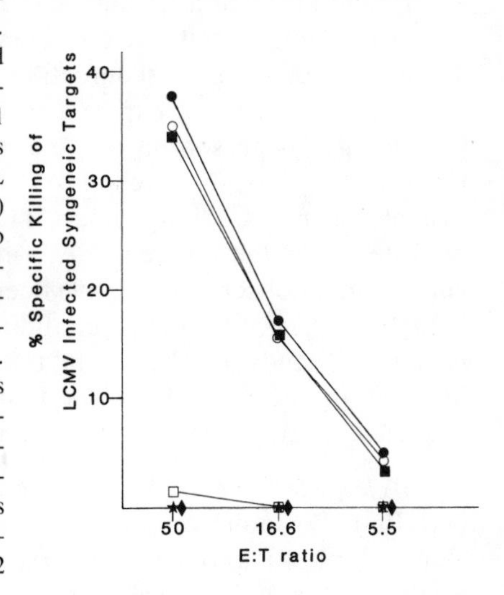

Figure 2. Generation of host-derived LCMV-specific CTL response by cured carriers. Spleen and lymph node cells from cured C57BL/6 carrier mice (Thy-1.2) were stimulated in vivo (in irradiated [650 rads] Thy-1.1 mice) with LCMV, and the CTL response was checked 8 days postinfection. Before the CTL activity was checked, the effector cells (E) were treated with the indicated antisera to characterize the cell-mediating LCMV-specific cytotoxicity. The specificity of the antisera used was checked as described previously (B. D. Jamieson and R. Ahmed, *Proc. Natl. Acad. Sci. USA*, in press). Target cells (T) were infected with the same plaque-purified isolate of LCMV ARM-13 that the carrier mice were exposed to in utero or neonatally. Symbols: ○, untreated cells; cells treated with (●) complement only, (■) anti-Thy-1.1 plus complement, (□) anti-Thy-1.2 plus complement, (★) anti-Thy-1.1 plus anti-Thy-1.2 plus complement, or (◆) anti-Lyt2.2 plus complement.

for the treatment of chronic virus infections. Such infections now constitute a major health problem. It is worth noting that exposure to virus during infancy increases the likelihood of developing a chronic infection. For example, in parts of Africa and Asia, where mother-to-offspring transmission of hepatitis B virus is common, up to 10 to 15% of the population are carriers (7, 14). Perinatal infections with the human immunodeficiency virus now account for a substantial number of acquired immunodeficiency syndrome cases (8). Using the LCMV model, we have shown that a chronic virus infection acquired at birth or in utero can be eliminated and that the host T-cell response then becomes functional and provides protection against reinfection. The immunotherapy protocol used in our studies is not directly applicable to the treatment of chronic virus infections of humans. However, the point I would like to emphasize is that even when an infection has been acquired congenitally or neonatally, reduction of the virus load (by any therapeutic protocol) may restore the potential of the host to become immunocompetent (i.e., the unresponsive state is reversible).

The findings of this study have significance to our understanding of tolerance mechanisms. Chronic infection of mice with LCMV has long been a classic model of immunological tolerance. The prediction of Burnet and Fenner (5) that an antigen introduced into the body during embryonic life would be mistaken for self and the Burnet (4) theory of clonal deletion of self-reactive clones were based partly on studies by Traub (19) of congenital infection of mice with LCMV. Congenitally infected mice are first exposed to the virus during embryonic life, and viral antigen is present in the fetal thymus. Following elimination of the virus, this previously unresponsive host becomes immunocompetent and capable of an LCMV-specific CTL response. These results show that persistence of the antigen is essential in maintaining the unresponsive state. Similar observations have been made by others studying neonatally induced tolerance to transplantation antigens (15, 16, 21). However, ours is the first study to show that tolerance is not permanent even when induced in utero, a situation that more closely mimics the natural sequence of events involved in the development of tolerance to self.

Acknowledgments. I thank the members of my laboratory, in particular Beth D. Jamieson and Rita J. Concepcion, and David D. Porter, Department of Pathology, University of California-Los Angeles School of Medicine, for their contributions to this work.

This research was supported by Public Health Service grant NS-21496 from the National Institutes of Health and an award (H860623) from the University of California Systemwide Task Force on AIDS. I am a Harry Weaver Neuroscience Scholar of the National Multiple Sclerosis Society.

Literature Cited

1. **Ahmed, R., B. D. Jamieson, and D. D. Porter.** 1987. Immune therapy of a persistent and disseminated viral infection. *J. Virol.* **61:**3920–3929.
2. **Ahmed, R., A. Salmi, L. D. Butler, J. M. Chiller, and M. B. A. Oldstone.** 1984. Selection of genetic variants of lymphocytic choriomeningitis virus in spleens of persistently infected mice: role in suppression of cytotoxic T lymphocyte response and viral persistence. *J. Exp. Med.* **60:**521–540.
3. **Buchmeier, M. J., R. M. Welsh, F. J. Dutko, and M. B. A. Oldstone.** 1980. The virology and immunobiology of lymphocytic choriomeningitis virus infection. *Adv. Immunol.* **30:**275–331.
4. **Burnet, F. M.** 1957. A modification of Jerne's theory of antibody production using the concept of clonal selection. *Aust. J. Sci.* **20:**67–69.
5. **Burnet, F. M., and F. Fenner.** 1949. *The Production of Antibodies.* Macmillan Publishing Co., Inc., New York.
6. **Cantor, H., and E. A. Boyse.** 1975. Functional subclasses of T lymphocytes bearing different Ly antigens. I. The generation of functionally distinct T-cell subclasses is a differentiative process independent of antigen. *J. Exp. Med.* **141:**1376–1399.
7. **Chen, D. S., N. H. Hsu, J. L. Sung, T. C. Hsu, S. T. Hsu, Y. T. Kuo, K. J. Lo, and Y. T. Shih.** 1987. A mass vaccination program in Taiwan against hepatitis B virus infection in infants of hepatitis B surface antigen-carrier mothers. *J. Am. Med. Assoc.* **257:**2597–2603.
8. **Curran, J. W., W. M. Morgan, A. M. Hardy, H. W. Jaffe, W. W. Darrow, and W. R.**

Dowdie. 1985. The epidemiology of AIDS: current status and future prospects. *Science* **229:**1352–1357.

9. **Gilden, D. H., G. A. Cole, and N. Nathanson.** 1972. Immunopathogenesis of acute central nervous system disease produced by lymphocytic choriomeningitis virus. II. Adoptive immunization of virus carriers. *J. Exp. Med.* **135:**874–889.
10. **Haywood, A. M.** 1986. Patterns of persistent viral infections. *N. Engl. J. Med.* **315:**939–948.
11. **Jamieson, B. D., L. D. Butler, and R. Ahmed.** 1987. Effective clearance of a persistent viral infection requires cooperation between virus-specific $Lyt2^+$ T cells and nonspecific bone marrow-derived cells. *J. Virol.* **61:**3930–3937.
12. **Ledbetter, J. A., R. V. Rouse, H. S. Micklem, and L. A. Herzenberg.** 1980. T cell subsets defined by expression of Lyt-1, 2, 3, and Thy-1 antigens: two-parameter immunofluorescence and cytotoxicity analysis with monoclonal antibodies modifies current views. *J. Exp. Med.* **152:**280–295.
13. **Lehmann-Grube, F., L. M. Peralta, M. Bruns, and J. Lohler.** 1983. Persistent infection of mice with the lymphocytic choriomeningitis virus, p. 43–103. *In* H. Fraenkel-Conrat and R. R. Wagner (ed.), *Comprehensive Virology*, vol. 18. *Virus-Host Interactions: Receptors, Persistence, and Neurological Diseases.* Plenum Publishing Corp., New York.
14. **Marion, P. L., and W. S. Robinson.** 1983. Hepadna viruses: hepatitis B and related viruses. *Curr. Top. Microbiol. Immunol.* **105:**99–121.
15. **Morecki, S., B. Leshem, A. Eid, and S. Slavin.** 1987. Alloantigen persistence in induction and maintenance of transplantation tolerance. *J. Exp. Med.* **165:**1468–1480.
16. **Nossal, G. J. V.** 1983. Cellular mechanisms of immunologic tolerance. *Annu. Rev. Immunol.* **1:**33–62.
17. **Oldstone, M. B. A., R. Ahmed, J. Byrne, M. J. Buchmeier, Y. Riviere, and P. Southern.** 1985. Virus and immune responses: lymphocytic choriomeningitis virus as a prototype model of viral pathogenesis. *Br. Med. Bull.* **41:**70–74.
18. **Oldstone, M. B. A., P. Blount, P. J. Southern, and P. W. Lampert.** 1986. Cytoimmunotherapy for persistent virus infection reveals a unique clearance pattern from the central nervous system. *Nature* (London) **321:**239–243.
19. **Traub, E.** 1936. Persistence of lymphocytic choriomeningitis virus in immune animals and its relation to immunity. *J. Exp. Med.* **63:**847–861.
20. **Volkert, M.** 1963. Studies on immunological tolerance to LCM virus. 2. Treatment of virus carrier mice by adoptive immunization. *Acta Pathol. Microbiol. Scand.* **57:** 465–487.
21. **Wood, P. J., P. G. Strome, and J. W. Streilein.** 1987. Characterization of cytotoxic cells in mice rendered neonatally tolerant of MHC alloantigens: evidence for repertoire modification. *J. Immunol.* **138:**3661–3668.

Part III.

HUMAN IMMUNODEFICIENCY VIRUS INFECTIONS

Chapter 13

The Human Immunodeficiency Virus Genome: Structure and Function

Flossie Wong-Staal

INTRODUCTION

The human immunodeficiency virus (HIV) genome holds the key to some of the unusual clinical features associated with HIV pathogenesis and to development of strategies for treatment of and vaccine for acquired immunodeficiency syndrome. HIV belongs to a rare group of retroviruses which are highly regulated. Infection by these viruses is characterized by a long and variable period for disease latency, and although virus expression for most retroviruses is governed strictly by *trans*-acting cellular genes, members of this group of viruses are endowed with their own set of regulatory elements. Only a handful of viruses share this property, and, interestingly, they include all of the human retroviruses identified to date, i.e., the leukemia viruses and the immunodeficiency viruses, as well as related animal viruses on both sides.

STRUCTURE OF THE HIV GENOME

Figure 1 shows the structure of a prototype HIV. It is first and foremost a retrovirus, since it encodes all of the structural proteins that are required for retrovirus replication: the four capsid proteins derived from the *gag* gene, the protease, reverse transcriptase, and endonuclease enzymes derived from the *pol* gene, and the major exterior glycoprotein and transmembrane protein derived from the *env* gene. However, in addition, this virus has at least five extra genes, *sor*, 3′*orf*, *tat*, *trs* (also referred to as *art*) and the recently identified *R* gene (1–3, 5, 6, 7, 10–12).

Flossie Wong-Staal • Laboratory of Tumor Cell Biology, National Cancer Institute, Bethesda, Maryland 20892.

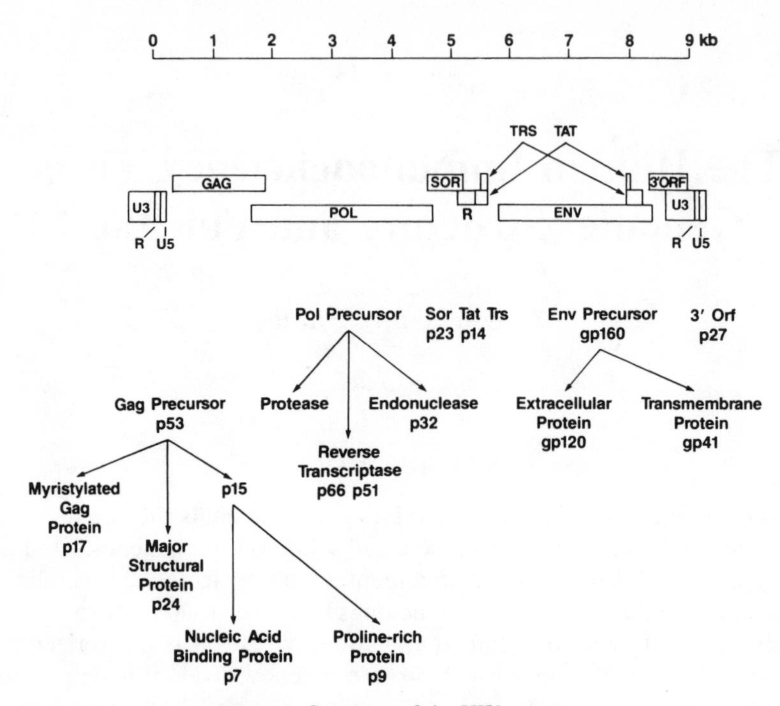

Figure 1. Structure of the HIV genome.

Some of the properties of these genes are listed in Table 1. All of them are immunogenic in vivo; i.e., some, if not all, of the people infected will have antibodies that react with peptides encoded by them. The *sor* product is a 23-kilodalton protein that is found intracellularly in the cytoplasm and membrane fractions of the infected cells. There is now evidence that the *sor* gene is critical for the formation of infectious particles (2a). The *tat* gene product migrates as a 14-kilodalton protein found predominantly in the nucleus but also in the cytoplasm at low levels. It appears to activate expression from the long terminal repeat (LTR), at both the transcriptional and posttranscriptional steps, and is absolutely required for virus production. The *trs* gene product migrates as an 18-kilodalton protein. It is localized in the nucleus and is also required for virus expression. However, it seems to be necessary only for expression of structural, but not regulatory, proteins. There is evidence that it modulates transcription. The 3′*orf* gene product is a 27-kilodalton cytoplasmic protein which appears to be totally dispensable for virus replication in vitro. In fact, mutants lacking this gene replicate better than the wild type, and so it

Table 1. Role of the HIV Accessory Genes for Virus Replication

Gene	Immunogenicity[a]	Size (kilodaltons)	Cellular localization	Function	Replication competence of mutants[b]
sor	+	23	Cytoplasm and inner membrane	Infectivity	±
tat	+	14	Nucleus and cytoplasm	Transcriptional and post-transcriptional activation	−
trs	+	18	Nucleus	Expression of structural proteins and modulation of transcription	−
3′*orf*	++	27	Cytoplasm	Negative regulation	++
R	+	—[c]	—	—	+

[a] ++, Highly immunogenic; +, moderately immunogenic.
[b] ++, Replicates to 3- to 10-fold higher titers than wild type; ±, can be transmitted at low levels by cell fusion only; −, no detectable virus replication.
[c] —, Not known.

seems to act as a negative regulator. The function of the *R* gene is still not known.

FUNCTION OF THE HIV GENOME

Figure 2 shows how we perceive the different viral genes to fit into the replication scheme. The left panel depicts events leading to the establishment of infection, and the right panel depicts events leading to the expression of a new round of infectious virus. For most retroviruses, the transition from left to right is usually automatic, and as mentioned above, virus expression is regulated solely by cellular factions. There are at least two obvious novel features about HIV. First, there is a complex regulatory pathway for virus expression that involves virus-encoded genes. Second, there is a unique viral protein, *sor*, which, together with the envelope proteins, determines the infectivity of the virus particles. I shall now briefly describe some experiments that led to the assignment of these functions for the accessory genes.

One of the approaches to take, and I think it is the most valid one, is to make small deletions, insertions, or even single-amino-acid substitutions in a gene, introduce this back into an infectious DNA clone of the virus, and transfect the clone into Cos cells, which are highly permissive for HIV expression, even though they cannot be reinfected, since they lack the virus receptor. It is possible to quantitate various parameters of virus expression in the Cos cells or to measure the infectivity and

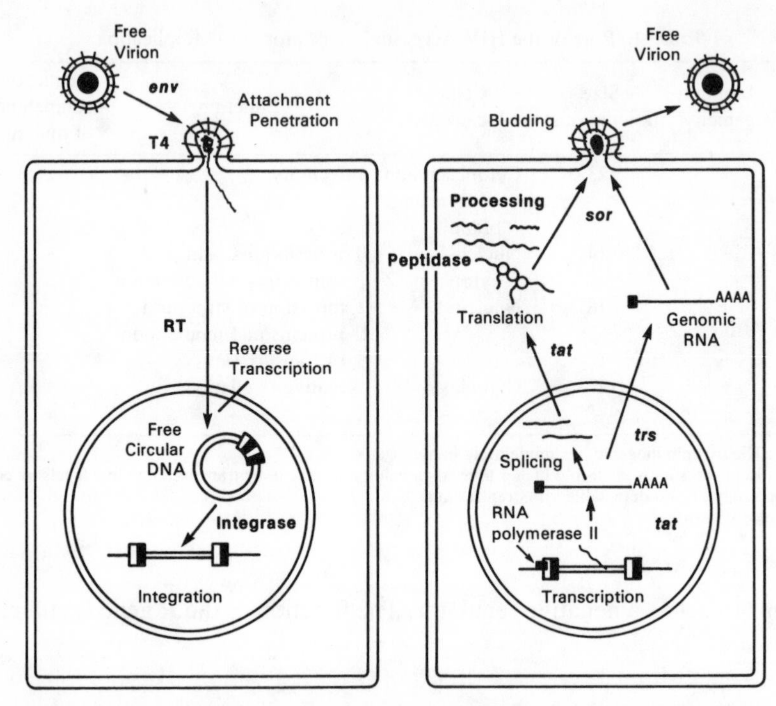

Figure 2. Virus replication cycle. Left panel, Early phase of infection (establishment); right panel, late phase of infection (expression).

cytopathicity of any virus that might be produced on T cells. By making minimal changes in the genome, it should be possible to truly limit the mutation to a single gene without significantly altering the structural configuration of the transcribed RNA or removing potential target sequences of the regulatory genes.

Mutants defective in the *sor* gene make the same number of virus particles as the wild type, but cell-free supernatant containing these particles failed to infect a variety of normal target T-cell lines. Nonetheless, the virus can be transmitted with reduced efficiency by cocultivation (2a).

The two regulatory genes that are crucial for virus expression are considered next. Mutants with a single-amino-acid change to a translational stop codon early in the *tat* or *trs* gene are totally incapacitated in making viral proteins (8a). However, their modes of action are completely different. If one analyzes the RNA of the transfected cells, one finds that the *tat* mutant makes no detectable RNA and the *trs* mutant makes a lot of RNA, which greatly deviates from wild-type RNA in its size distribution. Instead of at least four or five major species at 9.5, 5.5, 4.2, 2.0, and 1.8 kilobases (kb), only the lower, 1.8- to 2.0-kb, species are present at significant levels. This is simply degraded RNA, since these species contain sequences derived from the 5′ end as well as the 3′ end of the genome, and thus they represent fully spliced mature mRNA products. The 9.2-kb band codes for the *gag pol* proteins, the 5.5-kb band, usually a low-abundance species, codes for the *sor* protein, the 4.2-kb band codes for the *env* proteins, and the lower-molecular-weight species encode *tat*, *trs*, and 3′*orf*. Although *tat* is required for expression of all the molecular species, *trs* is required for expression of only the higher-molecular-weight forms. This result has very interesting implications, namely, that the virus has an ingenious way of disassociating the regulation of structural proteins from regulatory proteins. *tat* is a master switch that turns on expression of all viral RNAs, but *trs* is the control switch for the subset of structural proteins. Without *trs*, the expression of *tat* and 3′*orf* is not compromised. We also noticed in Northern (RNA) blot analysis that the absolute amount of viral RNA is increased 5- to 10-fold in the *trs* mutant. To determine whether these differences in steady-state viral mRNAs reflect differences in transcription rates or in mRNA stability, newly transcribed RNA in isolated nuclei was studied (8a). It should be emphasized that this assay does not measure the initiation of transcription, but rather the progression of transcription of preinitiated RNA. Consistent with the Northern blot data, there is a 5- to 10-fold reduction of nascent viral RNA in the *trs* mutant, suggesting that *trs* also down-regulates transcription. Therefore, at least two effects of *trs* have been defined, one at the transcriptional step and the other at the mRNA processing step. The *tat* gene is equally versatile. It definitely has a transcriptional effect. In addition, since the difference seen between the *tat* mutant and the wild type is more dramatic with steady-state mRNA than newly transcribed RNA, the *tat* gene also has the effect of mRNA stability. Furthermore, it has an enhancing effect on mRNA utilization. At least two *tat* mutants have been found which show very low levels of *tat* activity; these mutants make normal levels of RNA but very small amounts of viral proteins (8a). This suggests that although *tat* activates both transcription and posttranscriptional events, a lower threshold level of the protein is required for optimal transcription. The fact that it is a positive regulator at multiple levels of virus expression makes it an ideal target for antiviral strategies.

What does it take to make such a multifunction protein? Apparently, not very much. The *tat* gene is a very small gene with a coding capacity of only 86 amino acids. We showed previously that one can eliminate the last 28 amino acids without affecting its activity (9). There is some

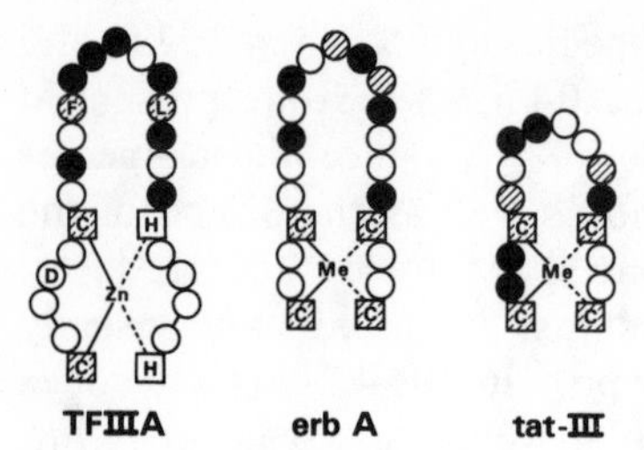

Figure 3. Structural analogy of *tat-3* (tat-III) to metal-binding finger proteins. Abbreviations: TFIIIA, transcription factor IIIA in *Xenopus* sp.; erb A, homolog of the thyroid hormone receptor.

indication that the amino-terminal portion may also be functionally irrelevant. We have obtained functional cDNA clones of the *tat* gene of the distantly related simian immunodeficiency virus (SIV). The *tat* genes of HIV and SIV are functionally homologous and can cross-*trans*-activate the other LTRs, at least in some cell types (S. Colombini and F. Wong-Staal, unpublished results). When we aligned the sequences of the SIV and HIV *tat* genes, we found very little homology on either end, but the central core region was highly conserved. Examination of this sequence revealed two potential nucleic acid-binding domains. The first domain has a helix-turn-helix structure, with a highly hydrophilic and basic stretch of arginine and lysine residues forming the second helix. Sadaie et al., using site-directed mutagenesis, showed that a single-amino-acid change which destroys the first helix structure also abolishes the function of this protein (R. Sadaie, unpublished results). The second potential nucleic acid-binding domain is rich in cysteine residues and can form a structure which resembles the metal-binding fingers of some DNA-binding proteins, such as transcription factor IIIA of *Xenopus* sp. and some of the hormone receptors. In *tat-3*, two pairs of cysteines form a tetrahedral core which presumably constitutes the metal-binding finger (Fig. 3). Again, mutations that disrupt this tetrahedral core would eliminate *tat* function. For transcription factor IIIA, it has been proposed that the protein not only regulates transcription, but may also bind to its own RNA and regulates translation. Therefore, there is precedent for a protein of this group to regulate both transcription and translation. Although there is no direct evidence that the *tat* protein does bind to either DNA or RNA, one observation makes a dual function for *tat* feasible. The target sequence for the *tat* protein has been mapped to a region of the LTR downstream from the initiation of transcription (8), which means that it is present both in the proviral DNA and in the transcribed mRNA. Therefore, *tat*, either by itself or in combination with cellular factors interacting with this region, can theoretically modulate both transcriptional and posttranscriptional processes.

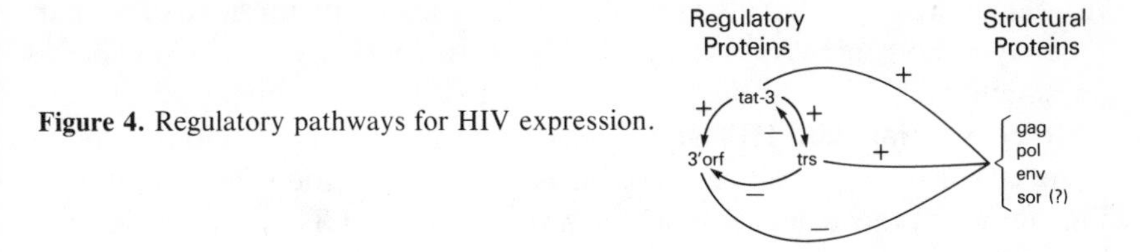

Figure 4. Regulatory pathways for HIV expression.

I mentioned above that *trs* may be a down-regulator for viral transcription. There is at least one other gene which down-regulates virus expression: 3′*orf*. Mutants defective in 3′*orf* replicate up to 10 times better than the wild type does (4). Paradoxically, 3′*orf* is well conserved, not only among different HIV-1 isolates, but also between HIV-1 and HIV-2. Figure 4 summarizes the network of positive and negative controls for virus expression. In an infected cell, there is first a basal level of transcription, but as soon as *tat* is expressed, the whole process is escalated; however, no structural proteins would be expressed until a certain threshold of Trs proteins is built up. However, *trs* is a double-edged sword. Although it activates expression of structural proteins, it also moderates overall virus expression. Similarly, 3′*orf* also lowers the level of virus expression without totally blocking it. Therefore, the virus can synthesize viral proteins rapidly but at moderate levels. Why would the virus evolve at least two genes that down-regulate virus expression? This relates directly to the pathogenic effect of the virus. The basis for the immunosuppressive effect of HIV is the depletion of T4 helper cells. Many pathways can contribute to this depletion: a direct cytopathic effect in the infected cell; killing of cells either expressing or bound to viral proteins, by antibody-dependent cell-mediated cytotoxicity or cytotoxic T lymphocytes; and, as recently proposed (W. Greene, personal communication), blocking of the inducibility of interleukin-2 by binding of HIV *env* to T4. Whatever the pathway, the magnitude of these effects is proportional to the amount of viral proteins made, in particular, the viral envelope protein. The negative regulatory elements for HIV, then, are used to avoid or at least to delay killing its host, which makes sense from an evolutionary point of view. The survival of the virus requires a symbiotic existence with the host, at least for some time. However, a number of environmental factors are known to enhance virus expression. First, following the initial observation by Zagury et al. (13) that HIV expression is turned on by antigen or mitogen activation, a number of groups have now provided molecular data that these agents actually stimulate transcription from the viral LTR (7b, 9a). Likewise, many groups have shown that a number of DNA viruses can *trans*-activate (6a,

7a). Siekevitz et al. (9a) also found that the *trans*-activator gene of human T-cell lymphotropic virus type I (HTLV-I), and possibly also of HTLV-II, *trans*-activates HIV transcription. This has great practical revelance, since coinfection with HIV and HTLV-I and HTLV-II is common among some risk groups and since these viruses home into the same target cell. Furthermore, these agents work on a region of the LTR which is distinct from the target region of *tat*, so that the effects of these enhancing agents and *tat-III* are independent and multiplicative. It may be concluded that disease progression may correlate with virus load, which also correlates with the cumulative exposure to environmental enhancing factors. Chronic antigen stimulation and infection with a variety of agents can be considered cofactors for disease progression.

Literature Cited

1. **Arya, S. K., C. Guo, S. F. Josephs, and F. Wong-Staal.** 1985. *Trans*-activator gene of human T-lymphotropic virus type III (HTLV-III). *Science* **229:**69–73.
2. **Feinberg, M. B., R. J. Jarrett, A. Aldovine, R. C. Gallo, and F. Wong-Staal.** 1986. HTLV-III expression and production involve complex regulation at the levels of splicing and translation of viral RNA. *Cell* **46:**807–817.

2a. **Fisher, A., B. Ensoli, L. Ivanoff, S. Petteway, M. Chamberlain, L. Ratner, R. C. Gallo, and F. Wong-Staal.** 1987. The *sor* gene of HIV is essential for efficient virus transmission *in vitro*. *Science* **237:**888–893.

3. **Fisher, A., M. B. Feinberg, S. F. Josephs, M. E. Harper, L. M. Marselle, G. Reyes, M. A. Gonda, A. Aldovini, C. Debouk, R. C. Gallo, and F. Wong-Staal.** 1986. The trans-activator gene of HTLV-III is essential for virus replication. *Nature* (London) **320:**367–373.
4. **Fisher, A. G., L. Ratner, H. Mitsuya, L. M. Marshelle, M. E. Harper, S. Broder, R. C. Gallo, and F. Wong-Staal.** 1986. Infectious mutants of HTLV-III with changes in the 3′ region and markedly reduced cytopathic effects. *Science* **233:**655–659.
5. **Franchini, G., M. Robert-Guroff, J. Ghrayeb, N. Chang, and F. Wong-Staal.** 1986. Cytoplasmic localization of the HTLV-III 3′ *orf* protein in cultured T-cells. *Virology* **155:**593–599.
6. **Franchini, G., M. Robert-Guroff, F. Wong-Staal, J. N. Ghrayeb, N. Kato, and N. Chang.** 1986. Expression of the 3′ open reading frame of the HTLV-III in bacteria: demonstration of its immunoreactivity with human sera. *Proc. Natl. Acad. Sci. USA* **83:**5282–5285.

6a. **Gendelman, H. E., W. Phelps, L. Feigenbaum, J. M. Ostrovo, A. Adachi, P. M. Howley, G. Khoury, H. S. Ginsburg, and M. A. Martin.** 1986. Transactivation of the human immunodeficiency virus long terminal repeat sequence by DNA viruses. *Proc. Natl. Acad. Sci. USA* **83:**9759–9763.

7. **Kan, N. C., G. Franchini, F. Wong-Staal, G. C. DuBois, W. G. Robey, J. A. Lautenberger, and T. S. Papas.** 1986. Identification of HTLV-III/LAV *sor* gene product and detection of antibodies in human sera. *Science* **231:**1553–1555.

7a. **Mosca, J. D., D. P. Bednarik, N. B. K. Raj, C. A. Rosen, J. G. Sodroski, W. A. Haseltine, and P. M. Pitha.** 1987. Herpes simplex virus type 1 can reactivate transcription of latent human immunodeficiency virus. *Nature* (London) **325:**67–70.

7b. **Nahel, G., and D. Baltimore.** 1987. An inducible transcription factor activates expression of human immunodeficiency virus in T-cells. *Nature* (London) **326:**711–713.

8. **Rosen, C. A., J. G. Sodroski, and W. A. Haseltine.** 1985. Location of cis-acting regulatory sequences in the human T-cell lymphotropic virus type III (HTLV-III/LAV) long terminal repeat. *Cell* **41:**813–823.

8a. **Sadaie, R., T. Benter, and F. Wong-Staal.** 1988. Site-directed mutagenesis of two trans-regulatory genes (tat-3, trs) of HIV. *Science* **239:**910–913.

9. **Seigel, L. J., L. Ratner, S. F. Josephs, D. Derse, M. Feinberg, G. Reyes, S. J. O'Brien, and F. Wong-Staal.** 1986. Transactivation induced by human T-lymphotropic virus type III (HTLV-III) maps to viral sequence encoding 58 amino acids and lacks tissue specificity. *Virology* **148:**226–231.

9a. **Siekevitz, M., S. F. Josephs, M. Dukovich, N. Peffer, F. Wong-Staal, and W. Greene.** 1987. Activation of the HIV LTR by T-cell mitogens and the transactivator protein of HTLV-I. *Science* **238:**1575–1578.

10. **Sodroski, J., W. C. Goh, C. Rosen, A. Tartar, D. Potetelle, A. Burny, and W. Haseltine.** 1986. A second post-transcriptional trans-activator gene required for HTLV-III replication. *Nature* (London) **321:**412–418.

11. **Sodroski, J., C. Rosen, F. Wong-Staal, S. Z. Salahuddin, M. Popovic, S. Arya, R. C. Gallo, and W. A. Haseltine.** 1985. *Trans*-acting transcriptional regulation of human T-cell leukemia virus type III long terminal repeat. *Science* **227:**171–173.

12. **Wong-Staal, F., P. K. Chanda, and J. Ghrayeb.** 1987. Human immunodeficiency virus type III: the eighth gene. *AIDS Res. Hum. Retroviruses* **3:**33–39.

13. **Zagury, D., J. Bernard, R. Leonard, R. Cheynier, M. Feldman, P. S. Sarin, and R. C. Gallo.** 1986. Long-term culture of HTLV-III infected T-cells: a model of cytopathology of T-cell depletion in AIDS. *Science* **231:**850–853.

Chapter 14

Natural History of Human Immunodeficiency Virus Infection

Scott D. Holmberg and James W. Curran

INTRODUCTION

Much is now known about the modes of transmission of human immunodeficiency virus (HIV), the agent that causes the acquired immunodeficiency syndrome (AIDS), in the United States. AIDS is a classic sexually transmitted disease; HIV may also be transmitted through blood-borne routes, such as by transfusion of blood and blood components, sharing of intravenous needles, and organ transplantation, or through perinatal transmission from mother to infant. Although much is known about the epidemiology of HIV, less is known about the natural history of infection (47). Do all, or just some, HIV-infected persons develop AIDS? If only some, how many, who, and why? What is the mean incubation period from HIV infection to HIV-associated diseases? What clinical or biologic parameters presage the development of AIDS or other manifestations of HIV infection?

Data on these critical questions have been unavoidably limited by the relatively short time since the first AIDS cases were recognized in 1982. In this report we briefly review our present understanding of the transmission of HIV as it may influence the course of the AIDS epidemic and our current knowledge of the natural history of HIV infection.

EPIDEMIOLOGY OF HIV

HIV is transmitted primarily in three ways: sexually, through blood-borne routes (usually through exposure to contaminated blood or nee-

Scott D. Holmberg and James W. Curran • AIDS Program, Center for Infectious Diseases, Centers for Disease Control, Atlanta, Georgia 30333.

dles), and, rarely, perinatally. Many epidemiologic factors can influence the spread of HIV from one person to another. In the United States, groups at increased risk for HIV infection include homosexual and bisexual men; intravenous-drug abusers (IVDAs) who share needles; persons who received transfusions or clotting factors (hemophiliacs), mainly before universal screening of blood and blood products for HIV antibody began in the United States in March 1985; heterosexuals who have sexual contacts with HIV-infected persons or with others at increased risk of HIV infection; and infants born to infected mothers (55). There is also a small but real occupational risk to health care workers.

Homosexual and Bisexual Men

As homosexual men have served as the sentinel population for AIDS in the United States (7–9, 48), they have also remained the most severely affected group. Of the over 50,000 AIDS patients reported to the Centers for Disease Control (CDC) as of January 1988, about 36,500 (73%) were homosexual or bisexual men (including homosexual and bisexual men who also have a history of intravenous-drug abuse).

In addition to the large cohort study conducted in San Francisco by the San Francisco Health Department and CDC (49; N. A. Hessol, G. W. Rutherford, P. M. O'Malley, L. S. Doll, W. W. Darrow, and H. W. Jaffe, *Abstr. III Int. Conf. AIDS*, M.3.1, p. 1, 1987), other ongoing studies of homosexual men have yielded important data about HIV transmission in homosexual men. The Multicenter AIDS Cohort Study, which involves over 5,000 homosexual men in four major U.S. cities (77), the University of California at San Francisco Study (57, 93), and studies of smaller cohorts of homosexual men in other cities (34, 53, 63, 86) indicate that HIV infection rates in homosexual men have declined steeply since 1983. This decline is attributed to changed sexual practices by homosexual men in response to the epidemic: fewer sexual partners, fewer anonymous partners, and safer sexual practices (24, 35). These changes have been reflected in the declining incidence of sexually transmitted infections other than those with HIV (11). The declining incidence of HIV infection in the San Francisco cohort study is depicted in Fig. 1.

AIDS cases among homosexual men continue to increase, but they reflect infections acquired several years ago. It will be some time before the declining incidence rate of HIV infection recently seen in some groups of homosexual and bisexual men is reflected in diminishing numbers of AIDS cases.

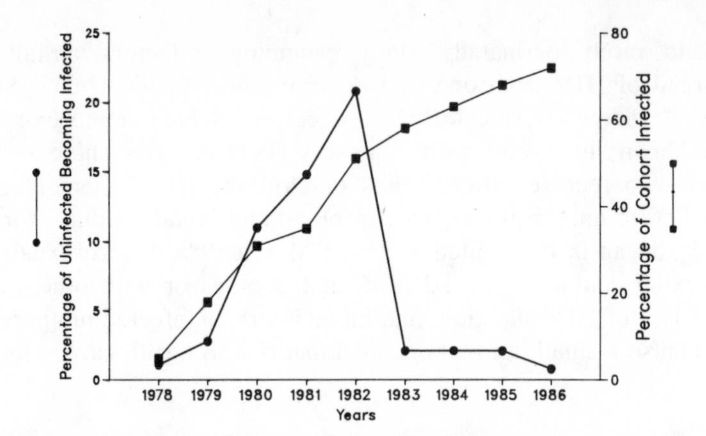

Figure 1. Incidence and prevalence of HIV infection in the San Francisco City Clinic/CDC study of homosexual and bisexual men. Incidence of HIV is calculated from a declining denominator each year as fewer men in the cohort remain uninfected.

IVDAs

IVDAs make up the second largest category of AIDS patients; they represent 16% of AIDS cases as of October 1987. In addition, IVDAs are reported to have a more rapidly progressive natural history of HIV infection than homosexuals (P. A. Selwyn, E. E. Schoenbaum, D. Hartel, T. Peterman, R. S. Klein, and G. H. Friedland, *Abstr. III Int. Conf. AIDS*, M.3.4, p. 2, 1987), perhaps because the latter group frequently present with Kaposi's sarcoma, a more slowly progressing manifestation of HIV infection (26).

Needle sharing is the obvious vehicle for HIV infection in IVDAs; in some "shooting galleries," needles may be shared up to 50 times. The extent of drug use since 1978, needle-sharing frequency, and time spent in shooting galleries have been shown to increase the risk for IVDAs of HIV infection (84). Conversely, the use of sterile needles and the length of time spent by IVDAs in treatment (methadone) programs are associated with a lower rate of infection. Geography also seems to play a role in the spread of HIV in IVDAs: about 80% of all IVDA AIDS patients are in New York City and northern New Jersey. Zones of decreasing HIV seropositivity with increasing distance from downtown Manhattan have been described (84).

There are an estimated 200,000 to 400,000 IVDAs in New York City and 750,000 to 1,000,000 nationwide; seropositivity in IVDAs studied recently in New York City is about 60% (W. R. Lange, B. J. Primm, F. S. Tennant, J. T. Payte, C. M. Luney, and J. H. Jaffe, *Abstr. III Int. Conf.*

AIDS, TP.54, p. 71, 1987). Of those attending methadone maintenance clinics in one study, about one-third were HIV seropositive (Selwyn et al., *Abstr. III Int. Conf. AIDS*, M.3.4, p. 2, 1987). Furthermore, unlike homosexual men whose self-reported changes in sexual behavior are reflected in actual decreases in HIV seroconversion rates, IVDAs, including those in treatment programs, continue to seroconvert at a high rate; in one study, about 10% of those attending methadone maintenance clinics are still seroconverting for HIV every year (P. A. Selwyn and E. E. Schoenbaum, personal communication).

Because HIV infection continues to spread in drug-abusing populations, it is anticipated that the course of the AIDS epidemic in the United States will be determined more by IVDAs and less by homosexual men as time progresses. Also, IVDAs are an important link from the reservoir of HIV-infected persons to other uninfected segments of the population. Women who acquire HIV infection from sexual contact with men in high-risk groups for AIDS much more frequently report sexual contact with IVDAs than with bisexual men, transfusion-infected men, or hemophiliacs (J. W. Ward, S. Kleinman, D. Douglas, A. Grindon, and S. Holmberg, *Transfusion*, in press). Also, female-to-male spread of HIV may occur from prostitutes, who are often IVDAs (16). Finally, about 80% of pediatric AIDS cases are in infants born to HIV-infected mothers (see below); most of these infected mothers or their sex partners are IVDAs (P. A. Selwyn, E. E. Schoenbaum, A. R. Feingold, M. Mayers, K. Davenny, and M. Rogers, *Abstr. III Int. Conf. AIDS*, TH.7.2, p. 157, 1987). Thus, the problems of the current high prevalence and continued incidence of HIV infection in IVDAs result in turn in increasing rates of heterosexual and perinatal transmission of HIV.

Heterosexual Men and Women

Although several early reports implicated male-to-female transmission of HIV (10, 39), there was controversy over whether transmission occurred in the reverse direction (76, 82, 94, 95). Since HIV survival outside CD4 helper cells is limited, and since vaginal and cervical secretions have few of these cells, some researchers were originally skeptical that female-to-male transmission could occur or, if so, could occur frequently. However, several reports now clearly document female-to-male transmission (6, 73). In-depth interviews with 65 men reported to CDC as having AIDS, but with no identified risk for AIDS, revealed that 17 (26%) reported sexual contact with prostitutes. Of these 17, 5 also reported having had contact with over 100 women in the previous 5 years (19). Furthermore, in Africa, where the ratio of male to

Table 1. Rates of Heterosexual Transmission of HIV as Reported from Various Studies

Partners of persons with AIDS (reference or abstract no.[a])	Male-to-female spread: fraction of exposed persons infected (%)	Female-to-male spread: fraction of exposed persons infected (%)
Partners of IVDAs (W.2.5)	41/88 (47)	7/12 (58)
Partners of IVDAs (28)	3/6 (50)	1/3 (33)
Partners of IVDAs (WP.40)	2/11 (16)	0/1 (0)
Partners of Haitian immigrants (28)	6/11 (55)	5/5 (100)
Partners of others (29)	5/11 (45)	6/9 (67)
Partners of hemophiliacs (51)	2/20 (10)	
Partners of hemophiliacs (56)	2/21 (10)	
Partners of hemophiliacs (W.2.6)	4/25 (16)	
Partners of bisexual men (72)	12/55 (22)	
Partners of transfusion-infected persons (75b)	10/55 (18)	2/25 (8)

[a] Abstract numbers from *III International Conference on AIDS*, Washington, D.C., 1 to 5 June 1987.

female AIDS cases is close to unity, HIV-infected men frequently reported many sexual contacts with female prostitutes (79).

Knowledge about heterosexual transmission is important in predicting the future of the AIDS epidemic (74), but it is difficult to find appropriate study populations (80) (Table 1). Sexual partners of IVDAs are often HIV infected, but some persons reporting acquisition of HIV from IVDA partners may themselves be drug users who share needles (28; B. R. Saltzman, G. H. Friedland, R. S. Klein, D. W. Maude, and N. H. Steigbigel, *Program Abstr. 26th Intersci. Conf. Antimicrob. Agents Chemother.*, abstr. no. 11, 1986; N. H. Steigbigel, D. W. Maude, C. J. Feiner, C. A. Harris, B. R. Saltzman, and R. S. Klein, *Abstr. III Int. Conf. AIDS*, W.2.5, p. 106, 1987) (Table 1). Useful information on rates of male-to-female spread has been obtained in studies of sexual partners of hemophiliacs (2, 51, 56; J. J. Goedert, M. E. Eyster, and R. J. Biggar, *Abstr. III Int. Conf. AIDS*, W.2.6, p. 106, 1987) (Table 1), but it is not possible to assess female-to-male spread in this population. The clearest studies of bidirectional heterosexual spread of HIV involve transfusion-infected persons and their partners. The date of infection (transfusion) is known for the index case, and so actual numbers of sexual encounters can be estimated for the index case and his or her sexual partner. Transfusion-infected persons studied were frequently monogamous and involved in stable heterosexual relationships, usually marriages. Although investigation of these persons yields more readily interpretable data, HIV-infected transfusion recipients and their partners tend to be older and less healthy than other heterosexuals in whom sexually transmitted infections occur.

In the only study of these persons to date, 2 (8%) of 25 husbands and 10 (18%) of 55 wives who had had sexual contact with infected spouses were seropositive for HIV (75b) (Table 1). Compared with seronegative wives, the seropositive wives were older (median ages, 54 and 62 years, respectively) and actually reported somewhat fewer sexual contacts with their infected husbands (means, 156 and 82 contacts, respectively). Although most husbands and wives remained uninfected despite repeated sexual contact without protection, some acquired infection after only a few contacts. This is consistent with an as yet unexplained biologic variation in transmissibility or host susceptibility and remains one of the most important current research issues in the epidemiology of HIV.

Blood Transfusion Recipients

With the availability and widespread use of an enzyme-linked immunosorbent assay to screen blood for HIV antibody, HIV infections transmitted via blood transfusion have been almost completely eliminated. However, before the widespread screening of the (estimated) 18 million blood units donated yearly, perhaps 12,000 persons may have been infected by transfused blood and survived the illness or procedure for which they received a transfusion (75, 75a). (About 60% of blood components are transfused to persons who do not survive hospitalization or the period immediately thereafter.) One epidemiologic study of the recipients of blood from HIV-infected donors showed that the likelihood of being infected with HIV was virtually 100%, irrespective of the type of component received (89).

The actual risk of becoming infected from a single unit of blood received before March 1985 was low, mostly because of deferral by potential donors who were advised not to donate if they belonged to known risk groups for AIDS (12). Donor self-deferral remains a cornerstone of reduction of the risk of HIV infection from blood transfusion.

Because a recently infected donor may not yet have developed antibodies to HIV, blood donations during this window period—about 3 months after infection—may not test positive on HIV antibody screening (C. Kenny, J. Parkin, G. Underhill, N. Shah, B. Burnell, E. Osborne, and D. J. Jeffries, Letter, *Lancet* **i**:565–566, 1987). It is rare for HIV-infected blood to slip past the barriers of donor self-deferral and HIV antibody screening, but it does occur (14). According to one estimate, about 460 transfused persons are infected with HIV from the more than 18 million units transfused yearly (90).

Hemophiliacs

Hemophiliacs receive clotting factors harvested from plasma products of blood from thousands of blood donors. Of the approximately 15,000 hemophiliacs in the United States, most (about 10,000) have severe hemophilia A and typically receive about 70,000 units of factor VIII each year. Since hemophiliacs are susceptible to a number of blood-borne viral infections, it is not surprising that hemophiliacs were infected by HIV early in the AIDS epidemic. Several investigations in different areas of the United States indicate a rise in seroprevalence of HIV antibody in hemophiliacs from about 10% in 1980 to about 80% in 1984 (52). The greatest incidence of HIV infection, as measured by HIV seroconversion, appears to have occurred from 1981 to 1983. More than 90% of frequent users of factor VIII concentrates during those years are infected (M. E. Eyster, personal communication). Hemophilia B patients were somewhat less likely to have become infected, because their clotting factor use was generally lower and because they were more likely to receive fresh frozen plasma or cryoprecipitate obtained from fewer donors. Still, by 1983, 39% of hemophilia B patients in one study were also HIV infected (M. E. Eyster, O. T. Preble, and J. J. Goedert, *Abstr. III Int. Conf. AIDS*, MP.70, p. 21, 1987).

The prognosis for HIV-infected hemophiliacs may be better or worse than that for other groups of infected persons (L. Smiley, G. C. White II, T. J. Matthews, K. J. Weinhold, and D. P. Bolognesi, *Program Abstr. 26th Intersci. Conf. Antimicrob. Agents Chemother.*, abstr no. 1018, 1986). One study found a low rate of progression to AIDS (2 to 4% per year) in HIV-infected hemophiliacs (M. A. Koerper and J. A. Levy, *Abstr. III Int. Conf. AIDS*, TH.10.3, p. 161, 1987), but another found frequent occurrence of lymphopenia (40%) and generalized lymphadenopathy (70%) in this population (Goedert et al., *Abstr. III Int. Conf. AIDS*).

In 1984, a method of heat treatment for factor concentrates was found to be effective in eliminating HIV that had been spiked into these concentrates before such heat treatment; the method was quickly and widely adopted. Also, since 1985, all donated blood and plasma used to produce clotting-factor concentrates have been screened for HIV antibody. These interventions have made infection of previously uninfected hemophiliacs exceedingly rare (1, 14).

Infants and Children: Perinatal Transmission

Children can acquire HIV infection through any of the mechanisms listed above, such as by transfusion of blood or antihemophilic clotting factors and (rarely) through sexual exposures. In addition, neonates born to infected mothers are at very high risk of acquiring infection (3, 83); 78% of children under 13 years of age who have AIDS have a parent with or at risk for HIV infection (21; M. J. Oxtoby, M. Rogers, P. Thomas, S. Manoff, and K. Winter, *Abstr. III Int. Conf. AIDS*, TH.7.1, p. 157, 1987). The first such AIDS-like illnesses in children were reported in 1982, and such illnesses have been retrospectively identified as early as 1980. Because most cases result from perinatal transmission (usually from IVDA mothers),. the demographic characteristics of pediatric AIDS patients reflect those of their parents: 80% are black or Hispanic (15), and 75% reside in New York, New Jersey, Florida, or California.

The actual mode of perinatal transmission has been difficult to elucidate (59). Does such transmission from mother to infant occur before or at birth? Most experts assume that both situations are possible.

The estimated risk of perinatal transmission has varied widely, but pooled data and larger studies suggest that infants born to HIV-infected mothers have a 40 to 60% risk of acquiring infection from their mothers (Oxtoby et al., *Abstr. III Int. Conf. AIDS*). Assessment of infection in infants is complicated by the fact that initial HIV antibodies in them may have been passively acquired from their mothers, so that HIV antibody tests do not reflect the infection status of the infant until he or she is able to elicit antibodies on his or her own (usually not until 9 months of age). When antigen tests become commercially available, they may be useful in early identification of perinatally infected infants.

It has been possible to assess the degree of pathogenicity in infected infants, and almost all available data indicate that infection in infants more quickly leads to AIDS and death than infection in older persons (Oxtoby et al., *Abstr. III Int. Conf. AIDS*). According to mathematical modeling of current statistics, the mean incubation from time of infection (birth) to AIDS for infants is 2 years, whereas the mean incubation from infection (transfusion) to AIDS for adults is 5 years or more.

Many children have been born to IVDA women in greater New York City alone (84). This presents a particular public health intervention problem, because it is difficult to get health intervention and educational messages to these inner-city minority women who also have very low rates of prenatal care (15). This emphasizes the wider problem that needle-sharing in IVDAs poses to persons other than the drug abusers themselves.

Health Care Workers

Health care workers are at increased risk for HIV infection through needle-stick accidents, lacerations sustained during surgery on an HIV-infected person, and other exposures to infected blood. Some early reports in the United States and Europe indicated that HIV infections were occurring in a few health care workers with such exposures (92). Two female health care workers seroconverted after accidental sticks from needles that had just been used on AIDS patients for whom they were caring (70, 71).

To evaluate and quantify the risk to health care workers, several studies have prospectively monitored health care workers caring for patients with HIV infection (65). For purposes of such surveillance, a definite case of occupationally acquired HIV infection should have the following features: (i) the health care worker should not belong to any other risk group for AIDS; (ii) he or she must be demonstrably seronegative at the time of the needle stick or other putative exposure; and (iii) seroconversion to HIV should occur within months of the documented exposure. An ongoing CDC study of such cases has found that of 351 health care workers with such exposures, 3 (<1%) have seroconverted after a needle-stick incident (17, 18). Previous studies, and reviews of collections of such studies, indicate that the risk of HIV infection from an accidental needle stick is low, apparently less than 1% per stick (31, 32, 41, 42). Furthermore, some of the needle-stick exposures resulting in HIV infection have been unusual, involving deep sticks with large amounts of injected blood (87).

Studies of seroconversion in health care workers after mucosal exposure to infectious blood were undertaken by various centers (31, 41), but none of 292 health care workers with such exposures were found to be HIV antibody positive (17). The first report of HIV transmission from mucosal exposures to infectious blood was of a mother who developed HIV antibody while tending her chronically ill child (13). In 1987, CDC reported the results of investigations of three incidents involving young female health professionals who apparently acquired HIV infection through contact with infected blood (17). These cases indicated that blood-to-mucocutaneous transmission of HIV, although very rare, may occur. Because of this possibility, CDC guidelines for health care workers have been reevaluated and continue to reinforce the importance of universal precautions to prevent exposure to potentially infectious clinical material (18).

NATURAL HISTORY OF HIV INFECTION

Few issues are considered as crucial to understanding the AIDS epidemic as the natural history of HIV infection in an HIV-seropositive person. What is the mean incubation period between HIV infection and the development of AIDS? What proportion of HIV-seropositive persons will develop AIDS, AIDS-related complex, or other clinical symptoms or remain asymptomatic? What is the long-term prognosis for someone who is seropositive for HIV antibody? These are key epidemiologic questions about AIDS and about the natural history of HIV infection (47, 85).

The major obstacle to knowing the mean incubation period between infection and disease is that such calculation requires that the peak of an epidemic be observed. The AIDS epidemic is unique in that it was identified and tracked shortly after it was recognized, so that we have not yet observed a peak in AIDS incidence and therefore cannot directly calculate mean incubation periods. Mathematical models to calculate the mean incubation period between HIV infection and disease (usually AIDS) cannot compensate for this unavoidable problem. Another problem in identifying a peak in the epidemic is that there is a lag in reporting cases from hospitals to state health departments and CDC; a peak in AIDS cases can be identified as such only about 12 months after it occurs (R. H. Byers, personal communication).

In turn, without knowing the distribution of incubation periods, it is not possible to precisely determine what proportion of HIV-infected persons will develop AIDS.

Acute Retroviral Infection

Within 4 to 5 weeks after inoculation with HIV, chimpanzees will produce detectable virus-specific immunoglobulin G antibodies; in half, immunoglobulin M antibody response will precede this immunoglobulin G response (29, 37). In humans, HIV antigen appears first, disappears, and then is followed by antibodies to *env* and *gag* proteins at about the time HIV antigen is no longer detectable (2, 20, 30, 36, 38; M. V. Ragni, T. A. O'Brien, J. A. Spero, and J. H. Lewis, *Abstr. III Int. Conf. AIDS*, TP.236, p. 101, 1987). When serologic responses following acute infection have been well characterized, HIV antibody is always detectable on standard enzyme-linked immunosorbent assays 8 to 14 weeks after infection (43; J. W. Ward, personal communication). Detection of antibody to HIV envelope glycoproteins gp120 and/or gp41 may precede or follow detection of antibody to core protein p24, but this is still not clearly determined.

Clinically, the acute infection may be associated with a distinctive mononucleosislike syndrome in the 2 months after infection (43; A. W. McLeod, M. T. Schechter, W. J. Boyko, K. J. P. Craib, B. Willoughby, and B. Douglas, *Abstr. III Int. Conf. AIDS*, MP.111, p. 28, 1987). The syndrome typically involves fever, rigors, arthralgias, myalgias, and a macular rash occurring 3 to 6 weeks after HIV infection. Preliminary data from studies of homosexual men and transfusion-infected persons suggest that as many as one-third of acutely infected persons may develop acute retroviral syndrome (McLeod et al., *Abstr. III Int. Conf. AIDS*; Ward, personal communication).

Persistence of HIV Infection

HIV infection is a persistent retroviral infection (22). HIV can be isolated from most persons with confirmed HIV antibody and from those with AIDS, generalized lymphadenopathy, and other HIV-associated conditions. In one study, the virus was isolated from peripheral blood lymphocytes of 22 (96%) of 23 asymptomatic HIV-infected blood donors who had apparently persistent infections for 1 to 4 years (27). A second study recovered virus from 8 (67%) of 12 homosexual men who were HIV seropositive for 2 to 6 years (50). Virus can still be recovered from 63 long-term (more than 3 years)-seropositive homosexual men in San Francisco (Hessol et al., *Abstr. III Int. Conf. AIDS*; A. R. Lifson, personal communication). No instances of nonpersistent HIV infection or definitive instances in which seropositive persons have become seronegative have been documented. However, even if such situations were shown to occur in a few individuals, HIV infection is persistent for virtually all infected persons.

Immunologic Parameters Preceding AIDS

Many investigators have attempted to define immunologic perturbations that may precede and prognosticate disease development in HIV-infected persons. A progressive decrease in helper T cells (CD4 antigen-positive lymphocytes) has been identified as a prognostic marker of disease development (25). Other studies have confirmed that a progressive decrease in helper T cells precedes the development of AIDS and other HIV-associated diseases (33; V. de Gruttola, J. Livartowski, W. Rozenbaum, P. Sette, B. Autran, and F. de Vathaire, *Abstr. III Int. Conf. AIDS*, THP.76, p. 176, 1987).

The importance of other markers of disease development has been less clear. The reappearance of p24 antigen just before or during AIDS may be a prognostic indicator of disease in HIV-infected persons (R. M. Hendry, K. R. Judkins, A. E. Wittek, H. C. Lane, and G. V. Quinnan, *Abstr. III Int. Conf. AIDS*, MP.125, p. 31, 1987; N. Kamani, L. R. Krilov, R. M. Hendry, A. E. Wittek, and G. B. Quinnan, *Abstr. III Int. Conf. AIDS*, THP.122, p. 183, 1987). Since the reappearance of this antigen is associated with the loss of p24 antibody, some researchers have hypothesized that reactivation of HIV infection, leading to disease progression, may accompany or even result from loss of inhibitory effect of p24 and other HIV antibodies. Neutralizing antibodies, usually found in low titer, generally appear to decrease further before development of disease in infected persons (58, 91; M. Arendrup, K. Ulrich, J. O. Nielsen, B. O. Lindhardt, C. Pedersen, and K. Krogsgaard, *Abstr. III Int. Conf. AIDS*, TP.81, p. 76, 1987).

The association of disease progression with other immunologic parameters is less well documented. Levels of alpha or gamma interferon (69; B. Bihari, F. Drury, V. Ragone, G. Ottomanelli, and E. Buimovici-Klein, *Abstr. III Int. Conf. AIDS*, THP.124, p. 184, 1987; Eyster et al., *Abstr. III Int. Conf. AIDS*), antilymphocyte antibody (B. Dorsett, W. Cronin, and H. L. Ioachim, *Abstr. III Int. Conf. AIDS*, MP.106, p. 27, 1987), and β-microglobulin (96; De Gruttola et al., *Abstr. III Int. Conf. AIDS*) have been correlated with the development of disease or severe prognosis in some HIV-infected persons. Because of the small numbers of persons studied, these associations should be considered preliminary.

Incubation Period

The incubation period, here defined as the time between HIV infection and onset of AIDS, and the percentage of persons progressing from HIV infection to AIDS are intertwined, and estimates of both have varied markedly (68, 78, 87, 88). Public Health Service estimates of the mean incubation period have necessarily increased as the epidemic has progressed and as persons with ever-longer incubation periods are identified. In 1984, the mean incubation period was estimated to be about 3 to 4 years (61); in 1987, with the same mathematical model, it is about 8 years (R. H. Byers, W. M. Morgan, W. W. Darrow, L. Doll, H. W. Jaffe, G. Rutherford, N. Hessol, and P. M. O'Malley, manuscript in preparation). Some researchers think (81) and dispute (S. Beal, Letter, *Nature* [London] **328:**673, 1987) that the average incubation period will be as long as 14 years. The average annual risk of developing AIDS after HIV infection has been estimated at anywhere between 2 to 4% (44) and 7% (60).

Variations in observations and estimates probably derive from the relative earliness of the prospective study of this epidemic of chronic infection, from the small sample sizes usually used, and from differences in populations studied (66). There may be unknown variations between risk groups, e.g., hemophiliacs versus homosexuals, because of yet-to-be-identified cofactors in pathogenesis. Geographic variations in rates of developing AIDS, e.g., homosexual men in San Francisco versus those in New York City (4), probably result because men in different cities are observed at different times relative to the time of infection.

Currently, over 20 cohorts of HIV-seropositive persons are being

Table 2. Progression to AIDS in Various Cohort Studies in the United States, 1983 to 1987

Type of cohort (abstract no.[a])	No. of seropositives in study	Approx. length of infection or follow-up (mo)	No. (%) developing AIDS
Heterosexuals, New York City (W.2.5)	48	>10	3 (6)
Homosexuals, Boston (TP.64)	57	>12	4 (7)
IVDAs, New York City (M.3.4)	184	>12	16 (9)
IVDAs, New York City (MP.90)	53	24	5 (9)
Homosexuals, four cities (TP.73)	1,828	24	164 (9)
Homosexuals, San Francisco (TP.53)	291	30	41 (14)
Homosexuals, San Francisco (M.3.2)	206	30	37 (18)
Homosexuals, San Francisco (TP.44)	35	30	10 (29)
Homosexuals, Atlanta (THP.100)	>75	42	21 (<28)
Homosexuals, Seattle (WP.76)	>178	42	25 (<14)
IVDAs, New York City (TP.87)	35	46	6 (17)
Homosexuals, Los Angeles (TP.72)	63	48	9 (14)
Homosexuals, San Francisco (TP.44)	>143	48	47 (<33)
Hemophiliacs, San Francisco (TH.10.3)	55	54	6 (11)
Transfusion-infected persons, San Francisco (M.3.5)	61	60	15 (25)
Hemophiliacs, Hershey, Pa. (TP.56)	50	60	7 (14)
Homosexuals, Washington, D.C. (TP.56)	42	60	13 (31)
Homosexuals, New York City (TP.56)	43	60	17 (40)
Homosexuals, San Francisco (M.3.1)	63	72	19 (30)

[a] From D. Cohn, Denver Disease Control Service, as derived from presentations and abstracts of the *III International Conference on AIDS*, Washington, D.C., 1 to 5 June 1987.

monitored in the United States. About 25% of HIV-infected persons observed in these cohorts develop AIDS within 5 years (Table 2). If one combines the results for the persons involved in these various studies (Table 2), there is a straight arithmetic progression in the percentage of HIV-infected persons developing AIDS, i.e., about 5% per year. However, the likelihood of progression of AIDS may not be the same each year after infection. In a few studies, the infection date of some persons can be approximated; such persons tend not to develop AIDS in the first 2 years after infection, but have progressively greater likelihood of developing AIDS each year thereafter (34, 89). The CDC/San Francisco City Clinic Cohort Study has the longest history and includes 63 men who have been seropositive for 3 years or more (Table 2); 30% of these have developed AIDS after a mean follow-up time of 72 months (Hessol et al., *Abstr. III Int. Conf. AIDS*).

Predictions beyond 5 or 6 years are necessarily tenuous, and, as described above, mathematical modeling has some unavoidable limitations. However, mathematical modeling of the incubation period is valuable in making minimum estimates of the impact of HIV infection. Also, it is useful to have models available to determine the kinds of epidemiologic data needed to predict future trends. Finally, and most practically, models do allow us to project into the immediate future and to plan for numbers of expected cases.

Infectivity and Spread to the General Population

Some researchers have applied models of transmission dynamics to predictions of the rate of spread from the reservoir of HIV-infected persons to the general populations of developed countries. The key parameter for the calculations of May and Anderson (64) is R_0, the basic reproductive rate of the infection. R_0 is the average number of secondary infections from an infected person; infection can maintain itself in a community only if this value is greater than 1.0. R_0 is determined by c (the average rate at which new sexual partners are acquired), by β (the average probability that infection is transmitted from an infected person to a susceptible one), and by D (the average duration of infectiousness). That is, $R_0 = \beta c D$. This model assumes that c is much lower for heterosexuals than for homosexuals, that β is probably between 0.05 and 0.5 per partner contact (45, 46), and that D is equivalent to the incubation period (assumed to be about 10 ± 4 to 5 years). According to these calculations, the AIDS epidemic in heterosexuals who have no other risk factor for AIDS, including having a sexual partner known to be at high risk for AIDS, would die out of its own accord.

This model may actually overestimate the risk of heterosexual spread. In two studies, β was about 0.001 per contact with an infected partner (72; see footnote to Table 2). Although some persons apparently acquired their infection after only one or a few sexual contacts with an infected person, indicating that host factors relative to transmissibility and infectivity are important, this value of β is still much lower than assumed in the model above. The rate at which heterosexuals acquire new partners, c, may have been or may be further reduced by yet-undocumented changes in sexual behavior in response to the AIDS epidemic itself. Finally, D, the proportion of time an infected person may be infective, may be less than the 100% assumed by the model. One research group has closely examined data from the wives of 25 male hemophiliacs. Four (16%) of these wives themselves became infected, but their seroconversion came after their husbands had been infected with HIV for 2 years or more and had had substantial decrements in helper T cells (T4 cells, CD4-positive cells) (Goedert et al., *Abstr. III Int. Conf. AIDS*). This interesting preliminary finding has led some to speculate that a drop in the level of T helper cells reflected a greater circulating virus titer, or some other biologic mechanism, that promoted the infectivity of the hemophiliac husbands. Thus, D may not be the whole period of HIV infection, since an infected person may be unlikely to transmit the virus during the 2 years after infection and may be less sexually active late in infection once he or she is sick with AIDS. For a number of reasons, then, the infection rate in heterosexuals, R_0, may actually be less than assumed in the model.

Current data indicate a low seroprevalence of HIV in the general heterosexual population. Although these groups are self-selected, the seroprevalence of HIV antibody in military recruits and blood donors in the United States is 0.15 and 0.02%, respectively. An ongoing study of sexually active heterosexuals who attend a sexually transmitted disease clinic in a highly endemic area for AIDS (New York City) and who deny being in a risk group for AIDS or having sexual contacts with AIDS patients or those in risk groups for AIDS indicates that of the first 205 of these persons interviewed and tested, only 1 (<0.5%) was infected; this person declined intensive interview (A. R. Lifson, personal communication). Thus, the ability of the virus to infect large numbers of heterosexual Americans apparently has been low.

Future of the AIDS Epidemic

Figure 2 shows the incidence of AIDS in the United States by quarter of diagnosis as projected from cases reported as of April 1986 (67). A polynomial model was fitted to the adjusted case counts as transformed by the Box-Cox method (5, 23). Reported cases in the past 2 years have so far neatly fit the model. These projected cases are close to the estimates made by public health, academic, and scientific officials at the Coolfont Conference in West Virginia in 1986 (62). These experts estimated that there will have been approximately 270,000 AIDS cases by the end of 1991, over half of whom will have died.

SUMMARY

In the United States, the spread of HIV has declined substantially in some populations of homosexual men, and the spread by transfused blood and blood products (clotting factors) has been virtually eliminated. However, because of the long incubation period between infection and disease, reported AIDS cases in these risk groups will continue to increase in the immediate future. In the longer term, because of continued seroconversions in IVDAs and transmission from them to sexual partners

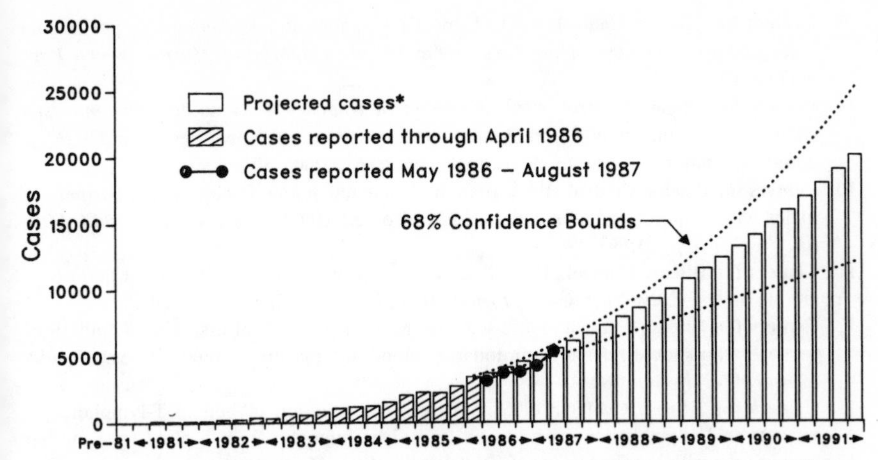

Figure 2. Incidence of AIDS in the United States: by quarter of diagnosis as projected from cases reported to CDC as of 30 April 1986, lagged 2 months to account for reporting delays (hatched bars); projected through 1991, by quarter of diagnosis (open bars); and reported to CDC between 1 May 1986 and 31 August 1987. (Adapted from references 5 and 23.)

and neonates, IVDAs will account directly and indirectly for an ever-increasing percentage of AIDS cases.

Until the peak of this epidemic occurs and is recognized, it will not be possible to answer some important questions about the natural history of HIV infection, such as the proportion of HIV-infected persons who will develop AIDS or the mean incubation period between HIV infection and AIDS.

Literature Cited

1. **Allain, J.-P.** 1986. Prevalence of HTLV-III/LAV antibodies in patients with hemophilia and in their sexual partners in France. *N. Engl. J. Med.* **315:**517.
2. **Allain, J. P., Y. Laurian, D. A. Paul, D. Senn, and Members of the AIDS-Haemophilia French Study Group.** 1986. Serologic markers in early stages of human immunodeficiency virus infection in haemophiliacs. *Lancet* **ii:**1233–1236.
3. **Ammann, A. J.** 1985. The acquired immunodeficiency syndrome in infants and children. *Ann. Intern. Med.* **103:**734–737.
4. **Blattner, W. A., R. J. Biggar, S. H. Weiss, M. Melbye, and J. J. Goedert.** 1985. Epidemiology of human T-lymphotropic virus type III and the risk of the acquired immunodeficiency syndrome. *Ann. Intern. Med.* **103:**665–670.
5. **Box, G. E. P., and D. R. Cox.** 1964. An analysis of transformations. *J. R. Stat. Soc. Ser. B* **26:**211–243.
6. **Calabrese, L. H., and K. V. Gopalakrishna.** 1986. Transmission of HTLV-III infection from man to woman to man. *N. Engl. J. Med.* **314:**987.
7. **Centers for Disease Control.** 1981. *Pneumocystis* pneumonia—Los Angeles. *Morbid. Mortal. Weekly Rep.* **30:**250–252.

8. **Centers for Disease Control.** 1981. Kaposi's sarcoma and *Pneumocystis* pneumonia among homosexual men—New York City and California. *Morbid. Mortal. Weekly Rep.* **30:**305–308.
9. **Centers for Disease Control.** 1982. A cluster of Kaposi's sarcoma and *Pneumocystis carinii* pneumonia among homosexual male residents of Los Angeles and Orange counties, California. *Morbid. Mortal. Weekly Rep.* **31:**305–307.
10. **Centers for Disease Control.** 1983. Immunodeficiency among female sexual partners of males with acquired immune deficiency syndrome (AIDS)—New York. *Morbid. Mortal. Weekly Rep.* **31:**697–698.
11. **Centers for Disease Control.** 1984. Declining rates of rectal and pharyngeal gonorrhea among males—New York City. *Morbid. Mortal. Weekly Rep.* **33:**295–297.
12. **Centers for Disease Control.** 1985. Update: revised public health service definition of persons who should refrain from donating blood and plasma—United States. *Morbid. Mortal. Weekly Rep.* **34:**547–548.
13. **Centers for Disease Control.** 1986. Apparent transmission of human T-lymphotropic virus type III/lymphadenopathy-associated virus infection from a child to a mother providing health care. *Morbid. Mortal. Weekly Rep.* **35:**76–79.
14. **Centers for Disease Control.** 1986. Transfusion-associated human T-lymphotropic virus type III/lymphadenopathy-associated virus infection from a seronegative donor. *Morbid. Mortal. Weekly Rep.* **35:**389–391.
15. **Centers for Disease Control.** 1986. Acquired immunodeficiency syndrome among blacks and Hispanics—United States. *Morbid. Mortal. Weekly Rep.* **35:**663–666.
16. **Centers for Disease Control.** 1987. Antibody to human immunodeficiency virus in female prostitutes. *Morbid. Mortal. Weekly Rep.* **36:**157–161.
17. **Centers for Disease Control.** 1987. Update: human immunodeficiency virus infections in health-care workers exposed to blood of infected patients. *Morbid. Mortal. Weekly Rep.* **36:**285–289.
18. **Centers for Disease Control.** 1987. Recommendations for prevention of HIV transmission in health-care settings. *Morbid. Mortal. Weekly Rep.* **36**(Suppl.)**:**1S–18S.
19. **Chamberland, M. E., K. G. Castro, H. W. Haverkos, B. I. Miller, P. A. Thomas, R. Reiss, J. Walker, T. J. Spira, H. W. Jaffe, and J. W. Curran.** 1984. Acquired immunodeficiency syndrome in the United States: an analysis of cases outside high-incidence groups. *Ann. Intern. Med.* **101:**617–623.
20. **Coutinho, R. A., J. Goudsmit, D. A. Paul, F. de Wolf, J. Lange, J. van der Noordaa, and the Dutch AIDS-Study Group.** 1986. The natural history of HIV infection in homosexual men, p. 141–148. *In* J. C. Gluckman and E. Vilmer (ed.), *Acquired Immunodeficiency Syndrome*. Elsevier, Paris.
21. **Cowan, M. J., D. Hellamen, D. Chudwin, D. W. Wara, R. S. Chang, and A. J. Ammann.** 1984. Maternal transmission of acquired immunodeficiency syndrome. *Pediatrics* **73:**382–386.
22. **Curran, J. W.** 1985. The epidemiology and prevention of the acquired immunodeficiency syndrome. *Ann. Intern. Med.* **103:**657–662.
23. **Curran, J. W., W. M. Morgan, A. M. Hardy, H. W. Jaffe, W. W. Darrow, and W. R. Dowdle.** 1985. The epidemiology of AIDS: current status and future prospects. *Science* **229:**1352–1357.
24. **Darrow, W. W., D. F. Echenberg, H. W. Jaffe, P. M. O'Malley, R. H. Byers, J. P. Gretchell, and J. W. Curran.** 1987. Risk factors for human immunodeficiency virus (HIV) infections in homosexual men. *Am. J. Public Health* **77:**479–483.
25. **Detels, R., J. L. Fahey, K. Schwartz, R. S. Greene, B. R. Visscher, and M. S. Gottlieb.**

1983. Relation between sexual practices and T-cell subsets in homosexually active men. *Lancet* **i**:609–611.

26. **Fauci, A. S., H. Masur, E. P. Gelmann, P. D. Markham, B. H. Hahn, and H. C. Lane.** 1985. The acquired immunodeficiency syndrome: an update. *Ann. Intern. Med.* **102:** 800–813.
27. **Feorino, P. M., H. W. Jaffe, E. Palmer, T. A. Peterman, D. P. Francis, V. S. Kalyanaraman, R. A. Weinstein, R. L. Stoneburner, W. J. Alexander, C. Raevsky, J. P. Getchell, D. Warfield, H. W. Haverkos, B. W. Kilbourne, J. K. A. Nicholson, and J. W. Curran.** 1985. Transfusion-acquired immunodeficiency syndrome: evidence for persistent infection in blood donors. *N. Engl. J. Med.* **312:**1293–1296.
28. **Fischl, M. A., G. M. Dickinson, G. B. Scott, N. Klimas, M. A. Fletcher, and W. Parks.** 1987. Evaluation of heterosexual partners, children, and household contacts of adults with AIDS. *J. Am. Med. Assoc.* **257:**640–644.
29. **Francis, D. P., H. W. Jaffe, P. M. Fultz, J. P. Getchell, J. S. MacDougal, and P. M. Feorino.** 1985. The natural history of infection with the lymphadenopathy-associated human T-lymphotropic virus type III. *Ann. Intern. Med.* **103:**719–722.
30. **Gaines, H., M. von Sydow, A. Sonnenberg, J. Albert, J. Czajkowski, P. Olov Pehrson, F. Chiodi, L. Moberg, E. M. Fenyo, B. Asjo, and M. Forsgren.** 1987. Antibody response in primary human immunodeficiency virus infection. *Lancet* **i:**1249–1253.
31. **Gerberding, J. L., C. E. Bryant-LeBlanc, K. Nelson, A. R. Moss, D. Osmond, H. F. Chambers, J. R. Carlson, W. L. Drew, J. A. Levy, and M. A. Sande.** 1987. Risk of transmitting the human immunodeficiency virus, cytomegalovirus, and hepatitis B virus to health care workers exposed to patients with AIDS and AIDS-related condition. *J. Infect. Dis.* **156:**1–8.
32. **Gerberding, J. L., P. C. Hopewell, L. S. Kaminsky, and M. A. Sande.** 1985. Transmission of hepatitis B without transmission of AIDS by accidental needlestick. *N. Engl. J. Med.* **312:**56–57.
33. **Goedert, J. J., R. J. Biggar, M. Melbye, D. L. Mann, S. Wilson, M. H. Gail, R. J. Grossman, R. A. DiGioia, W. C. Sanchez, S. H. Weiss, and W. A. Blattner.** 1987. Effect of T4 count and cofactors on the incidence of AIDS in homosexual men infected with human immunodeficiency virus. *J. Am. Med. Assoc.* **257**:331–334.
34. **Goedert, J. J., R. J. Biggar, S. H. Weiss, M. E. Eyster, M. Melbye, S. Wilson, H. M. Ginzburg, R. J. Grossman, R. A. DiGioia, W. C. Sanchez, J. A. Giron, P. Ebbesen, R. C. Gallo, and W. A. Blattner.** 1986. Three-year incidence of AIDS in five cohorts of HTLV-III-infected risk group members. *Science* **231:**992–995.
35. **Goedert, J. J., M. G. Sarngadharan, R. J. Biggar, S. H. Weiss, D. M. Winn, R. J. Grossman, M. H. Greene, A. J. Bodner, D. L. Mann, D. M. Strong, R. C. Gallo, and W. A. Blattner.** 1984. Determinants of retrovirus (HTLV-III) antibody and immunodeficiency conditions in homosexual men. *Lancet* **ii:**711–716.
36. **Goudsmit, J., F. de Wolf, D. A. Paul, L. G. Epstein, J. M. A. Lange, W. J. A. Krone, H. Speelman, E. C. Wolters, J. van der Noordaa, J. M. Oleske, H. J. van der Helm, and R. A. Coutinho.** 1986. Expression of human immunodeficiency virus antigen (HIV-Ag) in serum and cerebrospinal fluid during acute and chronic infection. *Lancet* **ii:**177–180.
37. **Goudsmit, J., L. Smit, W. J. A. Krone, M. Bakker, J. van der Noordaa, C. J. Gibbs, L. G. Epstein, and D. C. Gajdusek.** 1987. IgG response to human immunodeficiency virus in experimentally infected chimpanzees mimics the IgG response in humans. *J. Infect. Dis.* **155:**327–331.
38. **Groopman, J. E., F. W. Chen, J. A. Hope, J. M. Andrews, R. L. Swift, C. V. Bentos, J. L. Sullivan, P. A. Volberding, D. P. Sites, S. Landesman, J. Gold, L. Baker, D.**

Craven, and F. S. Boches. 1986. Serological characterization of HTLV-III infection in AIDS and related disorders. *J. Infect. Dis.* **153:**736–741.

39. **Harris, C., C. B. Small, R. S. Klein, G. H. Friedland, B. Moll, E. E. Emeson, I. Spigland, and N. Steigbigel.** 1983. Immunodeficiency in female sexual partners of men with the acquired immunodeficiency syndrome. *N. Engl. J. Med.* **308:**1181–1184.
40. **Haverkos, H. W., P. F. Pinsky, D. P. Drotman, and D. J. Bregman.** 1985. Disease manifestations among homosexual men with the acquired immunodeficiency syndrome (AIDS): a possible role of nitrites in Kaposi's sarcoma. *Sex. Transm. Dis.* **12:**203–208.
41. **Henderson, D. K., A. J. Saah, B. J. Zak, R. A. Kaslow, H. C. Lane, T. Folks, W. C. Blackwelder, J. Schmitt, D. J. LaCamera, H. Masur, and A. S. Fauci.** 1986. Risk of nosocomial infection with human T-cell lymphotropic virus type III/lymphadenopathy-associated virus in a large cohort of intensively exposed health care workers. *Ann. Intern. Med.* **104:**644–647.
42. **Hirsch, M. S., G. P. Wormser, R. T. Schooley, D. D. Ho, D. Felsenstein, C. C. Hopkins, C. Joline, F. Duncanson, M. G. Sarngadharan, C. Saxinger, and R. C. Gallo.** 1985. Risk of nosocomial infection with human T-cell lymphotropic virus (HTLV-III). *N. Engl. J. Med.* **312:**1–4.
43. **Ho, D. D., M. G. Sarngadharan, L. Resnick, F. Dimarzo-Veronese, T. R. Rota, and M. S. Hirsch.** 1985. Primary human T-lymphotropic virus type III infection. *Ann. Intern. Med.* **103:**880–883.
44. **Holmes, K. K.** 1985. Infectious diseases. *J. Am. Med. Assoc.* **254:**2254–2257.
45. **Holmes, K. K., D. W. Johnson, and H. J. Trostle.** 1970. An estimate of the risk of men acquiring gonorrhea by sexual contact with infected females. *Am. J. Epidemiol.* **91:**170–174.
46. **Hooper, R. R., G. H. Reynolds, O. G. Jones, A. Zaidi, P. J. Wiesner, K. P. Latimer, A. Lester, A. F. Campbell, W. O. Harrison, W. W. Karney, and K. K. Holmes.** 1978. Cohort study of venereal disease. I. The risk of gonorrhea transmission from infected women to men. *Am. J. Epidemiol.* **108:**136–144.
47. **Hopkins, D. R.** 1986. Key epidemiologic questions about AIDS and infection with HTLV-III/LAV. *Public Health Rep.* **101:**234–237.
48. **Jaffe, H. W., K. Choi, P. A. Thomas, H. H. Haverkos, D. M. Auerbach, M. E. Guinan, M. F. Rogers, T. J. Spira, W. W. Darrow, M. A. Kramer, S. M. Friedman, J. M. Monroe, A. E. Friedman-Kien, L. J. Laubenstein, M. Marmor, B. Safai, S. K. Dritz, S. J. Crispi, S. L. Fannin, J. P. Orkwis, A. Kelter, W. R. Rushing, S. B. Thacker, and J. W. Curran.** 1983. National case-control study of Kaposi's sarcoma and *Pneumocystis carinii* pneumonia in homosexual men. I. Epidemiologic results. *Ann. Intern. Med.* **99:**145–151.
49. **Jaffe, H. W., W. W. Darrow, D. F. Echenberg, P. M. O'Malley, J. P. Gretchell, V. S. Kalyanaraman, R. H. Byers, D. P. Drennan, E. H. Braff, J. W. Curran, and D. P. Francis.** 1985. The acquired immunodeficiency syndrome in a cohort of homosexual men. A six-year follow-up study. *Ann. Intern. Med.* **103:**210–214.
50. **Jaffe, H. W., P. M. Feorino, W. W. Darrow, P. M. O'Malley, J. P. Getchell, D. T. Warfield, B. M. Jones, D. F. Echenberg, D. P. Francis, and J. W. Curran.** 1985. Persistent infection with human T-lymphotropic virus type III/lymphadenopathy-associated virus in apparently healthy homosexual men. *Ann. Intern. Med.* **102:**627–628.
51. **Jason, J. M., J. S. McDougal, G. Dixon, D. N. Lawrence, M. S. Kennedy, M. Hilgartner, L. Aledort, and B. L. Evatt.** 1986. HTLV-III/LAV antibody and immune status of household contacts and sexual partners of persons with hemophilia. *J. Am. Med. Assoc.* **255:**212–215.
52. **Johnson, R. E., D. N. Lawrence, B. L. Evatt, D. J. Bregman, L. D. Zyla, J. W. Curran,**

L. M. Aledort, M. E. Eyster, A. P. Brownstein, and C. J. Carmen. 1985. Acquired immunodeficiency syndrome among patients attending hemophilia treatment centers (1978 to June 1984) and mortality experience of U.S. hemophiliacs (1968 to 1978). *Am. J. Epidemiol.* **121:**797–810.

53. **Kaplan, J. E., T. J. Spira, D. B. Fishbein, P. P. Pinsky, and L. B. Schonberger.** 1987. Lymphadenopathy syndrome in homosexual men. Evidence for continuing risk of developing the acquired immunodeficiency syndrome. *J. Am. Med. Assoc.* **257:**335–337.
54. **Kingsley, L. A., R. Detels, R. Kaslow, B. F. Polk, C. R. Rinaldo, Jr., J. Chimiel, K. Detre, S. F. Kelsey, N. Odaka, D. Ostrow, M. VanRaden, and B. Visscher.** 1987. Risk factors for seroconversion to human immunodeficiency virus among male homosexuals. *Lancet* **i:**345–348.
55. **Koop, C. E.** 1986. *Surgeon General's Report on Acquired Immunodeficiency Syndrome.* U.S. Department of Health and Human Services, Washington, D.C.
56. **Kreiss, J. K., L. W. Kitchen, H. E. Prince, C. K. Kaspar, and M. Essex.** 1985. Antibody to human T-lymphotropic virus type III in wives of hemophiliacs: evidence for heterosexual transmission. *Ann. Intern. Med.* **102:**623–626.
57. **Lang, W., R. E. Anderson, H. Perkins, R. M. Grant, D. Lyman, W. Winkelstein, R. Royce, and J. A. Levy.** 1987. Clinical, immunologic, and serologic findings in men at risk for acquired immunodeficiency syndrome. The San Francisco men's health study. *J. Am. Med. Assoc.* **257:**326–330.
58. **Lange, J. M. A., R. A. Coutinho, W. J. A. Krone, L. F. Verdonck, S. A. Danner, J. van der Noordaa, and J. Goudsmit.** 1986. Distinct IgG recognition patterns during progression of subclinical and clinical infection with lymphadenopathy associated virus/human T lymphotropic virus. *Br. Med. J.* **292:**228–230.
59. **Lapointe, N., J. Michaud, D. Pekovic, J. P. Chausseau, and J.-M. Dupuy.** 1985. Transplacental transmission of HTLV-III virus. *N. Engl. J. Med.* **312:**1325–1326.
60. **Laurence, J.** 1985. The immune system in AIDS. *Sci. Am.* **253:**84–93.
61. **Lui, K.-J., D. N. Lawrence, W. M. Morgan, T. A. Peterman, H. W. Haverkos, and D. J. Bregman.** 1986. A model-based approach for estimating the mean incubation period of transfusion-associated acquired immunodeficiency syndrome. *Proc. Natl. Acad. Sci. USA* **83:**3051–3055.
62. **MacDonald, D. I.** 1986. Coolfont report: a PHS plan for prevention and control of AIDS and the AIDS virus. *Public Health Rep.* **101:**341–348.
63. **Marmor, M., A. E. Friedman-Kien, S. Zolla-Pazner, R. E. Stahl, P. Rubinstein, L. Laubenstein, D. C. Williams, R. J. Klein, and I. Spigland.** 1983. Kaposi's sarcoma in homosexual men: a seroepidemiologic case-control study. *Ann. Intern. Med.* **100:**809–815.
64. **May, R. M., and R. M. Anderson.** 1987. Transmission dynamics of HIV infection. *Nature* (London) **326:**137–142.
65. **McCray, E., and The Cooperative Needlestick Surveillance Group.** 1986. Occupational risk of the acquired immunodeficiency syndrome among health care workers. *N. Engl. J. Med.* **314:**1127–1132.
66. **Medley, G. F., R. M. Anderson, D. R. Cox, and L. Billard.** 1987. Incubation period of AIDS in patients infected via blood transfusion. *Nature* (London) **328:**719–723.
67. **Morgan, W. M., and J. W. Curran.** 1986. Acquired immunodeficiency syndrome: current and future trends. *Public Health Rep.* **101:**459–465.
68. **Moss, A. R.** 1985. What proportion of HTLV-III antibody positives will proceed to AIDS? *Lancet* **ii:**223–224.
69. **Murray, H. W., J. K. Hillman, B. Y. Rubin, C. D. Kelly, J. L. Jacobs, L. W. Tyler, D. M. Donelly, S. M. Carriero, J. U. Godbold, and R. B. Roberts.** 1985. Patients at risk

for AIDS-related opportunistic infections: clinical manifestations and impaired gamma interferon production. *N. Engl. J. Med.* **313:**1504–1510.

70. **Neisson-Vernant, C., S. Arfi, D. Mathez, J. Leibowitz, and N. Monplaisir.** 1986. Needlestick HIV seroconversion in a nurse. *Lancet* **ii:**814.
71. **Oksenhelder, E., M. Harzic, J. M. LeRoux, C. Rabian, and J. P. Clavel.** 1986. HIV infection with seroconversion after a superficial needlestick injury to the finger. *N. Engl. J. Med.* **315:**582.
72. **Padian, N., L. Marquis, D. P. Francis, R. E. Anderson, G. W. Rutherford, P. M. O'Malley, and W. Winkelstein, Jr.** 1987. Male-to-female transmission of human immunodeficiency virus. *J. Am. Med. Assoc.* **258:**788–790.
73. **Peterman, T. A., and J. W. Curran.** 1986. Sexual transmission of human immunodeficiency virus. *J. Am. Med. Assoc.* **256:**2222–2226.
74. **Peterman, T. A., D. P. Drotman, and J. W. Curran.** 1985. Epidemiology of the acquired immunodeficiency syndrome (AIDS). *Epidemiol. Rev.* **7:**1–21.
75. **Peterman, T. A., H. W. Jaffe, P. M. Feorino, J. P. Getchell, D. T. Warfield, H. W. Haverkos, R. L. Stoneburner, and J. W. Curran.** 1985. Transfusion-acquired immunodeficiency syndrome in the United States. *J. Am. Med. Assoc.* **254:**2913–2917.

75a. **Peterman, T. A., K.-J. Lui, D. N. Lawrence, and J. R. Allen.** 1987. Estimating the risks of transfusion-associated acquired immune deficiency syndrome and human immunodeficiency virus infection. *Transfusion* **27:**371–374.

75b. **Peterman, T. A., R. L. Stoneburner, J. R. Allen, H. W. Jaffe, and J. W. Curran.** 1988. Risk of human immunodeficiency virus transmission from heterosexual adults with transfusion-associated infections. *J. Am. Med. Assoc.* **259:**55–58.

76. **Polk, B. F.** 1985. Female-to-male transmission of AIDS. *J. Am. Med. Assoc.* **254:**3177–3178.
77. **Polk, B. F., R. Fox, R. Brookmeyer, S. Kanchanaraksa, R. Kaslow, B. Visscher, C. Rinaldo, and J. Phair.** 1987. Predictors of the acquired immunodeficiency syndrome developing in a cohort of seropositive homosexual men. *N. Engl. J. Med.* **316:**61–66.
78. **Proudfoot, A.** 1986. Ratio between numbers of cases and carriers of AIDS. *Med. J. Aust.* **144:**614.
79. **Quinn, T. C., J. M. Mann, J. W. Curran, and P. Piot.** 1986. AIDS in Africa: an epidemiologic paradigm. *Science* **234:**955–963.
80. **Redfield, R. R., P. D. Markham, S. Z. Salahuddin, M. G. Sadgadharan, A. J. Bodner, T. M. Folks, W. R. Ballou, D. C. Wright, and R. C. Gallo.** 1985. Frequent transmission of HTLV-III among spouses of patients with AIDS-related complex and AIDS. *J. Am. Med. Assoc.* **253:**1571–1573.
81. **Rees, M.** 1987. The sombre view of AIDS. *Nature* (London) **326:**343–345.
82. **Schultz, S., J. A. Milberg, A. R. Kristal, and R. A. Stoneburner.** 1986. Female-to-male transmission of HTLV-III. *J. Am. Med. Assoc.* **255:**1703–1704.
83. **Scott, G. B., M. A. Fischl, N. Klimas, M. A. Fletcher, G. M. Dickinson, R. S. Levine, and W. P. Parks.** 1985. Mothers of infants with the acquired immunodeficiency syndrome. Evidence for both symptomatic and asymptomatic carriers. *J. Am. Med. Assoc.* **253:**363–366.
84. **Selwyn, P. A.** 1986. AIDS: what is now known. II. Epidemiology. *Hosp. Pract.* **21:**127–164.
85. **Sivak, S. L., and G. P. Wormser.** 1985. How common is HTLV-III infection in the United States? *N. Engl. J. Med.* **313:**1352.
86. **Stevens, C. E., P. E. Taylor, E. A. Zang, J. M. Morrison, E. J. Harley, S. Rodriguez de Cordoba, C. Bacino, R. C. Y. Ting, A. J. Bodner, M. G. Sarngadharan, R. C. Gallo, and**

P. Rubinstein. 1986. Human T-cell lymphotropic virus type III infection in a cohort of homosexual men in New York City. *J. Am. Med. Assoc.* **255:**2167–2172.

87. **Stricof, R. L., and D. L. Morse.** 1986. HTLV-III/LAV seroconversion following a deep intramuscular needlestick injury. *N. Engl. J. Med.* **314:**1115.
88. **Taylor, J. M. G., K. Schwartz, and R. Detels.** 1986. The time from infection with human immunodeficiency virus (HIV) to the onset of AIDS. *J. Infect. Dis.* **154:**694–697.
89. **Ward, J. W., D. A. Deppe, S. Samson, H. Perkins, P. Holland, L. Fernando, P. M. Feorino, P. Thompson, S. Kleinman, and J. R. Allen.** 1987. Risk of human immunodeficiency virus infection from blood donors who later developed the acquired immunodeficiency syndrome. *Ann. Intern. Med.* **106:**61–62.
90. **Ward, J. W., S. D. Holmberg, J. R. Allen, D. L. Cohn, S. E. Critchley, S. H. Kleinman, B. A. Lenes, O. Ravenholt, J. R. Davis, M. G. Quinn, and H. W. Jaffe.** 1988. Transmission of human immunodeficiency virus (HIV) by blood transfusions screened as negative for HIV antibodies. *N. Engl. J. Med.* **318:**473–478.
91. **Weber, J. N., P. R. Clapham, R. A. Weiss, D. Parker, C. Roberts, J. Duncan, I. Weller, C. Carne, R. S. Tedder, A. J. Pinching, and R. Cheingsong-Popov.** 1987. Human immunodeficiency virus infection in two cohorts of homosexual men: neutralizing sera and association of anti-gag antibody with prognosis. *Lancet* **i:**119–122.
92. **Weiss, S. H., W. C. Saxinger, D. Rechtman, M. H. Grieco, J. Nadler, S. Holman, H. M. Ginzburg, J. E. Groopman, J. J. Goedert, and P. D. Markham.** 1985. HTLV-III infection among health care workers: association with needle-stick injuries. *J. Am. Med. Assoc.* **254:**2089–2093.
93. **Winkelstein, W., Jr., D. M. Lyman, N. Padian, R. Grant, M. Samuel, J. A. Wiley, R. E. Anderson, W. Lang, J. Riggs, and J. A. Levy.** 1987. Sexual practices and risk of infection by the human immunodeficiency virus. The San Francisco men's health study. *J. Am. Med. Assoc.* **257:**321–325.
94. **Wykoff, R. F.** 1985. Female-to-male transmission of AIDS agent. *Lancet* **ii:**1017–1018.
95. **Wykoff, R. F.** 1986. Female-to-male transmission of HTLV-III. *J. Am. Med. Assoc.* **255:**1704–1705.
96. **Zolla-Pazner, S., D. William, W. El-Sadr, M. Marmor, and R. Stahl.** 1984. Quantitation of β_2-microglobulin and other immune characteristics in a prospective study of men at risk of acquired immune deficiency syndrome. *J. Am. Med. Assoc.* **251:**2951–2955.

Chapter 15

The Acquired Immunodeficiency Syndrome Epidemic in Africa: Present Status and Implications for the Future

Thomas C. Quinn

INTRODUCTION

The acquired immunodeficiency syndrome (AIDS) has become a global pandemic. Presently, 134 countries on five continents have officially reported over 80,000 cases of AIDS to the World Health Organization. However, this number is an underestimate of the actual number of cases, since thousands of additional AIDS cases remain unrecognized and unreported, particularly in developing countries such as those in Africa (23, 34). Even though most of the cases reported to the World Health Organization have been from the Americas, particularly the United States, the number of cases is increasing at the same logarithmic rate on nearly all other continents. Limited surveillance and serologic surveys have provided sufficient data to estimate that there are over 150,000 cases of AIDS worldwide, 500,000 cases of symptomatic people infected with the human immunodeficiency virus (HIV; the etiologic agent of AIDS), and approximately 5 million to 10 million people with asymptomatic HIV infection (23, 34). The majority of these asymptomatic carriers, approximately 3 to 5 million, are located predominantly in countries of Central Africa (34), and an additional 1.5 million cases are located within the United States (46, 51). It is from this human reservoir of asymptomatic people that HIV will be transmitted to other individuals and from whom more cases will occur. Unfortunately, many of these cases of AIDS will

---

Thomas C. Quinn • Laboratory of Immunoregulation, National Institute of Allergy and Infectious Diseases, Bethesda, Maryland 20892, and Johns Hopkins University School of Medicine, Baltimore, Maryland 21205.

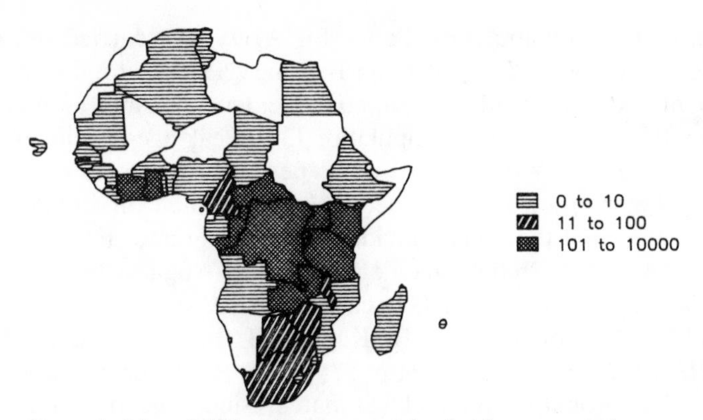

Figure 1. Map of Africa showing countries that have reported cases of AIDS to the World Health Organization as of December 1987.

occur among people in developing countries, which will undoubtedly create an enormous social, political, and economic strain on these countries. Without the present availability of an effective vaccine or curative antiviral drug, and with a case fatality rate of over 90% 4 years from the time of diagnosis, AIDS and HIV infection have been and will continue to become one of the most dreaded epidemics of our time.

THE AIDS EPIDEMIC IN AFRICA

The overall scope and intensity of AIDS and HIV infection in Africa are difficult to assess precisely. Although 40 of 45 African countries have reported nearly 8,000 cases of AIDS as of December 1987 (Fig. 1), it is generally accepted that thousands of additional cases have gone unrecognized or unreported in many areas of Central Africa where the prevalence rates of viral infection are extremely high, and the number of hospitalized patients who test positive for HIV ranges from 20 to 40% in some surveys (46). Unfortunately, surveillance for AIDS cases has been limited in Africa primarily because of the weaknesses in the health infrastructure and the lack of adequate resources. Diagnostic services for opportunistic infections and for serodiagnosis of HIV infection are not generally available. Furthermore, serodiagnosis is complicated and requires confirmatory testing because of the presence of cross-reacting antibodies (4, 31, 56). However, governments of African countries, in collaboration with international health agencies, have been attempting to address these issues, so that a clear picture of the descriptive epidemiology of AIDS in Africa is gradually emerging.

The first systematic surveillance for AIDS was carried out in Kinshasa, Zaire, between July 1984 and February 1985. A total of 332 cases were identified for an adjusted annual incidence of approximately 176 cases per 10^6 members of the population (38). Peak age-specific incidence rates of 786 cases per 10^6 and 601 cases per 10^6 were found among 30- to 39-year-old men and women, respectively. Continued surveillance data in 1986 indicated that the annual incidence is now greater than 1,000 cases per 10^6 adults (46). Preliminary data from 1987 suggest that these rates may now actually exceed 2,000 cases per 10^6 adults (R. Ryder, W. Bertrand, R. L. Colebunders, B. Kapita, H. Francis, and M. Lubaki, *Abstr. III Int. Conf. AIDS*, T7.6, p. 57, 1987). However, these disturbing numbers are probably minimal estimates, since the data reflect only recognized and reported cases of AIDS in several hospitals within Kinshasa.

Surveillance data can be used to reflect the basic epidemiologic trends of AIDS in Africa. In 1986, the male-to-female ratio of AIDS cases in Zaire was approximately 1:1.3 (46). As in developed countries, AIDS in Africa affects primarily young and middle-aged persons. The sex and age distribution of AIDS cases reflects patterns seen with other sexually transmitted diseases in developing countries in which the incidence and morbidity rates are higher among younger women and slightly older men (Fig. 2) (46). Women with AIDS were more likely than men to be unmarried (61% versus 36%), and nearly one-third of the married AIDS patients had at least one previous marriage. One-third of AIDS patients reported having at least one sexually transmitted disease during the 3 years preceding their illness (38). A total of 29% of patients had traditional medical practitioners, and 80% reported receiving medical injections; 9% of the patients received a blood transfusion during the 3-year period before the onset of illness. These data obviously do not allow for the direct assessment of high-risk activity associated with AIDS, since no information was provided from a control population without HIV infection. However, they do provide comparable information to that collected in the United States and Europe.

Although serologic studies often have their limitations, they better indicate the extent of HIV infection in selected populations in certain geographical areas. Rates of HIV infection found by serologic surveys carried out in over 25 different African countries demonstrate a wide range of infection among healthy populations, from 0.03% in Chad to as high as 18% in Rwanda and Uganda (42, 53; F. Mhalu, E. Mbena, U. Bredberg-Raden, J. Kiango, K. Nuamuryekunge, and G. Biberfeld, *Abstr. III Int. Conf. AIDS*, TP86, p. 76, 1987). Seroprevalence rates among high-risk groups such as men attending sexually transmitted

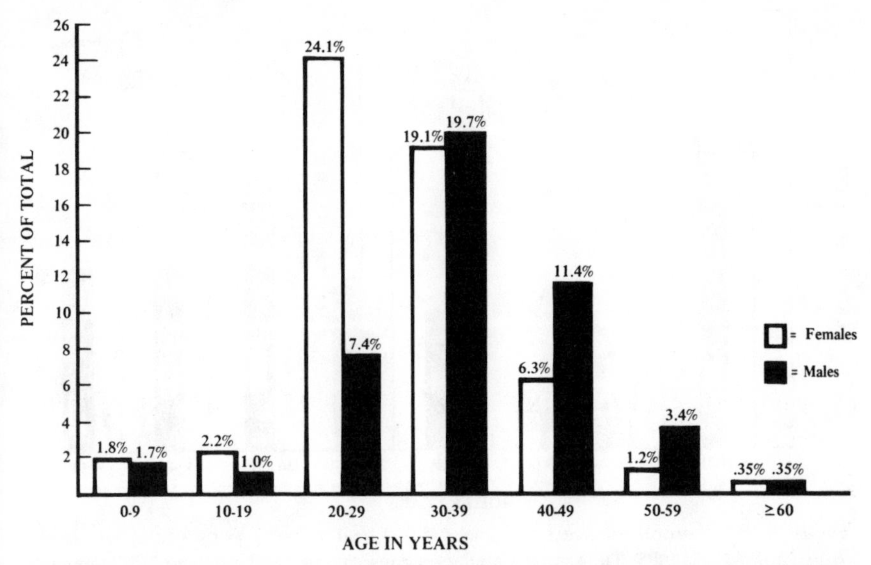

Figure 2. Distribution by age and sex of the first 500 AIDS cases diagnosed in Kinshasa, Zaire, between August 1985 and December 1985. Females were significantly younger than males. Women accounted for the majority of AIDS cases in the 20- to 29-year age group, whereas men accounted for the majority of cases in the 40- to 49-year and 50- to 59-year age groups. (From reference 46.)

disease clinics ranged from 15 to 30%; the rates for female prostitutes have ranged from 27 to 88%, depending upon selection, socioeconomic status, and geographic location (12, 30, 52).

Similar to AIDS surveillance data in Africa, the age- and sex-specific incidence rates of HIV infection in healthy individuals as measured by serologic surveys demonstrate a 1:1.4 male-to-female ratio (37, 46). In Kinshasa, Zaire, data on HIV seroprevalence from 5,099 healthy people (2,982 men and 2,117 women) show a bimodal curve, with peak prevalence under 1 year of age and among young adults aged 16 to 29 years (Fig. 3) (46). Although other factors may be influencing this distribution of HIV infection, this pattern suggests a sexually transmitted disease with higher prevalence rates among young, sexually active women. Combination of passive antibody transfer and transmission of virus from mother to infant is probably responsible for the high seroprevalence in children under age 2 (36). In a study in Lusaka, Zambia, which was primarily a hospital-based survey including outpatient clinics and prenatal clinics, the prevalence of antibodies was low in subjects aged less than 20 or older than 60 years (42). The peak prevalence occurred in men aged 35 years

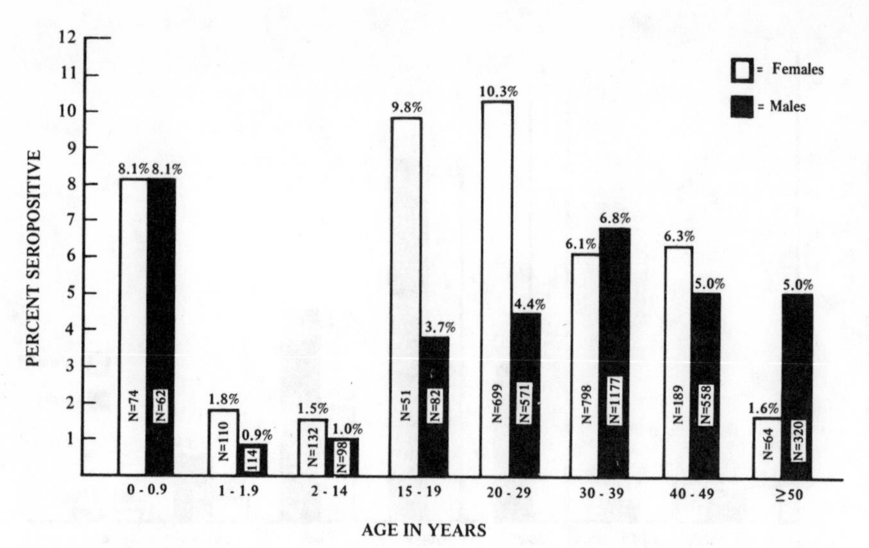

Figure 3. HIV seroprevalence rates among 5,099 healthy persons by age in Kinshasa, Zaire, in 1984 and 1985. The sample population consisted of 2,982 men and 2,117 women. Seropositive patients had serum samples that were repeatedly reactive on a commercially available enzyme-linked immunosorbent assay and were Western blot positive. In the age groups 16 to 19 years and 20 to 29 years, seroprevalence rates were significantly higher in women (10.3%) than men (4.3%). In the age group 50 to 59 years, seroprevalence rates were significantly higher in men (5.0%) than women (1.6%). (From reference 46.)

(32.9%) and in women aged 22 years (24.4%). There were no significant differences in prevalence by sex after adjusting for age. These seroprevalence data strongly suggest that HIV infection is common primarily for sexually active age groups and is transmitted predominantly heterosexually.

Longitudinal studies on HIV seroprevalence are useful in monitoring the spread of HIV infection within a given population. In a Kenyan study (44), serum was collected from 543 men and 123 women who had genital ulcer disease and who were attending a sexually transmitted disease clinic between 1980 and 1985, as part of studies on the etiology and epidemiology of chancroid. In addition, 535 serum samples from prostitutes seen between 1981 and 1985 in a special prostitute research clinic in Nairobi were tested, and serum samples were collected from 1,100 pregnant women selected on the basis of residence in eastern Nairobi between 1981 and 1985. Prevalence of antibodies to HIV in the prostitutes rose from 4% in 1981 to 61% in 1985. Between 1981 and 1982, 4% and 8%, respectively, of women with sexually transmitted diseases were seropositive. Antibodies to HIV in men with genital ulcers were first detected in 1981, and the prevalence of antibodies rose to 15% by 1985. In 1981, none of 111 pregnant women were seropositive, but by 1985, 2% of 1,100 pregnant women undergoing delivery at Pumwani Maternity Hospital had antibody to HIV. Thus, the relatively low but steady increase in seroprevalence among pregnant women in Nairobi, from 0 to 2% in 5 years, and among pregnant women in Kinshasa, from 0.25% to 8% within a 16-year period (46), indicates the potential rate of spread of HIV infection to the general population in some of these areas.

Consistent with this apparent rate of increase in the general population, HIV seroprevalence rates increased from 6.4% in 1984 to 7.8% in 1986 among 2,300 health care workers at a major hospital in Kinshasa (37; B. Ngaly, R. W. Ryder, B. Kapita, H. Francis, T. C. Quinn, and J. M. Mann, *Abstr. III Int. Conf. AIDS*, M3.6, p. 2, 1987). Between 1984 and 1986 there were 56 new infections, for a 2-year incidence rate of 3.7%. In summary, these results demonstrate that HIV can spread as rapidly in a high-risk population of heterosexuals such as female prostitutes as has been documented for homosexual men in San Francisco, with large numbers of sexual partners as the common risk factor (24). In contrast, there is a slow but significant evolution of infection in non-high-risk groups for AIDS such as pregnant women and health care workers in Central Africa (K. M. DeCock, N. Nzilambi, D. Forthal, R. Ryder, P. Piot, and J. B. McCormick, *Abstr. III Int. Conf. AIDS*, WP43, p. 117, 1987; Ngaly et al., *Abstr. III Int. Conf. AIDS*).

Seroprevalence data from rural and urban areas are significantly different (5, 13, 45, 46, 53; DeCock et al., *Abstr. III Int. Conf. AIDS*). In Burundi, Rwanda, and Zaire, seroprevalence rates appear higher in urban areas (5, 13, 42, 44, 45; Ryder et al., *Abstr. III Int. Conf. AIDS*). For example, although seroprevalence rates range from 6 to 8% in people residing in urban Kinshasa (46; Ryder et al., *Abstr. III Int. Conf. AIDS*), a serosurvey in 1986 in a rural area of Northern Zaire, using a cluster sampling technique, showed a prevalence rate of HIV antibody of 0.8% in 389 healthy adult villagers (DeCock et al., *Abstr. III. Int. Conf. AIDS*). For 136 pregnant women, the seroprevalence rate was 2%, compared with 8% in pregnant women in Kinshasa (DeCock et al., *Abstr. III Int. Conf. AIDS*; N. Nzila, R. W. Ryder, F. Behets, H. Francis, E. Bayende, A. Nelson, and J. M. Mann, *Abstr. III Int. Conf. AIDS*, TH7.6, p. 158, 1987). Similarly, the seroprevalence rate was 11% for 283 prostitutes in the rural region, but 27% for 377 urban prostitutes (DeCock et al., *Abstr. III Int. Conf. AIDS*; J. M. Mann, T. C. Quinn, H. Francis, M. Miatudila, P. Piot, and J. Curran, *Abstr. II Int. Conf. AIDS*, S17e, p. 105, 1986). The rate of infection in the healthy adult villagers in 1986 was similar to that documented in a retrospective serologic study which demonstrated a

0.8% infection rate of 659 serum samples collected in 1976 following an epidemic of Ebola hemorrhagic fever (DeCock et al., *Abstr. III Int. Conf. AIDS*). Thus, although HIV infection rates may increase rapidly in some selected populations, as demonstrated for the prostitutes of Nairobi, it can remain relatively stable in other populations such as those in rural areas.

MODES OF TRANSMISSION

Regardless of the geographic region, HIV appears to be transmitted primarily via three routes: sexual, parenteral, and perinatal (17). However, important regional variations exist within each of these transmission categories. For example, in North America and Europe the predominant sexual mode of transmission occurs among homosexual men, involving anal intercourse, whereas in developing countries such as Central Africa, heterosexual transmission is far more common than homosexual transmission. Similarly, in developed countries transmission among intravenous-drug abusers is quite common, but in developing countries HIV may be transmitted by needles not used for drug abuse, but primarily for medicinal purposes, scarification, or rituals. Thus, available data suggest that heterosexual transmission, frequent exposure to unsterilized needles, blood transfusions, and vertical transmission from mother to infant account for the spread of HIV infection and AIDS in Africa (46).

Heterosexual Transmission

In addition to the 1:1 male-to-female ratio among AIDS cases, the younger age and single marital status for female cases, serologic studies also confirm an overall equal HIV infection rate between the sexes, with slightly higher rates among younger women and older men (42, 46). These age-sex patterns of infection are those commonly seen with other sexually transmissible diseases, such as gonorrhea, chlamydia, and syphilis (1). In case-control studies in Rwanda (12, 13, 52), AIDS patients had a significantly higher number of heterosexual partners than controls (mean of 32 versus 3), and male patients had sexual intercourse significantly more often with female prostitutes than controls did (81% versus 34%); the risk of seropositivity increased significantly with the number of different sexual partners per year and with a history of other sexually transmitted diseases. African urban prostitutes have a high rate of infection and may have played an important role in the dissemination of HIV in some areas such as Nairobi (30), since HIV antibody prevalence was initially higher in prostitutes in that city than in men attending sexually transmitted disease clinics. HIV infection rates are also elevated in patients attending these sexually transmitted disease clinics, with rates of 15 to 30% in Nairobi and Lusaka (30, 42, 44). Seropositivity has also been documented to be high among male clients of female prostitutes (R. M. Greenblatt, S. L. Lukehart, F. A. Plummer, T. C. Quinn, C. W. Critchlow, and L. J. D'Conta, *AIDS*, in press; F. A. Plummer, J. N. Simonsen, E. N. Ngugi, D. W. Cameron, P. Piot, and J. O. Ndinya-Achola, *Abstr. III Int. Conf. AIDS*, M8.4, p. 6, 1987). However, in nearly all studies, HIV seropositivity among men and women was correlated with the number of sexual partners and frequently with a history of genital ulcerations (13, 30, 44, 46; Greenblatt et al., in press; Mann et al., *Abstr. II Int. Conf. AIDS*; M. Melbye, E. K. Njelesani, and A. Bayley, Letter, *Lancet* **ii:**1113–1115, 1986).

Even though evidence for bidirectional transmission is limited in the United States and Europe, there is sufficient biological and clinical evidence to support bidirectional transmission of HIV. First, it may be possible for HIV to be transmitted bidirectionally among heterosexual contacts, since the virus can be isolated from semen as well as cervicovaginal secretions (22, 55, 57). Second, African male AIDS patients, as well as expatriate males with AIDS who previously lived in Africa, frequently report a history of sexual intercourse with prostitutes, and the prevalence of HIV in male heterosexuals attending clinics for sexually transmitted diseases is increasing, suggesting female-to-male transmission (12, 30, 42; Plummer et al., *Abstr. III Int. Conf. AIDS*). Further evidence for such transmission comes from studies of households of AIDS patients in Zaire and Uganda, in which HIV antibody was significantly higher among spouses (61 to 68%) and infants of infected mothers than in other household members or controls (39, 49). Finally, several clusters of African AIDS cases have been identified in the chronology of events which suggest that both female-to-male and male-to-female transmission of HIV occur (25, 45).

Risk factors associated with HIV infection among heterosexuals include having a large number of sexual partners, having sexual intercourse with prostitutes, being a prostitute, or being a sexual partner of an infected person (46). Specific sexual activity, including anal intercourse, reported by only 4 to 8% of female AIDS patients and female prostitutes was not associated with HIV infection in surveys in Kenya, Zaire, Rwanda, or Zambia (12, 30, 42, 52; M. Merlin, R. Josse, E. Delaporte, J. P. Durand, C. Hengy, and A. J. Georges, *Abstr. III Int. Conf. AIDS*, M8.1, p. 5, 1987). Thus, receptive anal intercourse, which is a major risk factor for HIV infection in male homosexuals, may not necessarily be a major risk factor for heterosexual transmission.

Cofactors may play an important role in enhancing the sexual transmissibility of HIV infection. In studies among Nairobi prostitutes, HIV seropositivity was significantly associated with current sexually transmitted diseases such as gonorrhea, genital ulcers, and syphilis (30). In a study in Zambia, seropositivity in men was also correlated with the presence of genital ulcerations (42). In a prospective study of 116 initially seronegative prostitutes monitored for 2 years in Nairobi, 54 (76%) of 71 women with one or more episodes of genital ulcerative disease seroconverted, compared with 20 (44%) of 45 women without such disease (Plummer et al., *Abstr. III Int. Conf. AIDS*). In a prospective study of men with a sexually transmitted disease who had a recent history of sexual intercourse with a prostitute, 13 (14%) of 91 men with genital ulcerative disease seroconverted during a 2- to 3-month follow-up period, compared with 3 (3%) of 108 men with urethritis (D. W. Cameron, F. A. Plummer, J. N. Simonsen, J. O. Ndinya-Achola, L. J. D'Costa, and P. Piot, *Abstr. III Int. Conf. AIDS*, MP91, p. 25, 1987). In a study of female prostitutes in Zaire, HIV seropositivity, although not associated with a history of a sexually transmitted disease, was associated with placing of products within the vagina for hygenic purposes (Mann et al., *Abstr. II Int. Conf. AIDS*). All of these studies support the role of genital ulcerations, either secondary from trauma or from sexually transmitted diseases, in enhancing heterosexual HIV transmission.

Perinatal Transmission

With increasing prevalence rates of HIV infection among young, sexually active women, an increasing rate of infection among the children born to these women can be expected (17, 47). Perinatal transmission of HIV may occur in utero through transplacental passage of HIV, natally (at time of delivery, when there is a maternal-fetal blood exchange), or postnatally through breast feeding or other possible routes. The efficiency and risk factors associated with perinatal infection remain unknown, and prospective perinatal studies are currently under way in several African countries. Preliminary data from Kinshasa, Zaire, and Nairobi, Kenya, suggest that approximately 40 to 50% of children born to HIV antibody-positive mothers have either immunoglobulin M (IgM) antibody to HIV in cord blood or at a 3-month follow-up visit (M. Braddick, J. K. Kreiss, T. C. Quinn, J. O. Ndinya-Achola, G. Vercauteren, and F. A. Plummer, *Abstr. III Int. Conf. AIDS*, TH7.5, p. 158, 1987; Nzila et al., *Abstr. III Int. Conf. AIDS*, TH7.6, p. 158, 1987). In Kinshasa, Zaire, 39 (11%) of 351 children born to HIV-positive mothers died within 3 months of birth, compared with only 2 (0.6%) of 351 control children born to seronegative mothers ($P < 0.001$) (Nzila et al., *Abstr. III Int. Conf. AIDS*, TH7.6, p. 158, 1987). Of the infected children, 5% died within the first week of life, and the remaining 6% died after the first week. The mean birth weight of children dying within the first week of life was 1,651 g compared with 2,509 g for those dying after the first week of life. Immunologic studies performed in concert with studies of IgM antibodies in cord blood demonstrated a marked association between a small number of T4 cells in the mother and the probability of an infected child. The number of T4 cells in mothers of IgM-positive children was $278/mm^3$, compared with $510/mm^3$ in seropositive mothers of seronegative children and $703/mm^3$ in seronegative mothers of seronegative children ($P < 0.001$) (H. Francis, N. Lubaki, M. P. Duma, R. W. Ryder, J. Mann, and T. C. Quinn, *Abstr. III Int. Conf. AIDS*, F9.6, p. 214, 1987).

It is thus apparent that HIV infection in pregnant women appears to be an important cause of premature birth and perinatal death. Prospective studies already under way will be able to better define the consequences of HIV infection in pregnant African women, to determine the incidence of congenital HIV infection in children born to HIV-infected African women, to better identify the risk factors for congenital and perinatal HIV infection, and to describe the natural history of congenital and perinatal HIV infection. It is evident from seroprevalence studies of pregnant women in other areas that perinatal transmission may contribute significantly to the spread of HIV to the next generation of Africans. In areas other than Zaire, HIV seropositivity was confirmed in 0.3% of 768 pregnant women and new mothers in Gabon, 2.6% of 2,389 pregnant women in Nairobi, 2% of 200 consecutively tested women attending a prenatal clinic in Malawi, 14% of 1,011 women attending a prenatal clinic in Uganda, and 8.7% of 184 women giving birth in Lusaka, Zambia (33, 42, 46; Braddick et al., *Abstr. III Int. Conf. AIDS*; Nzila et al., *Abstr. III Int. Conf. AIDS*).

Transmission via Blood Transfusions

The importance of blood transfusions in HIV transmission in Africa is exemplified by the 6 to 18% seroprevalence rate of HIV infection among blood donors in Uganda, Rwanda, Zambia, and Zaire (33, 42, 46, 53; F. Mhalu, E. Mbena, U. Bredberg-Raden, J. Kiango, K. Nuamuryekunge, and G. Biberfeld, *Abstr. III Int. Conf. AIDS*, TP86, p. 76, 1987; Ryder et al., *Abstr. III Int. Conf. AIDS*). In Kinshasa, 9% of 295 AIDS patients interviewed between 1984 and early 1985 reported receiving at least one blood transfusion during the previous 5 years (38). In a large case-control study of seropositivity in healthy persons in Kinshasa, 9.3% of HIV seropositives, compared with 4.8% of HIV seronegatives, reported receiving blood transfusions (37). Similarly, 31% of hospitalized seropositive children, aged 2 to 24 months, with seronegative mothers had received blood transfusions prior to the present hospitalization, compared with 7% of seronegative control children of the same age (36). Of 50 patients with sickle-cell anemia in Zaire, 17% had confirmed antibody to HIV; the majority of these had also received multiple blood transfusions (N. Nzila, R. L. Colebunders, J. M. Mann, H. Francis, K. Nseka, and J. W. Curran, *Abstr. III Int. Conf. AIDS*, W4.6, p. 108, 1987).

In a study to examine the role of blood transfusions and the transmission of HIV to African children, 147 (14.1%) of 1,046 pediatric patients at Mama Yemo Hospital, Kinshasa, Zaire, had a history of previous blood transfusion (A. E. Greenberg, P. Nguyen-Dinh, J. M. Mann, N. Kabote, R. L. Colebunders, and T. C. Quinn, *J. Am. Med. Assoc.*, in press). Of these 1,046 pediatric patients, 40 (3.8%) were HIV seropositive, and there was a strong dose-response association between history of blood transfusion and HIV seropositivity. The odds of being HIV seropositive increased to 43 in those that received three or more blood transfusions. In a follow-up study of 167 hospitalized children with acute malaria, 21 (12.6%) were HIV seropositive (Greenberg et al., in press; P. Nguyen-Dinh, A. E. Greenberg, R. W. Ryder, J. M. Mann, N. Kabote, and H. Francie, *Bull. W.H.O.*, in press). Of 11 HIV seropositive acute malaria patients, 10 had received blood transfusions during the current hospitalization, and 4 of these children were documented to have been seronegative prior to blood transfusion. Follow-up of these four children demonstrated the development of IgM antibodies and persistence of IgG antibodies to HIV 6 months after these transfusions had been given, documenting the high risk of HIV infection from transfusion. The impact of this mode of HIV transmission may be substantial. In 1986, approximately 8,900 blood transfusions were given to children with malaria at Mama Yemo Hospital. Since the HIV-seropositive rate among blood donors was recently shown to be 6.3%, we can estimate that 561 HIV-seropositive blood donations were given to children with malaria in this hospital setting alone (Greenberg et al., in press; Ryder et al., *Abstr. III Int. Conf. AIDS*).

With such high seroprevalence rates among blood donors in some areas of Africa, it is obvious that blood screening for HIV should be instituted to prevent further transmission via this route. However, blood bank screening has been a difficult problem, since adequate facilities for blood banking and the financial resources necessary to support routine HIV screening are not available. The development and introduction of rapid diagnostic tests, such as a rapid latex agglutination slide test for HIV, may have interrupted this transmission route by allowing physicians to routinely screen blood prior to transfusion (6; T. C. Quinn, H. Francis, R. L. Kline, M. P. Duma, M. Sension, and C. Riggin, *Abstr. III Int.*

Conf. AIDS, T7.6, p. 57, 1987). In one preliminary survey at Mama Yemo Hospital, the latex agglutination slide test was shown to be 99% sensitive and 100% specific compared with the enzyme-linked immunosorbent assay and Western immunoblot, and required only 2 min to perform (Quinn et al., *Abstr. III Int. Conf. AIDS*). The implementation of such tests is urgently needed to prevent the transmission of HIV via blood transfusions.

Transmission via Injections

Use of unsterilized needles or other skin-piercing instruments for medical or ritual purposes has potential for HIV transmission similar to that seen with the reuse of needles for intravenous drug abuse in developed countries (Editorial, *Lancet* **ii:**1376–1377, 1984). In a seroprevalence study of 2,384 hospital workers in Kinshasa, significantly more HIV-seropositive than seronegative workers reported receiving medical injections during the previous 3 years; among those reporting injections, seroprevalence was nearly twice as high in those receiving five or more injections than in those receiving fewer than five injections (37). In a study of hospitalized seropositive children aged 1 to 24 months in Kinshasa who had similar medical problems, 16 born to seronegative mothers received significantly more injections than 222 seronegative children born to seronegative mothers (36). In addition, among adult patients with tuberculosis, HIV-seropositive patients reported being given significantly more injections during the 5-year period prior to hospitalization than seronegative patients did (35). These data suggest that injections and scarifications are associated with HIV infection, but it is difficult to distinguish whether the association is truly causal, i.e., providing a means of exposure to HIV, or secondary, i.e., is indirectly associated with treatment for early symptoms of HIV infection or other illnesses, such as sexually transmitted diseases. Further studies are required to fully define the role of injections in the transmission of HIV, but in the meantime, programs should be implemented to introduce proper sterilization of used needles and syringes, and, wherever financial resources are available, to introduce the use of disposable needles.

Transmission via Other Routes

As in Europe and in the United States (20, 21), there is no evidence to support casual or household transmission of HIV in Africa. For example, the rate of HIV seropositivity did not differ significantly between 186 nonspousal household contacts of 46 AIDS cases and 128 nonspousal household contacts of 43 seronegative controls in a study in Kinshasa (39). Similarly, in Uganda, 10 (71%) of sexual partners of 14 index AIDS cases were found to be seropositive, whereas only 2 (2%) of 100 nonsexual household contacts had evidence of infection (49). These seropositive nonsexual household contacts had their own independent risk factors for acquisition of HIV, including perinatal transmission, and evidence of household transmission other than through sexual exposure was not present. No association has been observed between any measure of patient or blood contact and HIV seroprevalence in hospital workers (37). In addition, the absence of HIV infection among expatriates who lack recognized risk factors for HIV infection, despite living in close proximity to possibly infected individuals, reaffirms the apparent lack of casual transmission of HIV in Africa.

There is no direct evidence for arthropod-borne transmission of HIV in Africa or in other countries where HIV infection exists, despite substantial arthropod densities (60). In Africa, the sex- and age-specific AIDS incidence and HIV seroprevalence data are not consistent with a vector-borne disease. Low seroprevalence among children 1 to 14 years old and among persons over 50 years old, and the significantly higher prevalence rate among women than men in the 20- to 39-year-old group, argue against vector-borne transmission (46). Malaria, a vector-borne disease, is particularly common in children between 1 and 24 months old in Kinshasa, but among 44 HIV-seropositive children in Kinshasa, 43 (98%) had other, known risk factors for AIDS, including birth to an HIV-seropositive mother, history of blood transfusion, and frequent exposure to unsterilized needles (36). In Burundi and Kinshasa, the geographic distribution of HIV infection and malaria appeared discordant (53; Greenberg et al., in press). A low titer of HIV in the blood of an infected person and the small amount of blood taken up by some insects emphasize the improbability of mechanical transmission of HIV by insect vectors.

CLINICAL MANIFESTATIONS

The clinical manifestations of HIV are diverse and range from asymptomatic infection to overwhelming illness due to HIV-induced encephalopathy, multiple opportunistic infections, and malignancies (7). The clinical manifestations may also vary in different populations according to the relative frequency of other endemic opportunistic infections. In tropical areas, such as Central Africa and the Caribbean, gastrointestinal and dermatologic manifestations are commonly observed (13, 43, 45, 46, 48, 54), whereas generalized lymphadenopathy and pulmonary symptoms, primarily due to *Pneumocystis carinii* infection, are frequently seen in AIDS patients in the United States (2, 17). In a recent review of 196

AIDS patients in Zaire (38), the mean duration of symptoms prior to diagnosis was 11.8 months (range, 10 to 78 months; median, 8 months). Symptoms consisted of profound weight loss (mean 29% loss in body weight; 99% of patients), fever (81% of patients), diarrhea (68%), cough (37%), dysphagia (35%), pruritis (30%), and dyspnea (23%). On physical examination, oral candidiasis was present in 47% of patients, and 61% of these also had dysphagia. A generalized pruritic macular eruption, frequently referred to as prurigo, was seen in 22%, and generalized lymphadenopathy was seen in 11% (15, 38). In Europe, where extensive diagnostic procedures have been performed with African AIDS patients, the most commonly observed opportunistic infections include oral esophageal candidiasis, central nervous system cryptococcosis, toxoplasmosis, tuberculosis, and cryptosporidiosis (11, 45). In contrast to American and European AIDS cases, of whom approximately 63% eventually develop *P. carinii* pneumonia, this opportunistic infection was found in only 14% of African patients diagnosed in Europe (2, 11). In Africa, where the diagnostic procedures are limited, oral candidiasis, cryptococcal meningitis, probable cytomegalovirus coreoretinitis, cryptosporidiosis, mucocutaneous herpesvirus infection, and disseminated Kaposi's sarcoma are most commonly observed.

Other infections that are seen more frequently in Africa than in the United States include disseminated histoplasmosis, salmonellosis, disseminated strongyloidiasis, and mycobacterial disease. In most African surveys, 20 to 40% of confirmed pulmonary patients hospitalized with tuberculosis were HIV seropositive, and two-thirds of these had extrapulmonary tuberculosis (35, 42). HIV infections may therefore substantially complicate both the management of individual patients with tuberculosis and strategies for tuberculosis control in countries where HIV infection occurs. It is possible that patients living in tropical areas and immunosuppressed as a result of HIV will manifest other infections endemic in those areas, such as leishmaniasis, leprosy, malaria, filiariasis, and other parasitic and bacterial infections.

Perhaps most impressive of the clinical presentations of HIV infection in Africa is the appearance of enteropathic AIDS, frequently referred to as slim disease (48; Mann et al., *Abstr. II Int. Conf. AIDS*). Most clinical studies of HIV infection in Central Africa show that profound weight loss and unexplained diarrhea occur in 99% and 80% of cases, respectively (16, 38, 45). Seroprevalence studies among these patients demonstrate the presence of HIV antibody in all patients who present with clinical diarrhea and a weight loss of 10 kg or more. Microbiologic evaluations of these patients revealed that ca. 50% of them harbored pathogens including *Cryptosporidium* spp., *Isospora belli*, *Entamoeba histolytica*, *Giardia lamblia*, *Strongyloides* spp., and *Salmonella* spp. (16, 51). The World Health Organization has developed a provisional clinical case definition for AIDS in Africa, where sophisticated diagnostic equipment may not be available (58). This case definition depends primarily on the presentation of weight loss, chronic diarrhea, and unexplained fever in association with at least one other clinical sign consistent with systemic immunosuppression.

HIV-2 INFECTION IN WEST AFRICA

In 1985, Kanki et al. (28) found anti-HIV antibody in serum samples from prostitutes in Dakar, Senegal, an area where AIDS had not been previously reported. By radioimmunoprecipitation, these Senegalese serum samples reacted more strongly with a retrovirus which had been isolated from an African green monkey, and which is referred to as the simian immunodeficiency virus (SIV_{AGM}), than with HIV-1 (18, 32). These observations led to a search for a new human retrovirus in West Africa. Kanki et al. (26) isolated a virus provisionally named human T-cell lymphotropic virus type IV (HTLV-IV) from a Senegalese prostitute, and this virus was shown to be more closely related serologically and genetically to SIV_{AGM} than to HIV-1 (P. Kanki, S. M'Boup, F. Barin, D. Ricard, F. Denis, and M. Esses, *Abstr. III Int. Conf. AIDS*, F6.6, p. 211, 1987). Serologic surveys in several countries in West Africa have shown relatively modest rates of infection with this virus (27). Seropositivity appeared to be most common in the groups at risk for HIV-1, including female prostitutes, patients with tuberculosis, and patients attending sexually transmitted disease clinics. Kanki et al. (26) have speculated that this new human retrovirus may be nonpathogenic, by analogy to SIV_{AGM} in African green monkeys. More recent genetic analysis demonstrated that an HTLV-IV isolate was essentially identical to SIV_{AGM} and that HTLV-IV-seropositive individuals are infected with a virus which strongly cross-reacts with SIV_{AGM} at regions of shared epitopes, but which might be a different virus from SIV_{AGM}, HTLV-IV, and HIV-1 (29).

Simultaneous with the report of the isolation of HTLV-IV, Clavel et al. (8) also isolated a virus, referred to as lymphadenopathy-associated virus type 2 and now renamed HIV-2, from patients with AIDS-like symptoms in Guinea-Bissau and Cape Verde in West Africa. HIV-2, although serologically cross-reactive with HTLV-IV and SIV_{AGM}, has been shown to have distinct genetic differences from SIV_{AGM} and consequently from HTLV-IV (9). Genomic analysis has indicated that HIV-1 and HIV-2 share similar genomic organization, indicating a common evolutionary origin, but differ significantly in terms of nucleotide sequences (9). The more conserved *gag* and *pol* genes displayed a 56% and 66% nucleotide sequence homology, respectively, and had less than 60% amino acid identity. The calculation of nucleotide sequence homology for the other viral genes gives even lower values, making HIV-1 and HIV-2 only 42% homologous. These data confirm that these two viruses are distinct elements of the HIV family and cannot be considered strains of the same virus according to the recommendations of the International Taxonomy Committee (14).

Clavel et al. (10) recently reported on the clinical, immunologic, and virologic data for 30 patients with HIV-2 infection, of whom 17 had AIDS, 4 had AIDS-related complex, 1 had diffuse lymphadenopathy, 2 had other clinical problems, and 6 were asymptomatic. The most common clinical presentations of HIV-2 were chronic diarrhea (14 patients) and severe weight loss (13 patients). Common infections diagnosed in HIV-2 patients included *Isospora belli*, *Cryptosporidium*, and *Mycobacterium tuberculosis* infections, aspergillosis, and *P. carinii* pneumonia. Four HIV-2-infected patients had Kaposi's sarcoma, four had diffuse lymphadenopathy, one had cerebral lymphoma, and one had acute encephalitis. Thus, the clinical spectrum of HIV-2 infection appears to be quite similar to that observed for HIV-1 infection.

Unfortunately, very little is known about the epidemiology of these related human retroviruses. The serologic diagnosis of HIV-2 is difficult, since antibodies to HIV-2 occasionally cross-react with HIV-1 (19). Natural history studies of HTLV-IV and HIV-2 infections have been limited by the lack of effective serologic assays and confusion regarding the classification of these retroviruses (L. G. Gurtler, G. Zoulek, G. Frosner, and F. Deinhardt, *Abstr. III Int. Conf. AIDS*, THP93, p. 179, 1987). The development of better serologic assays and the precise extent of the spread of the virus toward other areas including Africa, Europe, and the United States, will have to be assessed by large-scale prospective seroepidemiologic studies. However, it is apparent that HIV-2, a virus related to, yet distinct from, HIV-1, is the cause of AIDS in some West Africans.

THE FUTURE

Control of HIV infection has become a public health priority in many countries. With estimates of nearly 2 million HIV-infected people in the United States, and probably over 5 million infected people in Africa, it is evident that HIV has created a major international health problem. With the lack of an effective vaccine or curative treatment within the near future, primary prevention is the only effective means of controlling HIV infection and AIDS. As discussed throughout this chapter, the major routes of spread in both the United States and Africa are sexual transmission, blood transfusions, blood-contaminated needles, and perinatal transmission, and each of these modes must be dealt with effectively in any control program (59). The challenge to control HIV infection is great for all countries, but perhaps greatest for those in Africa, where a greater proportion of the population has already been infected and where the general social, political, and economic context of modern Africa may limit the effectiveness of a control program. The rapid urbanization in many parts of Africa has resulted in economic and sociologic changes that have influenced behavior and affected the health infrastructure. As a result, public health officials cannot immediately upgrade blood transfusion services to prevent HIV infection if the proposed intervention is likely to cost, per person, 10 to 30 times the annual per capita public health budget. Similarly, reuse of disposable injection equipment cannot be prevented when many hospital budgets are insufficient for obtaining antibiotics.

Encouraging changes in sexual behavior which has long been part of the culture of a society will be an enormous challenge, but one that must be accomplished. Creative educational approaches, respecting cultural traditions, are necessary to make the population aware of the danger of HIV infection and AIDS and to encourage protective measures. Depending upon the area, this may in practice mean educating all sexually active men and women, but particularly persons with sexually transmitted diseases. Reduction of risk of sexual transmission is based primarily on limiting the number of sexual partners, avoiding unsafe sexual practices, and consistently using condoms. A mutually monogamous relationship between seronegative individuals, or the use of condoms between partners of unknown serologic status, should help prevent the further sexual transmission of HIV. With over 120 countries already affected by this disease and 5 million to 10 million people carrying the virus, only worldwide cooperation will be successful in limiting the further spread of this disease.

Literature Cited

1. **Aral, S. O., and K. K. Holmes.** 1984. Epidemiology of sexually transmitted diseases, p. 127–141. *In* K. K. Holmes, P. A. Mardh, P. F. Sparling, and P. J. Weisner (ed.), *Sexually Transmitted Diseases*. McGraw-Hill Book Co., New York.
2. **Armstrong, D.** 1987. Opportunistic infections in the acquired immune deficiency syndrome. *Semin. Oncol.* **14:**40–47.
3. **Biggar, R. J.** 1986. The AIDS problem in Africa. *Lancet* **ii:**1018–1022.
4. **Biggar, R. J., P. C. Gigase, M. Melbye, L. Kestens, P. S. Sarin, A. J. Bodner, P.**

Demendts, W. J. Stevens, L. Paluku, and C. Delacollette. 1985. ELISA HTLV-III retrovirus antibody reactivity associated with malaria and immune complexes in healthy Africans. *Lancet* **ii:**520–523.

5. **Brun-Vezinet, F., C. Rouzioux, L. Montagnier, S. Chamaret, J. Gruest, S. Mitchell, P. Piot, and T. C. Quinn.** 1984. Prevalence of antibodies to lymphadenopathy-associated retrovirus in African patients with AIDS. *Science* **226:**453–456.
6. **Carlson, J. R., S. C. Mertens, J. L. Yee, M. B. Gardner, E. J. Watson-Williams, J. Ghrayeb, M. B. Jennings, and R. J. Biggar.** 1987. Rapid, easy, and economical screening test for antibodies to human immunodeficiency virus. *Lancet* **i:**361–362.
7. **Centers for Disease Control.** 1983. Tuberculosis—United States. *Morbid. Mortal. Weekly Rep.* **32:**478–480.
8. **Clavel, F., D. Guetard, F. Brun-Vezinet, S. Chamaret, M. A. Rey, M. O. Santos-Ferreira, A. G. Laurent, C. Dauguet, C. Katlama, and C. Rouzioux.** 1986. Isolation of a new human retrovirus from West African patients with AIDS. *Science* **233:**343–346.
9. **Clavel, F., M. Guyader, D. Guetard, M. Salle, L. Montagnier, and M. Alizon.** 1986. Molecular cloning and polymorphism of the human immune virus type 2. *Nature* (London) **324:**691–695.
10. **Clavel, F., K. Mansinho, S. Chamaret, D. Guetard, V. Favier, J. Nina, and M. O. Santos-Ferreira.** 1987. Human immunodeficiency virus type 2 infection associated with AIDS in West Africa. *N. Engl. J. Med.* **316:**1180–1185.
11. **Clumeck, N., M. Robert-Guroff, P. Van de Perre, A. Jennings, J. Sibomana, P. Demol, S. Cran, and R. C. Gallo.** 1985. Seroepidemiological studies of HTLV-III antibody prevalence among selected groups of heterosexual Africans. *J. Am. Med. Assoc.* **253:**2599–2602.
12. **Clumeck, N., J. Sonnet, H. Taelman, F. Mascart-Lemone, M. De-Bruyers, P. Van de Perre, J. Dasnoy, L. Marcelis, M. Lamy, C. Jonas, L. Eyckmans, H. Noel, M. Vanhaeverbeck, and J. P. Butzler.** 1984. Acquired immune deficiency syndrome in African patients. *N. Engl. J. Med.* **310:**492–497.
13. **Clumeck, N., P. Van de Perre, M. Carael, D. Rouvroy, and D. Nzaramba.** 1985. Heterosexual promiscuity among African patients with AIDS. *N. Engl. J. Med.* **313:**182–183.
14. **Coffin, J., A. Haase, J. A. Levy, L. Montagnier, S. Oroszian, N. Teich, H. Temin, K. Toyoshima, H. Varmus, and P. Vogt.** 1986. What to call the AIDS virus. *Nature* (London) **321:**10–12.
15. **Colebunders, R., J. M. Mann, H. Francis, K. Bila, L. Izaley, N. Kokonde, K. Kabasele, L. Ifoto, N. Nzilambi, T. C. Quinn, G. VanderGroen, J. W. Curran, G. Vercauteren, and P. Piot.** 1987. Evaluation of a clinical case-definition of acquired immunodeficiency syndrome in Africa. *Lancet* **i:**492–494.
16. **Colebunders, R., J. M. Mann, H. Francis, K. M. Bila, L. Izaley, K. Ndangi, A. F. Hood, T. C. Quinn, P. Gigase, E. VanMarck, G. Vercauteren, J. W. Curran, and P. Piot.** 1987. Persistent diarrhea, strongly associated with HIV infection in Kinshasa, Zaire. *Am. J. Gastroenterol.* **82:**859–864.
17. **Curran, J. W., W. M. Morgan, A. M. Hardy, H. W. Jaffe, W. W. Darrow, and W. R. Dowdle.** 1985. The epidemiology of AIDS: current status and future prospects. *Science* **229:**1352–1357.
18. **Daniel, M., N. Letvin, N. King, M. Kannagi, P. K. Sehgal, R. D. Hunt, P. J. Kanki, M. Essex, and R. C. Desrosiers.** 1985. Isolation of T-cell tropic HTLV-III like retrovirus from macaques. *Science* **228:**1201–1204.
19. **Denis, F., F. Barin, G. Gershy-Damet, J. L. Rey, M. Lhuillier, M. Mounier, G. Leonard, A. Sangare, A. Goudeau, S. MBoup, and M. Essex.** 1987. Prevalence of human

T-lymphotrophic retroviruses type III (HIV) and type IV in Ivory Coast. *Lancet* **i**:408–411.

20. **Fischl, M. A., G. M. Dickinson, G. B. Scott, N. Klimas, M. A. Fletcher, and W. Parks.** 1987. Evaluation of heterosexual partners, children, and household contacts of adults with AIDS. *J. Am. Med. Assoc.* **257**:640–644.
21. **Friedland, G. H., B. R. Saltzman, M. F. Rogers, P. A. Kahl, M. L. Lesser, M. M. Mayers, and R. S. Klein.** 1986. Lack of transmission of HTLV-III/LAV infection to household contacts of patients with AIDS or AIDS-related complex with oral candidiasis. *N. Engl. J. Med.* **314**:344–349.
22. **Ho, D. D., R. T. Schooley, T. R. Rota, J. C. Kaplan, T. Flynn, S. Z. Salahuddin, M. A. Gonda, and M. S. Hirsch.** 1984. HTLV-III in the semen and blood of a healthy homosexual man. *Science* **226**:451–453.
23. **Institute of Medicine.** 1986. *Confronting AIDS: Direction for Public Health, Health Care and Research.* National Academy Press, Washington, D.C.
24. **Jaffe, H. W., W. W. Darrow, D. F. Echenberg, P. M. O'Malley, J. P. Getchell, V. S. Kalyanaraman, R. H. Byers, D. P. Drennan, E. H. Braff, and J. W. Curran.** 1985. The acquired immunodeficiency syndrome in a cohort of homosexual men. A six-year follow-up study. *Ann. Intern. Med.* **103**:210–214.
25. **Jonckheer, T., I. Dab, P. Van de Perre, P. Lepage, J. Dasnoy, and H. Taelman.** 1985. Cluster of HTLV-III/LAV infection in an African family. *Lancet* **i**:400–401.
26. **Kanki, P. J., F. Barin, S. MBoup, J. S. Allan, J. L. Romet-Lemonne, R. Marlink, M. F. McLane, T. H. Lee, B. Arbeille, F. Denis, and M. Essex.** 1986. New human T-lymphotropic retrovirus related to simian T-lymphotropic virus type III (STLV-III_{AGM}). *Science* **232**:238–243.
27. **Kanki, P. J., S. MBoup, D. Ricard, F. Barin, F. Denis, C. Boye, L. Sangare, K. Travers, M. Albaum, R. Marlink, and M. Essex.** 1987. Human T-lymphotropic virus type 4 and the human immunodeficiency virus in West Africa. *Science* **236**:827–831.
28. **Kanki, P. J., M. F. McLane, N. W. King, Jr., N. L. Letvin, R. D. Hunter, P. Sehgal, M. D. Daniel, R. C. Desrosiers, and M. Essex.** 1985. Serologic identification and characterization of a macaque T-lymphotropic retrovirus closely related to HTLV-III. *Science* **228**:1199–1201.
29. **Kornfeld, H., N. Riedel, G. A. Vilianti, V. Hirsch, and J. I. Mullins.** 1987. Cloning of HTLV-IV and its relation to simian and human immunodeficiency virus. *Nature* (London) **326**:610–614.
30. **Kreiss, J. K., D. Koech, F. A. Plummer, K. K. Holmes, M. Lightfoote, P. Piot, A. R. Ronald, J. O. Ndinya-Achola, L. J. D'Costa, P. Roberts, E. N. Ngugi, and T. C. Quinn.** 1986. AIDS virus infection in Nairobi prostitutes: spread of the epidemic to East Africa. *N. Engl. J. Med.* **314**:414–418.
31. **Kuhnl, P., S. Seidl, and G. Holzberger.** 1985. HLA DR4 antibodies cause positive HTLV-III antibody ELISA results. *Lancet* **i**:1222–1223.
32. **Letvin, N. L., M. D. Daniel, P. K. Sehgal, R. C. Desrosiers, R. D. Hunt, L. M. Waldron, J. J. MacKey, D. K. Schmidt, L. V. Chalifoux, and N. W. King.** 1985. Induction of AIDS-like disease in macaque monkeys with T-cell tropic retrovirus STLV-III. *Science* **230**:71–73.
33. **Liskin, L., R. Blackburn, and J. H. Maier.** 1986. AIDS—a public health crisis. *Popul. Rep.* **14**:L193–L228.
34. **Mann, J.** 1987. The global AIDS situation. *World Health Stat. Q.* **40**:185–192.
35. **Mann, J., D. E. Snider, Jr., H. Francis, T. C. Quinn, R. L. Colebunders, P. Piot, J. W. Curran, N. Nzilambi, N. Bosenge, and M. Malonga.** 1986. Association between HTLV-III/LAV infection and tuberculosis in Zaire. *J. Am. Med. Assoc.* **256**:346–348.

36. **Mann, J. M., H. Francis, F. Davachi, P. Baudoux, T. C. Quinn, N. Nzilambi, N. Bosenge, R. L. Colebunders, P. Piot, N. Kabote, K. A. Pangu, M. Miatudila, and J. W. Curran.** 1986. Risk factors for human immunodeficiency virus seropositivity among children 1–24 months old in Kinshasa, Zaire. *Lancet* **ii:**654–657.
37. **Mann, J. M., H. Francis, T. C. Quinn, K. Bila, P. K. Asila, N. Bosenge, N. Nzilambi, L. Jansegers, P. Piot, K. Ruti, and J. W. Curran.** 1986. HTLV-III/LAV seroprevalence among hospital workers in Kinshasa, Zaire: lack of association with occupational exposure. *J. Am. Med. Assoc.* **256:**3099–3102.
38. **Mann, J. M., H. Francis, T. Quinn, K. A. Pangu, B. Ngaly, N. Nzila, B. Kapita, T. Muyembe, R. Kalisa, P. Piot, J. McCormick, and J. W. Curran.** 1986. Surveillance for AIDS in a Central African city: Kinshasa, Zaire. *J. Am. Med. Assoc.* **255:**3255–3259.
39. **Mann, J. M., T. C. Quinn, H. Francis, N. Nzilambi, N. Bosenge, K. Bila, J. B. McCormick, K. Ruti, P. K. Asila, and J. W. Curran.** 1986. Prevalence of HTLV-III/LAV in household contacts of patients with confirmed AIDS and controls in Kinshasa, Zaire. *J. Am. Med. Assoc.* **256:**721–724.
40. **Marlink, R. G., and M. Essex.** 1987. Africa and the biology of human immunodeficiency virus. *J. Am. Med. Assoc.* **257:**2632–2633.
41. **Marquart, K.-H., H. A. G. Muller, J. Sailer, and R. Moser.** 1985. Slim disease (AIDS). *Lancet* **ii:**186–187.
42. **Melbye, M., E. K. Njelesani, A. Bayley, K. Mukelabai, J. K. Manuwele, F. J. Bowa, S. A. Clayden, A. Levin, W. A. Blattner, R. A. Weiss, R. Tedder, and R. J. Biggar.** 1986. Evidence for heterosexual transmission and clinical manifestations of human immunodeficiency virus infection and related conditions in Lusaka, Zaire. *Lancet* **ii:**1113–1115.
43. **Pape, J. W., B. Liautaud, F. Thomas, J. R. Mathurin, M. M. St. Amand, M. Boncy, V. Pean, M. Pamphile, A. C. Laroche, J. Dehovitz, and W. Johnson.** 1985. The acquired immunodeficiency syndrome in Haiti. *Ann. Intern. Med.* **103:**674–678.
44. **Piot, P., F. A. Plummer, M. A. Rey, E. N. Ngugi, C. Rouzioux, J. O. Ndinya-Achola, G. Veracauteren, L. J. D'Costa, M. Laga, H. Nsanze, L. Fransen, D. Haase, G. Vande Groen, A. R. Ronald, and F. Brun-Vezinet.** 1987. Retrospective seroepidemiology of HIV infection in Nairobi populations. *J. Infect. Dis.* **155:**1108–1112.
45. **Piot, P., T. C. Quinn, H. Taelman, F. M. Feinsod, K. B. Minlangu, O. Wobin, N. Mbendi, P. Mazebo, K. Ndangi, W. Stevens, K. Kayembe, S. Mitchell, C. Bridts, and J. B. McCormick.** 1984. Acquired immunodeficiency syndrome in heterosexual population in Zaire. *Lancet* **ii:**65–69.
46. **Quinn, T. C., J. M. Mann, J. W. Curran, and P. Piot.** 1986. AIDS in Africa: an epidemiologic paradigm. *Science* **234:**955–963.
47. **Rogers, M. F.** 1985. AIDS in children: a review of the clinical, epidemiologic and public health aspects. *Pediatr. Infect. Dis.* **4:**230–236.
48. **Serwadda, D., R. D. Mugerwa, N. K. Sewankambo, A. Lwegaba, J. W. Carswell, G. B. Kirya, A. C. Bayley, R. G. Downing, R. S. Tedder, S. A. Clayden, R. A. Weiss, and A. G. Dalgleish.** 1985. Slim disease: a new disease in Uganda and its association with HTLV-III infection. *Lancet* **ii:**849–852.
49. **Sewankambo, N. K., J. W. Carswell, R. D. Mugerwa, G. Lloyd, P. Kataaha, R. G. Downing, and S. Lucas.** 1987. HIV infection through normal heterosexual contact in Uganda. *AIDS* **i:**113–116.
50. **Sewankambo, N. K., R. D. Mugerwa, R. Goodgame, J. W. Carswell, A. Moody, G. Lloyd, and S. Lucas.** 1987. Enteropathic AIDS in Uganda. An endoscopic, histological and microbiological study. *AIDS* **1:**9–14.
51. **U.S. Public Health Service.** 1986. PHS plan for prevention and control of AIDS and the AIDS virus. *Public Health Rep.* **101:**341–348.

52. **Van de Perre, P., N. Clumeck, M. Carael, E. Nzabihimana, M. Robert-Guroff, P. DeMol, P. Freyens, J. P. Butzler, R. C. Gallo, and J. B. Kanyamupira.** 1985. Female prostitutes: a risk group for infection with human T-cell lymphotropic virus type III. *Lancet* **ii:**524–526.
53. **Van de Perre, P., D. Munyambuga, G. Zissis, J. P. Butlzer, D. Nzaramba, and N. Clumeck.** 1985. Antibody to HTLV-III in blood donors in Central Africa. *Lancet* **i:**336–337.
54. **Vieira, J., E. Frank, T. J. Spira, and S. H. Landesman.** 1983. Acquired immune deficiency in Haitians: opportunistic infections in previously healthy Haitian immigrants. *N. Engl. J. Med.* **308:**125–129.
55. **Vogt, M. W., B. J. Witt, D. E. Craven, and M. S. Hirsch.** 1986. Isolation of HTLV-III/LAV from cervical secretions of women at risk for AIDS. *Lancet* **i:**525–527.
56. **Weiss, S. H., D. L. Mann, C. Murray, and M. Popovic.** 1985. HLA-DR antibodies and HTLV-III antibody ELISA testing. *Lancet* **ii:**157–159.
57. **Wofsy, C. B., J. B. Cohen, L. B. Hauer, N. S. Padian, B. A. Michaelis, L. A. Evans, and J. A. Levy.** 1986. Isolation of AIDS-associated retrovirus from genital secretions of women with antibodies to the virus. *Lancet* **i:**527–529.
58. **World Health Organization.** 1986. WHO workshop on AIDS in Central Africa. *Weekly Epidemiol. Rep.* **61:**72–73.
59. **World Health Organization.** 1986. Acquired immunodeficiency syndrome (AIDS). Plan of action for control of AIDS in Africa. *Weekly Epidemiol. Rep.* **61:**93–94.
60. **Zuckerman, A. J.** 1986. AIDS and insects. *Br. Med. J.* **292:**1094–1095.

Chapter 16

Monocytes and Human Immunodeficiency Virus Infection

Hans Wigzell, Birgitta Asjö, Eva-Maria Fenyö, Karin Lundin, Shunji Matsuda, Kenneth Nilsson, Urban Ramstedt, and Magnus Gidlund

INTRODUCTION

The virus causing acquired immunodeficiency syndrome, human immunodeficiency virus (HIV), displays a known tropism for $CD4^+$ cells because the CD4 molecules serve as cellular receptors for the virus (3, 7). CD4 is expressed in high concentrations on a majority of mature T lymphocytes (14) and at lower concentrations on perhaps all cells of the monocyte lineage (13). CD4 is also found on cells of the central nervous system with a characteristic distribution, but the association at the cellular subset level is still somewhat controversial (11). HIV binds to CD4 via its outer envelope protein, gp120 (3, 7, 10). This binding is an essential early step leading to syncytium formation and virus fusion with the cell membrane, where the inner part of the envelope protein, gp41, is presumed to play the part of a fusion protein. The ultimate consequence of viral infection may be the premature death of the cell. Immunohistopathology indicates that monocytes, and especially the germinal center follicular dendritic cells, could constitute a major reservoir for HIV, perhaps exceeding the $CD4^+$ T cells in importance (2).

We have compared the same HIV isolate for its capacity to infect normal human T cells and monocytes or cell lines of T or monocyte origin

Hans Wigzell, Karin Lundin, Shunji Matsuda, Urban Ramstedt, and Magnus Gidlund • Department of Immunology, Karolinska Institute, 10401 Stockholm, Sweden. **Birgitta Asjö, Eva-Maria Fenyö, and Shunji Matsuda** • Department of Virology, Karolinska Institute, 10401 Stockholm, Sweden. **Kenneth Nilsson** • Department of Pathology, University Hospital, 75185 Uppsala, Sweden.

(1). Experiments were carried out to compare various subsets of peripheral blood mononuclear cells for their sensitivity to HIV in short-term cultures. Monocytes with markers indicating different stages of maturation have been analyzed for their sensitivity to infection. Likewise, T lymphocytes obtained from normal blood donors and separated according to size have been screened for susceptibility to HIV infection. Monocytes have been reported to allow the propagation of HIV in a manner distinct from propagation in T cells; in monocytes, virus production occurs for a prolonged period and with different kinetics as compared with HIV replication in T lymphocytes (1, 4). It is not known whether this represents a difference in the physiology of T cells and monocytes or whether it could be caused by differences in CD4 expression in the two cell types. We have used clones from a human monocytoid cell line with different profiles of surface markers and CD4 density to evaluate the HIV infection characteristics of the various clones.

Finally, some preliminary experiments have been performed to test the impact of anti-HIV antibodies on the growth of HIV in human monocytes or peripheral blood lymphocytes. The rationale behind these studies is the well-known phenomenon of viral enhancement: with several viruses, antibodies at certain concentrations and of certain classes and specificities may enhance virus proliferation in macrophages (5).

SENSITIVITY OF NORMAL T CELLS AND MONOCYTES FROM PERIPHERAL BLOOD TO HIV INFECTION

HIV is known to be able to infect T cells and monocytes, although significant differences are known to exist not only between different virus isolates but also according to the physiological status of the cells (4, 12). Infection with HIV was carried out by using the human T-cell lymphotropic virus type IIIB isolate with reverse transcriptase activity of about 1.5×10^5 cpm/10^6 cells (1). We first wanted to explore the ability of human T-cell lymphotropic virus type IIIB to infect normal T lymphocytes fractionated according to size (and thus indicating a difference in activation) and to compare this with infection of monocytes with different morphological or functional phenotypes. The infection was carried out in the absence of interleukin-2 and Polybrene to more closely mimic natural situations. T lymphocytes were separated from peripheral blood by using sheep erythrocyte (SRBC) rosetting and Percoll gradient centrifugation; established protocols were followed (16). Monocytes were prepared either by adherence behavior on plastic petri dishes or by first depleting T cells by SRBC rosetting followed by sorting into $M3^+$ or $M3^-$

Table 1. Subsets of Human Blood Mononuclear Cells and Sensitivity to Infection with a Given HIV Isolate

Cell subset	% Fluorescent cells with marker:					ConA[a] (cpm)	HIV-RT[b] (cpm)
	M3	CD3	CD4	CD8	HLA-DR		
Unseparated	8	75	50	28	21	234,040	13,500
Large T	<1	96	59	54	<1	7,420	18,300
Medium T	<1	98	70	48	2	6,436	6,800
Small T	<1	99	74	53	<1	420	8,700
Adherent	40	28	ND[c]	ND	ND	47,620	44,900
$M3^+$	>99	ND	ND	ND	ND	1,440	12,900
$M3^-$	2	ND	ND	ND	ND	49,856	11,900

[a] [^{3}H]thymidine uptake with concanavalin A (ConA; 5 μg/ml given on day 3). The background was not subtracted; its average value is 1,372 without concanavalin A.
[b] HIV-RT, Reverse transcriptase activity 5 days after culture when using human T-cell lymphotropic virus type IIIB without Polybrene or interleukin-2. Results shown are the mean of three culture values.
[c] ND, Not done owing to low cell numbers.

monocytes with OKM3 antibodies and a fluorescence-activated cell sorter.

Table 1 shows the results obtained from such experiments. Concanavalin A stimulation was included as a marker for the presence of both monocytes and T cells, a dual requirement for mitogenesis to occur with this lectin. In the absence of interleukin-2, large preactivated T cells will support HIV infection more readily than medium or small T lymphocytes. These data indicate that activation of T cells is a normal prerequisite for HIV infection in the adherent population. This population consists of a mixture of monocytes and adherent large T cells as indicated by surface markers and concanavalin A responsiveness. The presence of the M3 marker, which is normally considered a marker of more mature monocytes, is strongly linked to the phagocytic behavior of the monocytes. However, both $M3^+$ and $M3^-$ monocytes were readily infected by HIV to a very similar degree. The present data would therefore indicate that under conditions excluding interleukin-2, the sensitivity of monocytes to HIV infection may exceed that of the average T lymphocyte.

In a subsequent series of studies, clones of the human monocytoid leukemia cell line U937 were analyzed for susceptibility to HIV (1). When such clones were characterized by phagocytic behavior, distinct and stable differences in phagocytic activity were noted. When clones with phagocytic and nonphagocytic behavior were analyzed for relative susceptibility to HIV, no correlation was found (Table 2). This study included an additional experiment with $M3^+$ or $M3^-$ peripheral blood monocytes as controls. We could thus firmly conclude that phagocytic

Table 2. Lack of Correlation between Phagocytic and Nonphagocytic Capacity in Human Monocytes and Susceptibility to HIV in the Absence of Anti-HIV Antibodies

Cells	% Phago-cytosis[a]	% ADCC[b]	M3 marker[c]	ConA response[d]	HIV prolif-eration[e]
Blood monocytes					
M3$^+$	11.5 (0.8)	8	>99%	1	12.4 (9.4–19.0)
M3$^-$	1.0 (1.0)	45	3%	26	11.9 (9.3–14.3)
U937 subclones					
Parental line	0	71	0%		0.7
Clone 1	31	0	100%		0.9
Clone 2	25	2	100%		223.5
Clone 16	0	81	0%		729.6

[a] % Phagocytosis, Percent uptake of ^{51}Cr-SRBC coated with IgG–anti-SRBC. Numbers in parentheses represent data obtained without antibodies.
[b] % ADCC, Percent specific lysis of ^{51}Cr-SRBC coated with IgG–anti-SRBC.
[c] M3 marker, Percent positive cells when using the monoclonal antibody OKM3.
[d] Concanavalin A (ConA) response, Stimulation index, i.e., response in thymidine uptake compared with background values.
[e] HIV proliferation, Reverse transcriptase values at day 7 (for monocytes) or day 10 (for U937). Numbers in parentheses show the range of results from four different cultures.

behavior in the absence of anti-HIV antibodies does not influence the sensitivity of human monocytes to HIV infection.

STRIKING POSITIVE CORRELATION BETWEEN CD4 DENSITY AND HIV SENSITIVITY IN U937 CLONES

Although variability in M3 and phagocytic behavior failed to correlate with HIV sensitivity, the CD4 molecule is known to play a decisive role in HIV infection of cells. Although T lymphocytes will frequently rapidly undergo cytopathic events and cell death after infection with HIV in vitro, U937 cells, as representatives of pure cells of the human monocyte family, usually allow a more prolonged production of HIV over time in vitro (1). In our studies with the U937 parental line and subclones, we observed a very great variability among clones in relation to sensitivity to HIV (Table 3). A clear-cut correlation was observed between CD4 density on the target cells and rapid infection, severe cytopathic effects, and cell death. In our experiments, clone 16 of U937, with a CD4 density approaching that present on CD4$^+$ T cells, behaved more like T cells with regard to sensitivity to HIV contact in vitro. Other clones and the parental line, which have CD4 concentrations more comparable to those of freshly obtained monocytes, were more sluggish in developing cytopathic effects

Table 3. Phenotype and HIV Sensitivity of U937 Parental Line and Clones

Cell	% Highly positive cells with:			Lysosomal enzyme level	Day postinfection when HIV sensitivity[a] detected
	HLA-DR	CD4	FcR		
Parental	10–20	<10	35	Low	48
Clone 1	>90	50–60	100	High	20
Clone 2	>90	60–70	100	High	10
Clone 16	10–20	>95	15	Low	6

[a] HIV sensitivity indicated by significant reverse transcriptase and/or cytopathic effects.

and could support productive HIV infection for prolonged periods. Our data would therefore support the view that a major factor in determining the fate of an HIV-infected cell is the density of CD4 on its surface rather than whether it happens to be of the T-cell or monocyte lineage.

UPTAKE OF HIV gp120 ON CD4$^+$ AND CD4$^-$ CELLS: IMPACT OF ANTIBODIES AGAINST gp120

For the following studies, gp120 from HIV (human T-cell lymphotropic virus type IIIB) was purified and labeled with ^{125}I by following protocols previously described (1). Binding of gp120 to CD4 is a major factor in the selective cellular tropism of HIV. There exist additional possibilities of uptake or binding of virus and virus proteins, for instance, by the combination of viral antigens with antibodies in the presence or absence of complement. Table 4 shows the results of using ^{125}I-gp120 and its binding to CD4$^+$ or CD4$^-$ cells in the presence or absence of human antibodies from HIV-seropositive individuals. The uptake of gp120 is influenced in the absence of antibodies by one single parameter: the presence or absence of CD4 on the cells tested (Table 4). However, when antibodies against gp120 were included in the assay, binding to the CD4$^+$ cells was inhibited. In fact, this was repeatedly found with samples from every HIV-seropositive individual analyzed to date. In contrast, with certain cells, an increase in the uptake of HIV gp120 was observed, namely, in cells with known expression of the Fc receptor for immunoglobulin G (IgG) on their surfaces. For instance, in our experiment K562 cells displayed a pronounced increase in the uptake of gp120 in the presence of antibodies, whereas insignificant binding occurred in the absence of antibodies.

Since the binding of immunocomplexes to Fc receptors for many cell types (15) is heavily influenced by competing monomeric immunoglobulin, the influence of carrying out the binding tests in the presence of

Table 4. Uptake of HIV gp120 on CD4$^+$ or CD4$^-$ Cells as Measured by Impact of Antibodies against gp120 and IgG Levels[a]

Serum donor	Serum dilution in PBS[b]	Uptake of ^{125}I-gp120 (cpm)[c]	
		CD4$^+$ FcR$^-$ cells (Jurcat)	CD4$^-$ FcR$^+$ cells (K562)
Expt A			
HIV$^+$	1:5	224	631
HIV$^+$	1:50	307	1,623
HIV$^-$	1:5	1,038	348
HIV$^-$	1:50	1,103	451
Expt B			
HIV$^+$	1:250	706	10,078
HIV$^+$	1:250 plus 10 mg of IgG/ml	698	1,759
HIV$^+$	1:250 plus 3 mg of IgG/ml	701	3,167
HIV$^-$	1:250	2,305	987
HIV$^-$	1:250 plus 10 mg of IgG/ml	1,943	942
HIV$^-$	1:250 plus 3 mg of IgG/ml	2,137	978

[a] Two experiments were done. Experiment A was done to show evidence of prozone effects of anti-HIV antibodies. Experiment B was done to show that increases in normal IgG levels only partially block uptake on FcR$^+$ cells. The same HIV$^+$ donor serum was used in both experiments.
[b] PBS, Phosphate-buffered saline.
[c] Nonspecific uptake was not subtracted. Underlined numbers show a significantly increased uptake.

physiological concentrations of normal human IgG was analyzed. This led to a significant reduction in the cellular binding of HIV gp120 as compared with the result of allowing the antibody reaction to occur in more dilute concentrations of IgG (Table 4). However, even at normal physiological concentrations of immunoglobulin, a highly significant increase in binding to cellular Fc was observed when serum samples from HIV-seropositive individuals were used.

IMPACT OF ANTI-HIV ANTIBODIES ON HIV REPLICATION

HIV antibodies from HIV-seropositive individuals frequently display inhibitory activity against HIV in vitro; that is, they exhibit neutralizing activity, albeit at low titers. We next analyzed the capacity of antisera from HIV-seropositive donors to influence HIV replication in Fc receptor-positive or receptor-negative cells. Some of these cells were CD4$^+$, and others were CD4$^-$.

We first analyzed whether Fc receptor-positive, CD4-negative cells which do not display phagocytic activities could be infected by HIV in the presence of anti-HIV antibodies. Table 5 shows examples of two such studies with K562 and Raji cells. Not shown are data in parallel demon-

Table 5. Preliminary Studies on the Impact of HIV-Seropositive Sera on HIV Replication in Various Cell Types

Cell type[a]	Virus dose	HIV serum dilution in 10% NHS[b]	HIV-RT and CPE on day[c]:							
			4	6	7	11	13	14	17	21
PBL	1	10% NHS		11.5			4.0			
PBL	1	10% HIV$^+$ serum		11.0			4.6			
PBL	1	1% HIV$^+$ in 10% NHS		11.4			9.6			
PBL	1	0.1% HIV$^+$ in 10% NHS		59.1			All dead			
PBL	1:10	10% NHS		<1			4.4			
PBL	1:10	10% HIV$^+$ serum		<1			7.7			
PBL	1:10	1% HIV$^+$ in 10% NHS		<1			21.6			
PBL	1:10	0.1% HIV$^+$ in 10% NHS		1.4			All dead			
K562	1	10% NHS		2.2			<1			
K562	1	10% HIV$^+$ serum		<1			<1			
K562	1	1% HIV$^+$ in 10% NHS		1.1			<1			
K562	1	0.1% HIV$^+$ in 10% NHS		1.4			<1			
Raji	1	10% NHS		2.8			<1			
Raji	1	10% HIV$^+$ serum		4.3			3.1			
Raji	1	1% HIV$^+$ in 10% NHS		3.1			1.0			
Raji	1	0.1% HIV$^+$ in 10% NHS		5.2			1.4			
U937, clone 2		10% NHS	<1, −		2, ±	21, +		16, ++	34, ++	45, ++
U937, clone 2		1:5	<1, −		<1, ±	<1, +		<1, ±	1, −	<1, −
U937, clone 2		1:25	<1, −		<1, ±	10, ±		18, ++	41, ++	65, ++
U937, clone 2		1:125	<1, −		2, +	15, +		12, ++	131, ++	109, ++
U937, clone 2		1:615	<1, ±		3, +	20, +		15, ++	90, ++	140, ++
U937, clone 2		1:3,125	<1, ±		9, +	25, +		16, ++	56, ++	106, ++
U937, clone 2		1:15,625	<1, −		6, +	14, +		4, ++	98, ++	89, ++

[a] Both K562 and Raji cell lines are FcR$^+$ CD4$^-$ nonphagocytic cells.
[b] NHS, Normal human serum.
[c] HIV-RT, Mean of triplicate reverse transcriptase values at indicated days after infection (10^3 cpm per ml of culture). CPE, cytopathic effects read in a double-blind manner: −, no abnormal changes; ±, few giant cells; +, many giant cells and some balloons; ++, most cells form clumps, many giant cells, many

strating that the serum concentrations used indeed induce a highly significant enhancement in the binding of HIV gp120 to the cells, similar to the results shown in Table 4. No infection of K562 or Raji cells could be induced by any dilution of the sera used at any date studied. Fc receptors on nonphagocytic cells therefore do not seem to function to allow HIV to replicate in $CD4^-$ cells. We subsequently attempted to induce HIV-enhanced proliferation in $CD4^+$ or $CD4^-$ phagocytic cells by using the same human anti-HIV sera at different concentrations. Our attempts to produce stable $CD4^-$ mutants from our U937 phagocytic clones have so far failed, and so we were unable to use any human phagocytic cells in this assay. However, murine cells have been reported to be able to support the growth of HIV if transfected (9). We accordingly used murine macrophages in these tests as $CD4^-$ cells. The murine CD4 analog, L3T4, does not in itself display measurable binding to HIV gp120. The results of using murine macrophages were all negative; that is, over a 4-week period no concentration of HIV antibodies tested resulted in significant increases in reverse transcriptase activity or in detectable cytopathic effects.

In contrast, the results obtained when using human monocytes or peripheral blood lymphocytes, although to be regarded as preliminary, suggest that serum samples from HIV-seropositive individuals may contain enhancing antibodies, allowing more rapid replication of HIV. Table 5 shows results of using either peripheral blood lymphocytes from normal donors or U937 cells as target cells for HIV infection. High-titer seropositive serum samples were diluted in normal human serum and analyzed for their effect on HIV replication. The serum samples from HIV-seropositive individuals allowed more rapid and efficient replication of HIV, as indicated by reverse transcriptase activity and cytopathic effects, whereas higher concentrations of serum had an inhibitory action. Serum samples from certain HIV-seropositive individuals appear, however, to have this enhancing capacity. We would therefore conclude that serum samples from HIV-seropositive individuals may contain elements allowing more efficient viral replication, but that the nature and relevance of these preliminary findings must be refined and extended.

CONCLUSIONS

HIV displays its selective tropism as a result of the high affinity of the envelope protein gp120 for the cell surface glycoprotein CD4. It is likely that CD4, besides serving as a point of attachment and route of entry for HIV, may have additional functions in relation to HIV infections. Reactions with CD4 are normally essential when T lymphocytes make primary contact with antigens in the context of HLA-D molecules (14), with the result frequently being T-cell activation. Since activation of T cells is known to be a feature in assisting the proliferation of HIV, CD4 may accordingly play a similar dual role when HIV reacts with it. Whether CD4 on monocytes has a similar activating role as a signal transducer is at present unknown.

In infected individuals, T lymphocytes expressing virus infection are very rare. In contrast, in the enlarged lymph nodes, the cells most frequently found to express HIV are the follicular dendritic cells (2). Likewise, in the central nervous system, HIV production and presence are intimately linked to cells with monocytoid features (4, 8). Langerhans cells in the skin are also known to be able to harbor HIV in infected individuals. In vitro, monocyte cell lines such as U937 have been found to be excellent producers of HIV for prolonged periods after infection (1). It is therefore plausible that a major part of HIV proliferation in vivo occurs within the monocyte family of cells. One reason for this may reside at the level of CD4 expression, where the average $CD4^+$ T lymphocyte may express 10-times-higher amounts of CD4 than the average $CD4^+$ monocyte does (13). As described above and previously (1), this variation in receptor density has a profound consequence for the survival of the infected cell, with a striking, positive correlation existing between CD4 density and rapid virus proliferation and cell death. In fact, when U937 clones having the same density of CD4 as the average $CD4^+$ T cell were used, such monocytoid cells also died rapidly, like T cells.

In infected individuals, the most common cell expressing HIV may therefore be a monocyte-derived cell rather than a T lymphocyte. If the ordinary route of clinical transmission of HIV via sexual intercourse occurs predominantly via cell-bound virus and not via free virus particles (which is an assumption to be proven or disproven), it is likely that monocytes are more relevant than T lymphocytes to this process. It is not difficult to assume, then, that venereal diseases associated with enhanced numbers of mononuclear cells, predominantly monocytes, in the secretions may automatically enhance the spread of the virus, both in the recipient and in the donor of HIV. Treatment of venereal disease, in fact, may be the most cost-beneficial way of reducing the rate of transmission of HIV at present.

In infected individuals a continuous battle occurs for various periods until, in most individuals, acquired immunodeficiency syndrome and other diseases finally appear. In this battle, humoral and cellular immune reactions play hitherto poorly defined roles in infection. It is not in fact known which aspect(s) of the immune reactions against HIV is good, that is, inhibits the virus, or, alternatively, which reaction(s) may allow the virus to grow at an enhanced rate. It is well known that under certain conditions, humoral antibodies against viruses which can grow in monocytes allow that virus to grow faster (5). In this report we present some very preliminary findings showing that serum samples from HIV-seropositive individuals may allow a more rapid proliferation of HIV in vitro under certain defined conditions.

The data, as they stand, have not delineated the molecules responsible for this effect to be HIV-specific antibodies, and obviously much more work must be done to address this problem. Low titers of HIV envelope antibodies were detected in chimpanzees which had undergone attempted HIV vaccination followed by challenge with live virus (6). Although no or only weak protective reactions may be present in these animals, no evidence of the opposite effect, i.e., viral enhancement, was observed.

In sum, there exists at present no evidence to suggest that viral enhancement is relevant in the spread of HIV in vivo, although preliminary in vitro data indicate the need for further studies. Nevertheless, the fact that HIV infection may occur in virtually all monocyte-derived cell types in addition to the $CD4^+$ T cells has had a profound impact on our understanding of the immunopathology of the disease. It also has significant implications for drug therapy, since monocytes and T lymphocytes are known to have significant differences in metabolism and also handle HIV in quantitatively different manners.

Acknowledgments. This work was supported by the Swedish Medical Research Council, by the U.S. Army Medical Research Acquisition Activity, Bethesda, Md., and by the Karolinska Institute, Stockholm, Sweden.

Literature Cited

1. **Asjö, B., I. Ivhed, M. Gidlund, S. Fuerstenberg, E. M. Fenyö, K. Nilsson, and H. Wigzell.** 1987. Susceptibility to infection by the human immunodeficiency virus (HIV) correlates with T4 expression in a parental monocytoid cell line and its subclones. *Virology* **157**:359–365.
2. **Biberfeldt, P., K. J. Chyat, L. M. Marcel, G. Biberfeldt, R. C. Gallo, and M. E. Harper.** 1986. HTLV-III expression in infected lymph nodes and relevance to pathogenesis of lymphadenopathy. *Am. J. Pathol.* **125**:436–442.
3. **Dalgleish, A. G., P. C. L. Beverley, P. R. Clapham, D. H. Crawford, M. F. Greaves, and R. A. Weiss.** 1984. The CD4(T4) antigen is an essential component of the receptor for the AIDS retrovirus. *Nature* (London) **312**:763–766.
4. **Gartner, S., P. Markovits, D. M. Marleovitz, M. H. Kaplan, R. C. Gallo, and M. Popovic.** 1986. The role of mononuclear phagocytes in HTLV-III/LAV infection. *Science* **223**:215–219.
5. **Halstead, S. B.** 1982. Immune enhancement of viral infection. *Prog. Allergy* **31**:301–364.
6. **Hu, S. L., P. Fultz, H. M. McClure, J. W. Eichberg, E. K. Thomas, J. Zarling, M. C.**

Singhal, S. G. Kosowski, R. B. Swenson, D. C. Anderson, and G. Todaro. 1987. Effect of immunization with a vaccinia-HIV env recombinant on HIV infection of chimpanzees. *Nature* (London) **328:**721–723.

7. **Klatzman, D., E. Champagne, S. Chamaret, J. Gruest, D. Guetard, T. Hercend, J.-C. Gluckman, and L. Montagnier.** 1984. T-lymphocyte T4 molecule behaves as the receptor for the human retrovirus LAV. *Nature* (London) **312:**767–768.
8. **Koenig, S., H. E. Gendelman, J. M. Orenstein, M. C. DalCanto, G. H. Pezeshkpour, M. Yungbluth, F. Janotta, A. Aksamit, M. A. Martin, and A. S. Fauci.** 1986. Detection of AIDS virus in macrophages in brain tissue from AIDS patients with encephalopathy. *Science* **233:**1089–1093.
9. **Levy, J. A., C. Cheng-Mayer, D. Dina, and P. A. Luciw.** 1986. AIDS retrovirus (ARV-2) clone replicates in transfected human and animal fibroblast. *Science* **232:**998–1001.
10. **Lundin, K., A. Nygren, L. A. Arthur, W. G. Robey, B. Morein, U. Ramstedt, M. Gidlund, and H. Wigzell.** 1987. A specific assay measuring binding of 125-I-Gp120 from HIV to $T4^+/CD4^+$ cells. *J. Immunol. Methods* **97:**93–100.
11. **Maddon, P. J., A. G. Dalgleish, J. S. McDougal, P. R. Clapham, R. A. Weiss, and R. Axel.** 1986. The T4 gene encodes the AIDS virus receptor and is expressed in the immune system and in the brain. *Cell* **47:**333–348.
12. **McDougal, S. J., A. Mawle, S. P. Cort, J. K. A. Nicholson, G. D. Cross, J. A. Scheppler-Campbell, D. Hicks, and J. Sligh.** 1985. Cellular tropism of the human retrovirus HTLV-III/LAV. I. Role of T cell activation and expression of the T4 antigen. *J. Immunol.* **135:**3151–3162.
13. **Moscicki, R. A., E. P. Amento, S. M. Krane, J. T. Kunich, and R. R. Colvin.** 1983. Modulation of surface antigens of a human monocyte cell line, U937, during incubation with T lymphocyte-conditioned medium: detection of T4 antigen and its presence on normal blood monocytes. *J. Immunol.* **131:**743–748.
14. **Reinherz, E., and S. F. Schlossman.** 1980. The differentiation and function of human T cells. *Cell* **19:**821–827.
15. **Wigzell, H.** 1984. Antibody-dependent cell-mediated cytotoxicity, p. 1–13. *In 9th International Convocation on Immunology*. S. Karger, Basel.
16. **Wigzell, H., and U. Ramstedt.** 1986. Natural killer cells, p. 60.1–60.11. *In* D. M. Weir (ed.), *Handbook of Experimental Immunology*. Blackwell Scientific Publications, New York.

Chapter 17

CD4: the Human Immunodeficiency Virus Receptor

J. Steven McDougal

INTRODUCTION

The human immunodeficiency virus (HIV), also known as human T-cell lymphotropic virus type III (HTLV-III), lymphadenopathy-associated virus, or acquired immunodeficiency syndrome-related virus (2, 6, 16, 26), infects a subpopulation of thymus-derived T cells. This T-cell subpopulation is defined functionally as helper-inducer T cells or phenotypically as $CD4^+$ T cells because it performs critical recognition and induction functions in the immune response and expresses the cell surface protein CD4 (or T4). Clinically, infection results in gradual helper-inducer T-cell depletion, progressive immune unresponsiveness, and increasing susceptibility to opportunistic infections and malignancies—manifestations of the acquired immunodeficiency syndrome. The other major clinical consequence of HIV infection, subacute encephalopathy, is not necessarily related to the degree of $CD4^+$ T-cell depletion; rather, neurologic disease results from an apparent initial infection of monocyte-derived cells in the nervous system (15, 24, 47). Nevertheless, the two clinical sequelae share a common pathogenic mechanism: HIV has evolved an envelope with a diabolic avidity for the CD4 molecule, which is present at highest density on helper-inducer T cells but is also found at a relatively lower density on monocytes and monocyte-derived cells.

In this review, experimental evidence that the CD4 molecule functions as a receptor for HIV is outlined as a point of departure for examining relevant HIV-CD4 cell interactions. There are five types of experiments that, when taken together, provide compelling evidence that

---

J. Steven McDougal • Immunology Branch, Division of Host Factors, Center for Infectious Diseases, Centers for Disease Control, Atlanta, Georgia 30333.

CD4 functions as a receptor for HIV, specifically for the outer envelope glycoprotein gp110 (or gp120).

EXPERIMENTAL EVIDENCE FOR RECEPTOR FUNCTION OF CD4

First, the preferential infectivity or tropism of HIV for CD4-positive cells does not really constitute proof but would lead to the conclusion that CD4 is the HIV receptor. Although virtually all mammalian cells tested will replicate infectious HIV when HIV is introduced into the cells as integrated proviral DNA by transfection (12, 25, 42), infection with native, exogenous virus has been reproducibly and consistently demonstrated only in CD4-positive cells. Klatzmann et al. (22), examining peripheral blood cell subpopulations, first demonstrated the preferential replication of HIV in $CD4^+$ T cells and the selective isolation of HIV from $CD4^+$ T cells of infected patients. Numerous investigations with separated-cell populations, whole-cell populations monitored for phenotypic depletion, and cell lines have confirmed these findings and extended them to include cells with lesser degrees of CD4 expression, such as monocytes and some transformed B-cell lines (7, 8, 11, 16–18, 22, 27, 34, 36, 37, 40).

In cultures of normal human lymphocytes (stimulated with phytohemagglutinin, cultured in the presence of interleukin-2, and injected with HIV), cytoplasmic virus (detected by immunofluorescence) appears and then disappears in a proportion (1 to 10%) of cells (many of which form syncytia); this is followed by release of virus detected by assays for particulate reverse transcriptase activity, viral antigens, and infectivity (Fig. 1). Infection is associated with loss of the CD4 molecules from the surfaces of infected cells and, ultimately, loss of $CD4^+$ T cells from the culture (20, 34). Residual CD4-negative cells are not susceptible to a second round of infection (22, 34). Thus, in vitro, one can reproduce the essential immunologic feature of acquired immunodeficiency virus in a relatively short time.

It has been suggested, on the basis of in vitro and in vivo observations (7, 10, 22, 38, 39), that HIV preferentially infects a subset of $CD4^+$ T cells. However, virtually all $CD4^+$ T cells are potentially susceptible to HIV, because under optimal conditions of HIV exposure, activation, and culture, nearly all $CD4^+$ T cells are ultimately depleted, but there are clear differences in the amount of HIV replication and the rate of cell death between so-called resting $CD4^+$ T cells and activated or proliferating $CD4^+$ T cells. Activated $CD4^+$ T cells replicate more virus and die more quickly than nonactivated $CD4^+$ T cells (34, 48), and activation of T cells has been considered a reasonable mechanism by which postulated

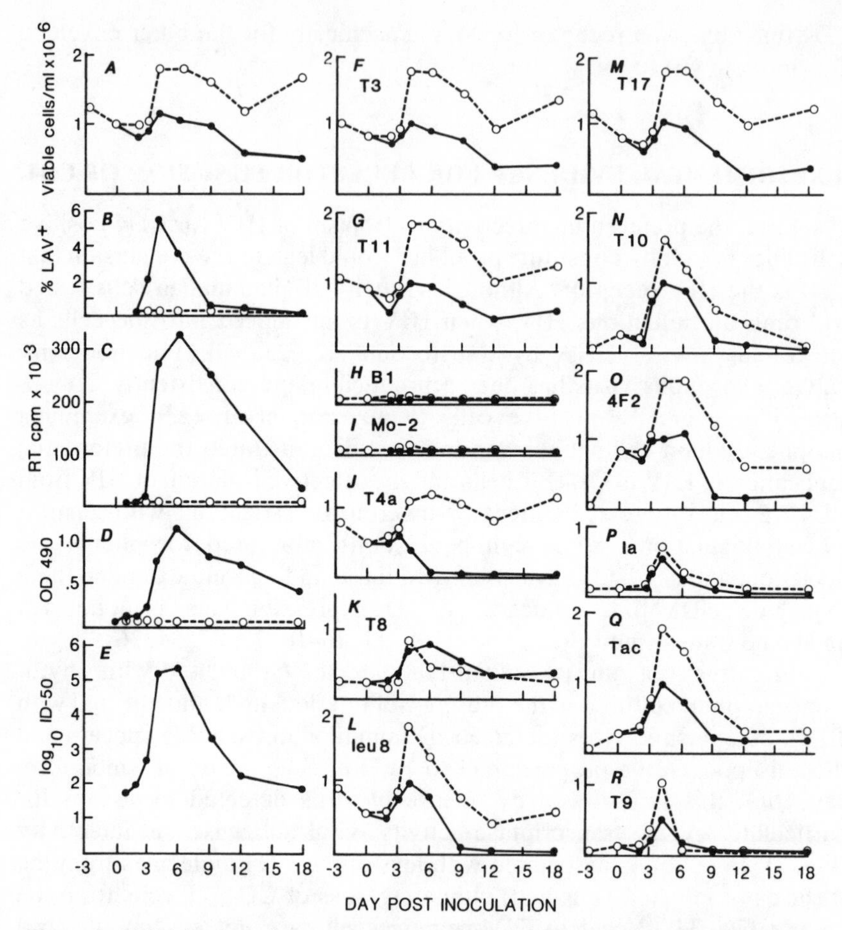

Figure 1. HIV infection of normal human lymphocytes. Phytohemagglutinin-stimulated lymphocytes were divided; half were inoculated with HIV (●) and the other half served as controls (○). Viable cell counts (A), cytoplasmic virus (B), supernatant reverse transcriptase (RT) activity (C), supernatant viral antigen (D), and infectivity titer of supernatant (E) were monitored over time, as were cell surface phenotypes (F to R). Reprinted from the *Journal of Immunology* (34) with permission from the American Association of Immunologists.

cofactors might act to influence the rate at which progressive $CD4^+$ T-cell depletion and immunodeficiency occur in infected persons.

In the second type of experiment, reported by Popovic et al. (M. Popovic, R. C. Gallo, and D. L. Mann, *Clin. Res.* **33:**560A, 1984), Klatzmann et al. (23), and Dalgleish et al. (7), the addition of certain anti-CD4 monoclonal antibodies (MAbs) to cultures of T cells prevents or substantially reduces the infectivity and replication of an inoculum of HIV. The inhibition is greater if CD4 MAb is added to cells before rather than after the virus and is maintained in media throughout the culture period; it is not increased by combining CD4 MAbs with MAbs directed at other cell surface components; and it is not due to effects of CD4 MAb on cellular proliferation or activation (34). In addition, Dalgleish et al. (7) were able to block penetration of a pseudotype virus formed between HIV and vesicular stomatitis virus and to inhibit fusion (syncytia) between HIV-infected and uninfected CD4 cells, providing more direct evidence that CD4 MAbs act early in infection and at the cell membrane.

In the third type of experiment, indirect immunofluorescence, HIV binds to the surfaces of $CD4^{+}$ cells but not to CD4-negative cells. The proportion of cells that bind HIV is similar to the proportion of cells that express CD4 whether one examines normal lymphocyte populations or CD4-enriched or -depleted populations or whether the cells are freshly obtained or examined after activation with mitogens (34). Cells with a high density of CD4, such as T cells, bind more HIV and register relatively higher fluorescence intensity than do cells with low-density CD4 expression, such as monocytes (Fig. 2) (34, 37). Binding is saturable, specifically inhibited by certain anti-CD4 MAbs (but not by a reasonably comprehensive panel of MAbs directed at other cell surface molecules), and reciprocal in that HIV and anti-CD4 MAbs inhibit binding of each other (34, 37).

Not all anti-CD4 MAbs inhibit HIV binding to CD4 cells equivalently. A panel of CD4 MAbs that reacted at similar titers in direct binding to CD4 differed when titrated for inhibition of HIV binding (Fig. 3). OKT4A, OKT4D, OKT4F, and Leu3a MAbs were potent inhibitors of HIV binding, whereas binding was not inhibited by OKT4 and OKT4C MAbs and was only partially inhibited by OKT4B and OKT4E (35). Similar results were reported by Sattentau et al. (43) for inhibition of syncytia. The CD4 epitopes detected by these MAbs map as clusters on the molecule (14), and the pattern of inhibition suggests that HIV binds to a restricted portion of CD4, most probably in the amino end of the molecule.

We considered the possibility that the antigen-combining site (or idiotope) of an anti-CD4 MAb shares some structural homology with the CD4-binding site of HIV—particularly if it binds to precisely the same epitope of CD4. If so, antiserum raised to the idiotope of an anti-CD4 MAb might react with the CD4-binding site of HIV. Accordingly, we prepared anti-idiotypic sera in rabbits to each of the four CD4 MAbs that inhibited HIV binding. The serum samples were anti-idiotypic by the criteria that they reacted uniquely with the respective CD4 MAbs used as

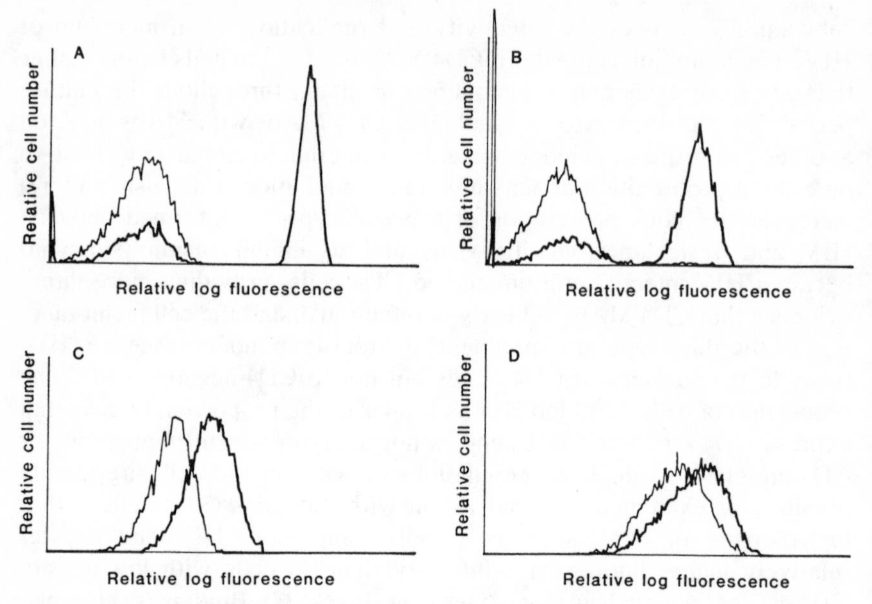

Figure 2. CD4 expression and HIV binding by lymphocytes and monocytes. Lymphocyte (A and B) or monocyte (C and D) reactivity with MAb OKT4A (——, panels A and C) and binding of HIV (——, panels B and D) were determined by indirect immunofluorescence. ——, Results with reagent controls. Reprinted from the *Journal of Immunology* (37) with permission from the American Association of Immunologists.

an immunogen and inhibited their binding to CD4. They did not react with HIV or inhibit HIV binding (35). Two other groups have generated an anti-idiotypic MAb to the Leu3a MAb that reacts with HIV (4, 46). The anti-idiotypic MAb reacts with both Leu3a and OKT4A, a cross-specificity not present in our anti-Leu3a (or anti-OKT4A) reagent, and this may explain the discrepancy. If Leu3a contains a partial or whole internal image of the CD4-binding site of HIV, Leu3a becomes a candidate vaccine if, as an immunogen, it reliably induces an anti-HIV response.

The fourth piece of evidence comes from radioimmunoprecipitation experiments demonstrating bimolecular complexes of the CD4 molecule and the gp110 envelope glycoprotein of HIV (33). $CD4^+$ T cells were surface labeled with ^{125}I, HIV was added, and the cells were washed and lysed. Immunoprecipitation of viral proteins coprecipitated a 57-kilodalton cell surface protein that by comigration, blocking, and preabsorption experiments was identified as the CD4 molecule. Conversely, if HIV is radiolabeled and the CD4 molecule is precipitated (by using an anti-CD4 MAb that does not interfere with HIV binding), an HIV protein

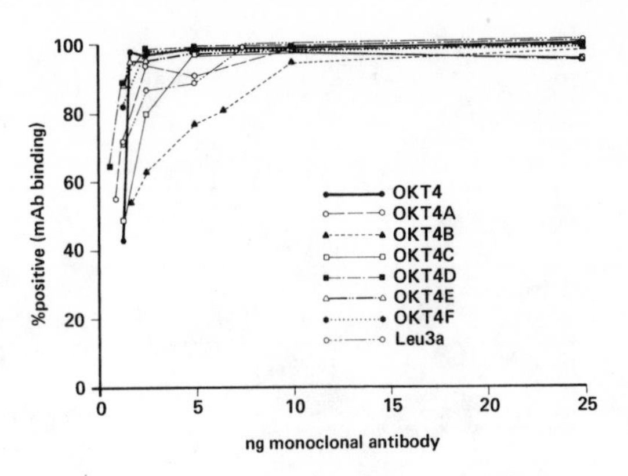

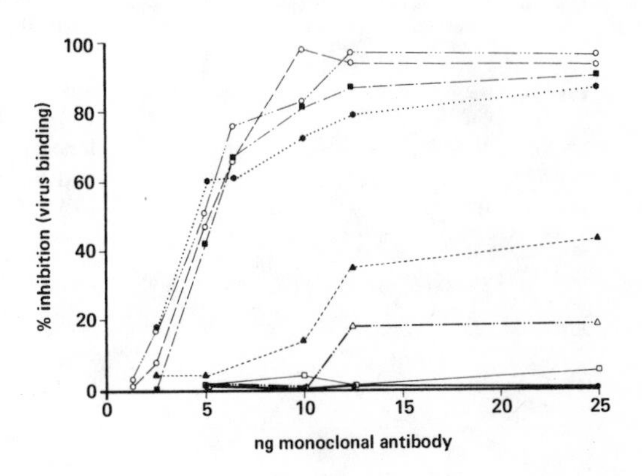

Figure 3. Binding and inhibition of HIV binding by anti-CD4 MAbs. The anti-CD4 MAbs, as indicated, were tested in indirect immunofluorescence for binding to the $CD4^+$ cell line CEM (top) or for inhibition of HIV binding to CEM cells (bottom). Reprinted from the *Journal of Immunology* (35) with permission from the American Association of Immunologists.

of 110 kilodaltons coprecipitates with CD4. Using a bifunctional chemical cross-linker, we could not identify any additional cell surface proteins that bound to HIV (33). Further evidence that gp110 is the principal viral protein that binds to CD4 has been provided by Bolognesi (3) and Lasky et al. (24a), who used highly purified native gp110 and recombinant gp110, respectively.

The preceding experiments were designed to detect the initial binding of gp110 to CD4 on the surfaces of uninfected cells. Recent studies by Hoxie et al. (20) have demonstrated that $CD4^+$ cells infected with HIV lose cell surface expression of CD4 and form intracellular complexes of gp110 and CD4. Results of such an experiment are shown in Fig. 4, in which HIV-infected $CD4^+$ cells were internally radiolabeled and lysed at a time when there was no detectable cell surface expression of CD4. Immunoprecipitation with anti-CD4 MAb precipitates the CD4 molecule

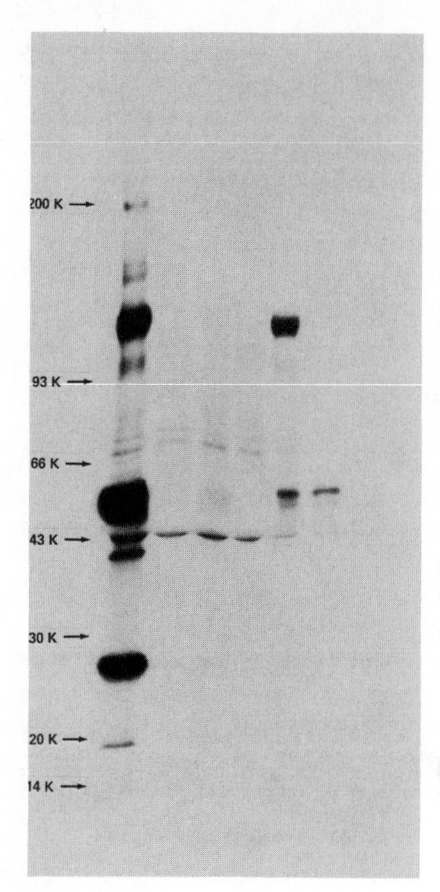

Figure 4. Coprecipitation of gp110 with CD4 in HIV-infected cells. HIV-infected (lanes 1, 3, 5, and 7) or uninfected (lanes 2, 4, 6, and 8) CEM cells were internally labeled with [^{35}S]methionine and [^{35}S]cysteine and lysed. Radioimmunoprecipitations were performed on the lysates. The lysates were reacted with the following antisera, and the precipitates were resolved by electrophoresis (32): lanes 1 and 2, anti-HIV; lanes 3 and 4, nonimmune serum; lanes 5 and 6, MAb OKT4; lanes 7 and 8, control MAb.

and viral gp110. Intracellular complexing of CD4 and gp110 may explain the curious observation that CD4$^+$ cells lose cell surface expression of CD4 after infection and before cell death (20, 34). More importantly, several recent investigations suggest that CD4 expression is a necessary requisite for cell death. One of the earliest, perhaps least elegant, yet very compelling, experiments was that of DeRossi et al. (9), who forced HIV into CD4-negative CD8$^+$ T cells and found that HIV replication was equivalent to that in CD4 cells, but the cells did not die. Asjö et al. (1) have shown a correlation between the degree of CD4 expression before infection and the cytopathic effect. Numerous studies of deletion and site-directed mutants of HIV or transfection studies with cloned HIV genes have failed to implicate any of the HIV regulatory genes in the cytopathic effect. However, Fisher et al. (13) produced a mutant of HIV with a deletion in the *env* gene that was replication competent but not cytopathic, and Sodroski et al. (44) and Lifson et al. (29) found that transfection of the *env* gene into CD4-positive cells results in syncytium formation and cell death. Transfection of the *env* gene into CD4-negative cells does not result in syncytia, although these cells will form syncytia with CD4-positive cells (29, 44). Thus, it seems quite likely that CD4 expression is both a determinant of tropism and a requisite for cell death.

The final piece of evidence comes from experiments involving cell lines transfected with the CD4 gene. If cell lines that ordinarily do not express CD4, do not bind HIV, and cannot be infected with native HIV are transfected with the human CD4 gene, they express CD4, bind HIV, and, in the case of human cell lines, replicate infectious HIV (30). Murine cell lines transfected with CD4 express CD4 and bind virus, but the virus does not replicate. Since murine cell lines replicate HIV when transfected with infectious HIV cDNA clones (12, 25, 42), it is possible that the human CD4 protein is not processed, cycled, or internalized appropriately for HIV infection in murine cells. Recombinant clones linking the extracellular (HIV-binding) domain of human CD4 with the transmembrane and cytoplasmic domains of the murine L3T4 homolog are being constructed to address this possibility. In addition, site-directed or deletion mutation of CD4 is being performed to further define the HIV-binding region of CD4.

Similar approaches are being used to define the binding site on gp110. Lasky et al. (24a) and Sodroski et al. (44) have described amino acid substitutions or deletions in the carboxy end of gp110 that render the molecule nonbinding. Within the 549-residue molecule, these mutations were in a relatively conserved portion of gp110 spanning residues 360 to 475. In a different approach, others have used antiserum directed at various broad regions of gp110 in neutralization assays, and these results also implicate the carboxy end of gp110, although a neutralizing epitope is not necessarily a binding epitope (19, 31, 41).

For reasons outlined below, the binding site on gp110 appears to require a proper three-dimensional orientation or conformation formed by approximating discontiguous amino acid sequences. We treated HIV with the physical, chemical, and enzymatic regimens listed in Table 1 and assessed their effect on HIV binding capacity. Specific details of the manipulations and controls to assure or deny authentic binding to CD4 are given elsewhere (35). Reduction of disulfide bonds with dithiothreitol and alkylation with iodoacetamide followed by dialysis abolished the capacity of HIV to bind. The effect probably occurs at intramolecular disulfide bonds, since there is no evidence that gp110 is covalently bound to other

Table 1. Effect of Physical, Chemical, and Enzymatic Treatment on HIV Envelope Binding to CD4$^+$ Cells

HIV binding site destroyed	HIV binding site preserved
100°C for 10 min	65°C for 30 min
	56°C for 30 min
	8 M urea
	1% Sodium dodecyl sulfate
	95% Alcohol
Dithiothreitol reduction and iodoacetamide alkylation	Iodoacetamide reduction
	Periodate oxidation and borohydride reduction
Trypsin	Neuraminidase
	Endoglycosidase H
	Endoglycosidase F

gp110 molecules or other viral proteins. If the tertiary structure conferred by covalent disulfide bonding is not disrupted, the proper secondary structure required for binding appears to be thermodynamically favored, because treatment with heat or denaturants, followed by their removal, did not affect binding. Proteolysis with trypsin destroyed binding, but derivatization of carbohydrate by periodate oxidation and borohydride reduction or glycolytic digestion did not. Our carbohydrate digestions were not complete, as assessed by reduction in apparent molecular weight, and this may explain differences in results from those of other groups who have found that deglycosylation or carbohydrate-binding lectins interfere with HIV-CD4 interactions (28, 31). Carbohydrate may function to maintain conformation or protect the molecule from degradation, but it seems unlikely (or at least unappealing) that the binding site is composed only of carbohydrate in view of the results with trypsin and because it would require that HIV (which has no glycosylation enzymes of its own) acquire its highly specific CD4-binding capacity from the host cell glycosylation system.

The above results indicate that the binding site requires a proper protein conformation and not simply the proper primary amino acid sequence; thus, it is unlikely that synthetic peptides matching portions of the gp110 molecule would bind to CD4 or inhibit virus binding, unless, of course, they were formulated in the proper tertiary structure. Nevertheless, it is possible that a receptor that apparently binds to a ligand formed by approximating discontiguous amino acid sequences would have some affinity for contiguous sequences that form part of the bond. A panel of 22 peptides corresponding to gp110 sequences spanning most portions of the molecule (Table 2) were tested for inhibition of HIV binding at a $\geq$10:1 ratio by weight of peptide to HIV, which is conservatively estimated to be

Table 2. Synthetic Peptides Containing Sequences of gp110

gp110 residue no.[a]	Sequence		
	NH_2 leader	Encoded sequence	COOH tail
54–83	Ace-Y	CASDAKAYDTEVHNVWATHACVPTDPNPQE	G-COOH
89–108	Ace	VTENFNMWKNDMVEQMHEDI	C-$CONH_2$
113–134	NH_2-Y	DQSLKPCVKLTPLCVSLKCTDL	$CONH_2$
166–190	Ace	ISTSIRGKVQKEYAFFYKLDIIPID	C-$CONH_2$
268–280	Ace	GSLAEEEVVIRSA	C-COOH
308–328	Ace	TRKSIRIQRGPGRAFVTIGKI	C-$CONH_2$
323–336	NH_2	VTIGKIGNMRQAHC	$CONH_2$
345–360	Ace	ATLKQIASKLREQFGN	C-$CONH_2$
362–382	Ace-C	KTIIFKQSSGGDPEIVTHSFN	G-COOH
398–423	Ace	STWSTEGSNNTEGSDTITLPC	COOH
424–450	NH_2	RIKQFINMWQEVGKAMYAPPISGQIRC	$CONH_2$
469–493	Ace	GSEIFRPGGGDMRDNWRSELYKYKV	C-$CONH_2$
494–517	NH_2-C	VKIEPLGVAPTKAKRRVVQREKRA	C-COOH
488–499	Ace	LYKYKVVKIEPL	C-$CONH_2$
252–266	NH_2-K	CTHGIRPVVSTQLLL	$CONH_2$
224–233	NH_2	APAGFAILKC	K-$CONH_2$
210–222	Ace-K	CPKVSFEPIPIHY	$CONH_2$
192–201	Ace	DTTSYTLTSC	E-COOH
147–162	Ace	SSSGEMMMEKGEIKNC	COOH

[a] Residue number designations are from reference 45.

at least a 1,000:1 molar ratio. We also tested peptide preparations that were covalently coupled to bovine serum albumin on the premise that the relative avidity of a low-affinity interaction between putative peptide inhibitor and $CD4^+$ cells would be enhanced by virtue of the potential for multipoint attachment. None of the peptides inhibited HIV binding. This panel was by no means comprehensive, and perhaps some peptide could bind to CD4 as a mimic of gp110. However, in view of the conformation requirements noted above, it seems unlikely that such a peptide would be constructed solely from a single primary sequence. On the other hand, antiserum to the peptides may be sufficiently reactive with native gp110 and may be useful for mapping the portion(s) of gp110 involved in binding. We have prepared antisera to the peptides, and two of them (antipeptides 34 and 36) inhibited HIV binding. Inhibition was not as complete as that obtained with antiserum raised to the whole molecule, and the pattern of inhibition with dilution of antibodies suggests that they inhibit sterically rather than by competing for binding to precisely the same site.

What are the prospects of developing an immunogen that would reliably elicit an antibody response that blocks binding and is protective?

Human anti-HIV serum inevitably contains antibodies that inhibit HIV binding quite potently, and binding inhibition titers correlate roughly with anti-gp110 titers (35). Furthermore, there is an association of lower anti-gp110 titers with disease progression and lower levels of $CD4^+$ cells, suggesting that gp110 antibody is involved in control of infection, although this relationship exists for antibody to a number of other proteins as well (32). The fact that disease progresses despite the presence of antibody does not necessarily mean that antibody would not be protective if induced (by vaccination) before infection (when the viral inoculum or load is presumably less than in established infection). On the premises that the CD4-binding site of gp110 is conserved between strains of HIV (which is likely), is immunogenic, and is accessible to antibody, anti-HIV serum samples should block binding and afford some protection. Unfortunately, there is some reason to be pessimistic about the last two premises. For instance, serum samples from patients infected with HIV apparently do not react with (or rarely react with) the envelope of lymphadenopathy-associated virus type 2 or HTLV-IV (5, 21). If the two viruses have identical CD4-binding sites, one must conclude that the precise site is not immunogenic in humans. Then there is the problem of strain variation in neutralization assays. Although exclusive neutralization of HIV isolates by serum samples from patients from whom the isolates were obtained or the development over time of resistance to neutralization has not been reported, some strains are clearly better indicators of neutralization than others in existent assays, and preferential neutralization of certain strains by individual serum samples has been reported (1, 19, 31, 41, 46). For instance, Matthews and co-workers (31, 41) reported that gp110 or a recombinant peptide derived from the HTLV-IIIB strain of HIV induced antibodies that neutralized the HTLV-IIIB strain but not the HTLV-III RF strain (which differ by about 20% in envelope amino acid sequence). They also demonstrated that human antibodies absorbed and eluted from immobilized gp110 of HTLV-IIIB preferentially neutralized HTLV-IIIB. However, antibodies from the same serum samples that did not adsorb to the gp110 column were broadly neutralizing. Some of this confusion might be resolved by postulating the existence of antibodies that detect conformational determinants on gp110. Such antibodies would not be elicited or absorbed by denatured immunogens or absorbants, respectively. There are differences in antibody binding to reduced versus native envelope (T. Lee, R. Redfield, M. J. Chou, J. Allan, D. Burke, and M. Essex, *Abstr. III Int. Conf. AIDS*, WP.17, p. 113, 1987; J. S. McDougal, J. K. A. Nicholson, G. D. Cross, S. P. Cort, M. S. Kennedy, and A. Mawle, *Abstr. III Int. Conf. AIDS*, TP.108, p. 80, 1987). We have found that purified antibodies reactive with conformational determinants compared with antibodies reactive with denatured determinants, on a weight basis, are much more potent inhibitors of HIV binding (McDougal et al., *Abstr. III Int. Conf. AIDS*). It remains to be determined whether protective immunity can be achieved and whether it will require a multicomponent vaccine (with respect to viral proteins and viral strains). However, research on the structure of the CD4-binding site may define immunogens that elicit protective anti-HIV responses superior to those obtained during natural infection.

Literature Cited

1. **Asjö, B., I. Ivhed, M. Gidlund, S. Fuerstenberg, E. M. Fenyö, K. Nilsson, and H. Wigzell.** 1987. Susceptibility to infection by the human immunodeficiency virus (HIV) correlates with T4 expression in a parental monocytoid cell line and its subclones. *Virology* **157:**359–365.
2. **Barre-Sinoussi, F., J. C. Chermann, F. Rey, M. T. Nugeyre, S. Chamarat, J. Gruest, C. Dauguet, C. Axler-Blin, F. Vezinet-Brun, C. Rouzioux, W. Rozenbaum, and L. Montagnier.** 1983. Isolation of a T-lymphotropic retrovirus from a patient at risk for the acquired immunodeficiency syndrome (AIDS). *Science* **220:**868–871.
3. **Bolognesi, D. P.** 1986. Immunobiological properties of the major envelope glycoprotein of HTLV-III_B (HIV), p. 138–141. *In* M. Girard, G. De The, and L. Vallette (ed.), *Colloque des "Cent Gardes."* Pasteur Vaccins, Paris.
4. **Chanh, T. C., G. R. Dreesman, and R. C. Kennedy.** 1987. Monoclonal anti-idiotypic antibody mimics the CD4 receptor and binds human immunodeficiency virus. *Proc. Natl. Acad. Sci. USA* **84:**3891–3895.
5. **Clavel, F., F. Guetard, F. Brun-Vezinet, S. Chamarat, M. Rey, M. O. Santos-Feureira, A. G. Laurent, C. Dauguet, C. Katlama, C. Rouzioux, D. Klatzmann, J. L. Champalimaud, and L. Montagnier.** 1986. Isolation of a new human retrovirus from West African patients with AIDS. *Science* **233:**343–346.
6. **Coffin, J., A. Haase, J. A. Levy, L. Montagnier, S. Oroszlan, N. Teich, H. Temin, K. Toyoshima, H. Varmus, P. Vogt, and R. Weiss.** 1986. Human immunodeficiency viruses. *Science* **232:**697.
7. **Dalgleish, A. G., P. C. L. Beverley, P. R. Clapham, D. H. Crawford, M. F. Greaves, and R. A. Weiss.** 1985. The CD4 (T4) antigen is an essential component of the receptor for the AIDS retrovirus. *Nature* (London) **312:**763–767.
8. **Dalgleish, A. G., and P. Clapham.** 1985. B cells in the pathogenesis of AIDS. *Immunol. Today* **6:**71.
9. **DeRossi, A., G. Franchini, A. Aldovini, A. Del Mistro, L. Chieco-Bianchi, R. C. Gallo, and F. Wong-Staal.** 1986. Differential response to the cytopathic effects of human T-cell lymphotropic virus type III (HTLV-III) superinfection in $T4^+$ (helper) and $T8^+$ (suppressor) T cell clones transformed by HTLV-I. *Proc. Natl. Acad. Sci. USA* **83:**4297–4301.
10. **Fauci, A. S.** 1984. Immunologic abnormalities in the acquired immunodeficiency syndrome (AIDS). *Clin. Res.* **32:**491–499.
11. **Fauci, A. S., H. Masur, E. P. Gelmann, P. D. Markham, B. H. Hahn, and H. C. Lane.** 1985. The acquired immunodeficiency syndrome: an update. *Ann. Intern. Med.* **102:**800–813.

12. **Fisher, A. G., E. Collalti, L. Ratner, R. C. Gallo, and F. Wong-Staal.** 1985. A molecular clone of HTLV-III with biological activity. *Nature* (London) **316:**262–265.
13. **Fisher, A. G., L. Ratner, H. Mitsuya, L. M. Marselle, M. E. Harper, S. Broder, R. C. Gallo, and F. Wong-Staal.** 1986. Infectious mutants of HTLV-III with changes in the 3′ region and markedly reduced cytopathic effects. *Science* **233:**655–659.
14. **Fuller, T. C., J. E. Trevithick, A. A. Fuller, R. B. Colvin, A. B. Cosimi, and P. C. Kung.** 1984. Antigenic polymorphism of the T4 differentiation antigen expressed on human T helper/inducer lymphocytes. *Hum. Immunol.* **9:**89–102.
15. **Gabuzda, D. H., D. D. Ho, S. M. de la Monte, M. S. Hirsch, T. R. Rota, and R. A. Sobel.** 1986. Immunohistochemical identification of HTLV-III antigen in brains of patients with AIDS. *Ann. Neurol.* **20:**289–295.
16. **Gallo, R. C., S. Z. Salahuddin, M. Popovic, A. M. Shearer, M. Kaplan, B. F. Haynes, T. J. Palker, R. Redfield, J. Oleske, B. Safai, G. White, P. Foster, and P. D. Markham.** 1984. Frequent detection and isolation of cytopathic retroviruses (HTLV-III) from patients with AIDS and at risk for AIDS. *Science* **224:**500–502.
17. **Gartner, S., P. Markovits, D. M. Markovitz, M. H. Kaplan, R. C. Gallo, and M. Popovic.** 1986. The role of mononuclear phagocytes in HTLV-III/LAV infection. *Science* **233:**215–219.
18. **Ho, D. D., T. R. Rota, and M. S. Hirsch.** 1986. Infection of monocyte/macrophages by human T lymphotropic virus type III. *J. Clin. Invest.* **77:**1712–1715.
19. **Ho, D. D., M. G. Sarngadharan, M. S. Hirsch, R. T. Schooley, T. R. Rota, R. C. Kennedy, T. C. Chanh, and V. L. Sato.** 1987. Human immunodeficiency virus neutralizing antibodies recognize several conserved domains on the envelope glycoproteins. *J. Virol.* **61:**2024–2028.
20. **Hoxie, J. A., J. D. Alpers, J. Rackowski, K. Huebner, B. S. Haggarty, A. J. Cedarbaum, and J. C. Reed.** 1986. Alterations in T4 (CD4) protein and mRNA synthesis in cells infected with HIV. *Science* **234:**1123–1127.
21. **Kanki, P. J., F. Barin, S. M'Boup, J. S. Allan, J. C. Romet-Lemonne, R. Marlink, M. F. McLane, T. Lee, B. Arbeille, F. Denis, and M. Essex.** 1986. New human T-lymphotropic retrovirus related to simian-T-lymphotropic virus type III (STLV-III AGM). *Science* **232:**238–243.
22. **Klatzmann, D., F. Barre-Sinoussi, M. T. Nugeyre, C. Dauguet, E. Vilmer, C. Griscelli, F. Brun-Vezinet, C. Rouzioux, J. C. Gluckman, J. C. Chermann, and L. Montagnier.** 1984. Selective tropism of lymphadenopathy associated virus (LAV) for helper-inducer T lymphocytes. *Science* **225:**59–62.
23. **Klatzmann, D., E. Chapagne, S. Chamarat, J. Gruest, D. Guetard, T. Hercend, J. C. Gluckman, and L. Montagnier.** 1985. T-lymphocyte T4 molecule behaves as the receptor for human retrovirus LAV. *Nature* (London) **312:**767–768.
24. **Koenig, S., H. E. Gendelman, J. M. Orenstein, M. C. Dal Canto, G. H. Pezeshkpour, M. Yungbluth, F. Janotta, A. Aksamit, M. Martin, and A. Fauci.** 1986. Detection of AIDS virus in macrophages in brain tissue from AIDS patients with encephalopathy. *Science* **233:**1089–1093.
24a. **Lasky, L. A., G. Nakamura, D. H. Smith, C. Fennie, C. Shimasaki, E. Patzer, P. Berman, T. Gregory, and D. J. Capon.** 1987. Delineation of a region of the human immunodeficiency virus type 1 gp120 glycoprotein critical for interaction with the CD4 receptor. *Cell* **50:**975–985.
25. **Levy, J. A., C. Cheng-Mayer, D. Dina, and P. A. Luciw.** 1986. AIDS retrovirus (ARV-2) clone replicates in transfected human and animal fibroblasts. *Science* **232:**998–1000.
26. **Levy, J. A., A. D. Hoffman, S. M. Kramer, J. A. Landis, J. M. Shimabukuro, and L. S.**

Oschiro. 1984. Isolation of lymphocytopathic retroviruses from San Francisco patients with AIDS. *Science* **225:**840–842.

27. **Levy, J. A., J. Shimabukuro, T. McHugh, C. Casavant, D. P. Stites, and L. S. Oschiro.** 1985. AIDS-associated retroviruses (ARV) can productively infect other cells besides human T helper cells. *Virology* **147:**441–448.
28. **Lifson, J., S. Coutre, E. Huang, and E. Engleman.** 1986. Role of envelope glycoprotein carbohydrate in human immunodeficiency virus (HIV) infectivity and virus-induced cell fusion. *J. Exp. Med.* **164:**2101–2106.
29. **Lifson, J. D., M. B. Feinberg, G. R. Reyes, L. Rabin, B. Banapour, S. Chakrabarti, B. Moss, F. Wong-Staal, K. S. Steimer, and E. G. Engleman.** 1986. Induction of CD4-dependent cell fusion by the HTLV-III/LAV envelope glycoprotein. *Nature* (London) **323:**725–728.
30. **Maddon, P. J., A. G. Dalgleish, J. S. McDougal, P. R. Clapham, R. A. Weiss, and R. Axel.** 1986. The T4 gene encodes the AIDS virus receptor and is expressed in the immune system and the brain. *Cell* **47:**333–348.
31. **Matthews, T. J., A. J. Langlois, W. G. Robey, N. T. Chang, R. C. Gallo, P. J. Fischinger, and D. P. Bolognesi.** 1986. Restricted neutralization of divergent human T-lymphotropic virus type III isolates by antibodies to the major envelope glycoprotein. *Proc. Natl. Acad. Sci. USA* **83:**9709–9713.
32. **McDougal, J. S., M. S. Kennedy, J. K. A. Nicholson, T. J. Spira, H. W. Jaffe, J. E. Kaplan, D. B. Fishbein, P. O'Malley, C. H. Aloisio, C. M. Black, M. Hubbard, and C. B. Reimer.** 1987. Antibody response to human immunodeficiency virus in homosexual men. Relation of antibody specificity, titer, and isotype to clinical status, severity of immunodeficiency, and disease progression. *J. Clin. Invest.* **80:**316–324.
33. **McDougal, J. S., M. S. Kennedy, J. M. Sligh, S. P. Cort, A. Mawle, and J. K. A. Nicholson.** 1986. Binding of HTLV-III/LAV to T4^{+} T cells by a complex of the 110k viral protein and the T4 molecule. *Science* **231:**382–385.
34. **McDougal, J. S., A. Mawle, S. P. Cort, J. K. A. Nicholson, G. D. Cross, J. A. Scheppler-Campbell, D. Hicks, and J. Sligh.** 1985. Cellular tropism of the human retrovirus HTLV-III/LAV. I. Role of T cell activation and expression of the T4 molecule. *J. Immunol.* **135:**3151–3162.
35. **McDougal, J. S., J. K. A. Nicholson, G. D. Cross, M. S. Kennedy, S. P. Cort, and A. Mawle.** 1986. Binding of the human retrovirus HTLV-III/LAV/ARV/HIV to the CD4 (T4) molecule: conformation dependence, epitope mapping, antibody inhibition, and potential for idiotypic mimicry. *J. Immunol.* **137:**2937–2944.
36. **Montagnier, L., J. Gruest, S. Chamaret, C. Dauget, C. Axler, D. Guetard, M. T. Nugeyre, F. Barre-Sinoussi, J. C. Chermann, J. B. Brunet, D. Klatzmann, and J. C. Gluckman.** 1984. Adaptation of lymphadenopathy associated virus (LAV) to replication in EBV-transformed B-lymphoblastoid cell lines. *Science* **225:**63–67.
37. **Nicholson, J. K. A., G. D. Cross, C. S. Callaway, and J. S. McDougal.** 1986. In vitro infection of human monocytes with human T-lymphotropic virus type III/lymphadenopathy-associated virus (HTLV-III/LAV). *J. Immunol.* **137:**232–329.
38. **Nicholson, J. K. A., J. S. McDougal, and T. J. Spira.** 1985. Alterations of functional subsets of T helper and T suppressor cell populations in acquired immunodeficiency syndrome (AIDS) and chronic unexplained lymphadenopathy *J. Clin. Immunol.* **5:**269–274.
39. **Nicholson, J. K. A., J. S. McDougal, T. J. Spira, G. D. Cross, B. M. Jones, and E. L. Reinherz.** 1984. Immunoregulatory subsets of the T helper and T suppressor cell population in homosexual men with chronic unexplained lymphadenopathy. *J. Clin. Invest.* **73:**191–201.

40. **Popovic, M., E. Read-Connole, and R. C. Gallo.** 1984. T4 positive human neoplastic cell lines susceptible to and permissive for HTLV-III. *Lancet* **ii:**1472.
41. **Putney, S. D., T. J. Matthews, W. G. Robey, D. L. Lynn, M. Robert-Guroff, W. T. Mueller, A. J. Langlois, J. Ghrayeb, S. R. Petteway, K. J. Weinhold, P. J. Fischinger, F. Wong-Staal, R. C. Gallo, and D. P. Bolognesi.** 1986. HTLV-III/LAV neutralizing antibodies to an E. coli-produced fragment of the virus envelope. *Science* **234:**1392–1395.
42. **Rosen, C. A., J. G. Sodroski, and W. A. Haseltine.** 1985. The location of cis-acting regulatory sequences in the human T cell lymphotropic virus type III (HTLV-III/LAV) long terminal repeat. *Cell* **41:**813–823.
43. **Sattentau, Q. J., A. G. Dalgleish, R. A. Weiss, and P. C. L. Beverley.** 1986. Epitopes of the CD4 antigen and HIV infection. *Science* **234:**1120–1123.
44. **Sodroski, J., W. C. Goh, C. Rosen, K. Campbell, and W. A. Haseltine.** 1986. Role of the HTLV-III/LAV envelope in syncytium formation and cytopathicity. *Nature* (London) **322:**470–474.
45. **Wain-Hobson, S., P. Sonigo, O. Danos, S. Coles, and M. Alizon.** 1985. Nucleotide sequence of the AIDS virus, LAV. *Cell* **40:**9–17.
46. **Weiss, R. A., P. R. Clapham, A. G. Dalgleish, P. C. L. Beverley, Q. J. Sattentau, L. A. Lasky, P. W. Berman, P. Maddon, and R. Axel.** 1986. HIV-I: neutralizing antibodies and cellular receptors, p. 134–135. *In* M. Girard, G. De The, and L. Vallette (ed.), *Colloque des "Cent Gardes."* Pasteur Vaccins, Paris.
47. **Wiley, C. A., R. D. Schrier, J. A. Nelson, P. W. Lampert, and M. B. A. Oldstone.** 1986. Cellular localization of human immunodeficiency virus infection within the brains of acquired immune deficiency syndrome patients. *Proc. Natl. Acad. Sci. USA* **83:**7089–7093.
48. **Zagury, D., J. Bernard, R. Leonard, R. Cheynier, M. Feldman, P. S. Sarin, and R. C. Gallo.** 1986. Long term cultures of HTLV-III infected T cells: a model of cytopathology of T cell depletion in AIDS. *Science* **231:**850–853.

Chapter 18

Immunobiology of Human Immunodeficiency Virus Infection

Janet K. A. Nicholson

INTRODUCTION

Infection of the human host with the human immunodeficiency virus (HIV) affects many different aspects of the immune response. First, and perhaps primarily, the $CD4^+$ T cells which induce many of the cellular and humoral responses are the target of HIV infection (7, 9, 13). Depletion of these cells, resulting from the cytolytic properties of virus infection, is the hallmark of the disease acquired immunodeficiency syndrome (AIDS) (2, 7, 8, 28). Another common response to HIV infection is an increase of $CD8^+$ cells, which regulate immunoglobulin production and mediate cell-mediated cytotoxic responses (6, 21, 23). Lastly, monocytes and macrophages, which process and present antigen to T and B cells, also appear to be targets of HIV and may actually harbor the virus (10, 12, 22). The evidence that monocytes are targets of HIV and the characterization of various lymphocyte populations in patients infected with HIV are the subject of this report.

INFECTION OF HUMAN MONOCYTES WITH HIV IN VITRO

Monocytes and their tissue counterparts, macrophages, are one of the first lines of defense against foreign material, including microorganisms. They migrate to the site of entry, phagocytose the material, and, in most cases, degrade and inactivate it. Monocytes are also involved in cellular immune responses by ingesting, processing, and presenting antigens to T and B cells. Processed antigens are presented to and

Janet K. A. Nicholson • Division of Host Factors, Center for Infectious Diseases, Centers for Disease Control, Atlanta, Georgia 30333.

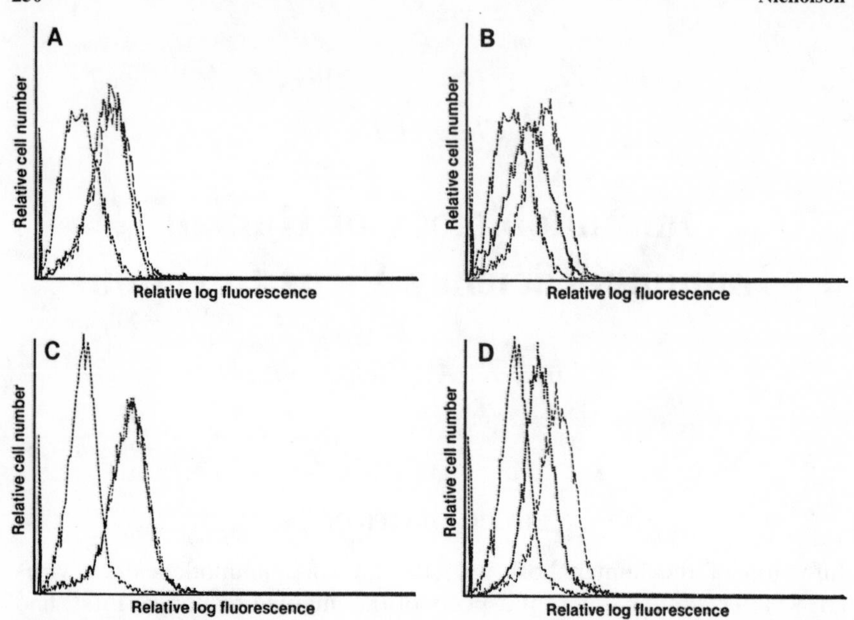

Figure 1. Inhibition of HIV binding to monocytes by anti-OKT4A and vice versa. (A and B) HIV binding (------). Inhibition of this binding by anti-OKT4 is shown in panel A (······) and inhibition by anti-OKT4A is shown in panel B (······). (C) HIV inhibition (······) of anti-OKT4 (------). (D) HIV inhibition (······) of anti-OKT4A (------). (Reprinted from the *Journal of Immunology* [22] with permission from the American Association of Immunologists.)

recognized by immune cells in the context of major histocompatilibity (MHC) class II antigens on the monocyte cell surface.

Another structure expressed by monocytes is the CD4 molecule, although, compared with T cells, this antigen is found at low density (22, 30). The CD4 molecule functions as the receptor for HIV on T cells (5, 13, 19), and so it is possible that HIV binds to the CD4 molecules on monocytes. It is also possible that monocytes ingest HIV independent of an interaction between the virus and the CD4 molecule.

In a binding assay in which normal human peripheral blood monocytes, purified by adherence to gelatin-coated flasks, were incubated with HIV (LAV-1), we demonstrated inhibition of virus binding by preincubating the cells with OKT4A (the epitope of the CD4 antigen that binds to the envelope protein, gp110) (Fig. 1). Binding was not inhibited by OKT4, another epitope on the CD4 antigen that is not involved in virus binding. Conversely, OKT4A binding, but not OKT4 binding, to monocytes was inhibited by HIV (LAV-1) (Fig. 1). Unlike in the T cells, the binding was not totally inhibited. These data suggest that the T4A antigen on monocytes is involved in HIV binding, but do not rule out the

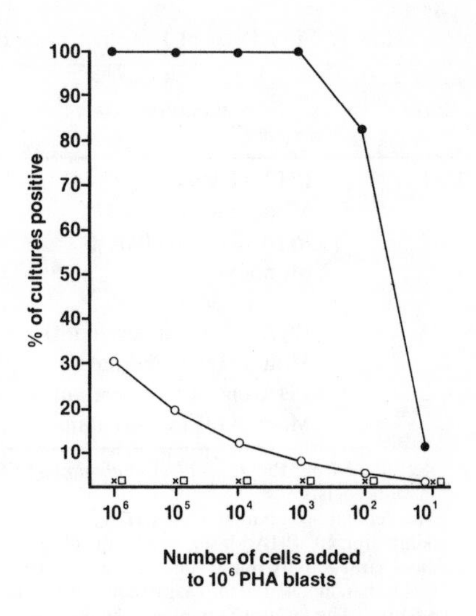

Figure 2. Inoculation of various cells with HIV. T cells (●), B cells (×), monocytes (○), and erythrocytes (□) were cultured with HIV (LAV-1) overnight and washed. Various numbers of cells were added to PHA blasts and distributed to microdilution wells (10 replicate wells were used). Culture supernatants were harvested 11 days later and assayed for the presence of virus by an enzyme-linked immunosorbent capture assay. (Reprinted from the *Journal of Immunology* [22] with permission from the American Association of Immunologists.)

participation of another antigen or receptor on the monocyte. Class II MHC antigen density is reportedly decreased in density on monocytes of AIDS patients (11). This suggests that HIV may be bound to class II antigens on monocytes or that HIV infection down-regulates class II antigen expression. I found that monoclonal antibodies to class II antigens on monocytes variably inhibited in vitro HIV binding; however, a combination of both T4A and class II antibodies did not inhibit HIV binding any more than did each antibody alone. If there is another antigen on monocytes besides the CD4 antigen that binds to HIV, it does not appear to be the class II MHC antigen.

We examined the ability of various cell types from normal human peripheral blood to replicate HIV by dividing peripheral blood cells into T-cell, B-cell, monocyte, and erythrocyte populations (22). We incubated the various cell types for 18 h with a culture supernatant containing HIV (LAV-1), washed the cells, and cocultured various numbers of cells with a constant number of phytohemagglutinin (PHA)-stimulated lymphocytes for 2 to 3 weeks. Culture supernatants were assayed for virus replication by an antigen-capture enzyme-linked immunosorbent assay (20) every 3 to 4 days. Virus was easily detected in culture supernatants from inoculated T cells, but not from B cells or erythrocytes (Fig. 2). The monocyte preparations contained detectable virus, but at much lower levels than the T cells did (Fig. 2).

We then set up bulk cultures of monocyte preparations that had less than 1% T-cell contamination to monitor virus production in both the cells

Table 1. Fifty Percent Infective Dose of HIV-Pulsed Cells and Their Supernatants[a]

Expt	Coculture	ID_{50} at day postinoculation[b]: 3	10	17
1	PHA blasts (cells)	2	12	ND[c]
	Monocytes (cells)	6,310	40	ND
2	PHA blasts (cells)	<1	1	ND
	Monocytes (cells)	>100,000	1,413	ND
1	PHA blasts (supernatant)	$10^{2.95}$	$10^{1.47}$	ND
	Monocytes (supernatant)	$<10^{0}$	$10^{2.07}$	ND
2	PHA blasts (supernatant)	$10^{5.43}$	$10^{2.22}$	$10^{2.36}$
	Monocytes (supernatant)	$<10^{0}$	$10^{0.35}$	$10^{2.93}$

[a] Reprinted from the *Journal of Immunology* (22) with permission from the American Association of Immunologists.
[b] The cellular 50% infective dose (ID_{50}) is the number of cells that, when inoculated into assay cultures containing 10^5 PHA blasts, results in 50% of the cultures being positive. The smaller the number, the more viral activity in the cell population. The supernatant ID_{50} is the reciprocal dilution of supernatant that, when inoculated into assay cultures containing 10^5 PHA blasts, results in 50% of the cultures being positive. The larger the number, the greater the viral activity.
ND, Not done.

and the cell supernatants over time. Because the level of virus in the monocyte cultures was so low, we expanded the virus through additional subculturing with PHA blasts. Virus inoculation had no effect on monocyte survival. This contrasts with T-cell survival, which is decreased in the inoculated cultures (18). We were unable to detect cytoplasmic labeling of monocytes with fluorescein-conjugated polyclonal antiserum to HIV because of the high background detected with the reagent alone on uninfected monocytes. Expansion of both the monocyte culture supernatants and the monocytes from virus-inoculated bulk cultures indicated that, with time, more virus appeared (Table 1). This contrasts with results from PHA-stimulated, HIV-inoculated lymphocyte cultures, in which virus production peaks at about 10 days postinoculation and then declines.

Detection of HIV in monocytes in both in vivo and in vitro systems has also been reported (10, 12, 22). Not all patients have detectable virus in their monocytes by the methods used (12), nor do all strains infect normal monocytes with the same efficiency (10). The differences in virus isolates that allow one strain but not another to effectively infect monocytes are unclear. Whether monocytes only harbor the virus in cytoplasmic vesicles or whether they actually replicate virus is also unknown; however, there is evidence that both may occur.

My colleagues and I have found that monocytes do bind HIV and that HIV binding is inhibitable by OKT4A. The kinetics of HIV infection in monocytes in vitro is considerably different from that of T cells, and the magnitude of virus production is considerably lower, but HIV production increases with time. On the basis of these results, we propose that monocytes do bind and replicate HIV and may serve as a reservoir of HIV in the host.

LYMPHOCYTE SUBPOPULATIONS IN HOMOSEXUAL MEN EXPOSED TO HIV

The most common laboratory finding for patients with AIDS has been a decline in the number of $CD4^+$ T cells (2, 7, 8, 28). Patients who have signs and symptoms of HIV infection, but do not have AIDS, often have an increase in the $CD8^+$ T-cell population (7, 23, 26). Since the $CD8^+$ cells are heterogeneous in terms of function, the increase in the number of $CD8^+$ cells in these patients may be due to increases in a particular subpopulation of $CD8^+$ cells. To examine this, we selected patients from the San Francisco area who were involved in a hepatitis B vaccine study in 1978 to 1980. These were asymptomatic homosexual men who had been tested for antibodies to HIV as well as for the presence of virus in their peripheral blood mononuclear cells. We randomly selected the men and then divided them into groups on the basis of antibody and viral culture status. We evaluated their lymphocyte subsets by two-color immunofluorescence and flow cytometer analysis. Compared with our heterosexual controls, all of the men had elevated numbers of lymphocytes, primarily because of increases in the number of CD8 (Leu2)-positive cells (Fig. 3). $CD4^+$ cell numbers were also elevated in the HIV antibody-negative men. Leu7, an antibody that identifies large granular lymphocytes (1), was present at elevated levels in all of the homosexual men.

Two-color immunofluorescence can be very valuable in subdividing functionally heterogeneous cell populations into functionally defined populations. One analysis of interest is the division of $CD8^+$ cells into $Leu7^+$ and $Leu7^-$ cells. An increase in $CD8^+$ $Leu7^+$ cells, which are believed to have cytotoxic function (4, 25), has been reported to occur in hemophiliac patients and in patients with AIDS or AIDS-related syndromes (17, 26, 27, 31). We found an increase in $CD8^+Leu7^+$ cells in all of the homosexual groups we studied here (Fig. 4). On the basis of these results and results with acute Epstein-Barr virus-infected mononucleosis patients (M. Berdicca and J. K. A. Nicholson, unpublished observations), we conclude that an elevation in the number of these cells is not specific for AIDS or HIV infection, but may be a response to other agents as well. An elevation in the converse population, the $CD8^+$ $Leu7^-$ cells, is found

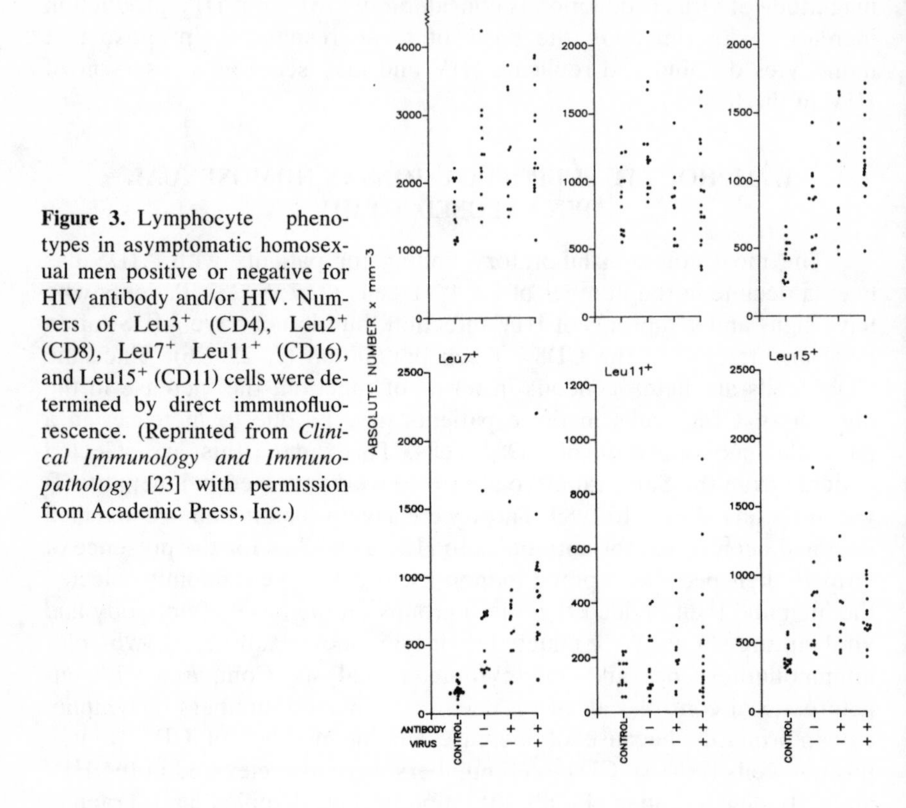

Figure 3. Lymphocyte phenotypes in asymptomatic homosexual men positive or negative for HIV antibody and/or HIV. Numbers of Leu3$^+$ (CD4), Leu2$^+$ (CD8), Leu7$^+$ Leu11$^+$ (CD16), and Leu15$^+$ (CD11) cells were determined by direct immunofluorescence. (Reprinted from *Clinical Immunology and Immunopathology* [23] with permission from Academic Press, Inc.)

only in HIV-infected men (Fig. 4); however, the function of this cell population is unknown.

Leu15 monoclonal antibody subdivides CD8 cells into suppressor (bright CD8$^+$ Leu15$^+$) (14), cytotoxic (Leu15$^-$) (3), and natural killer (dull CD8$^+$ Leu15$^+$) (L. Lanier, personal communication) cells. CD8$^+$ cytotoxic and suppressor cell numbers were elevated in all homosexual groups studied here (23); however, even greater increases in the cytotoxic cell numbers were observed in men who had been exposed to HIV (Fig. 4). There are recent reports of the presence of cells with cytotoxic activity against HIV-infected target cells in healthy HIV antibody-positive patients (29; S. Koenig, P. Earl, D. Powell, H. C. Lane, B. Moss, and A. S. Fauci, *Abstr. III Int. Conf. AIDS*, abstr. no. T.9.3, p. 59, 1987); however, cytotoxic function has not been easily demonstrated. The only reports of such functions were from experiments with Epstein-Barr virus-trans-

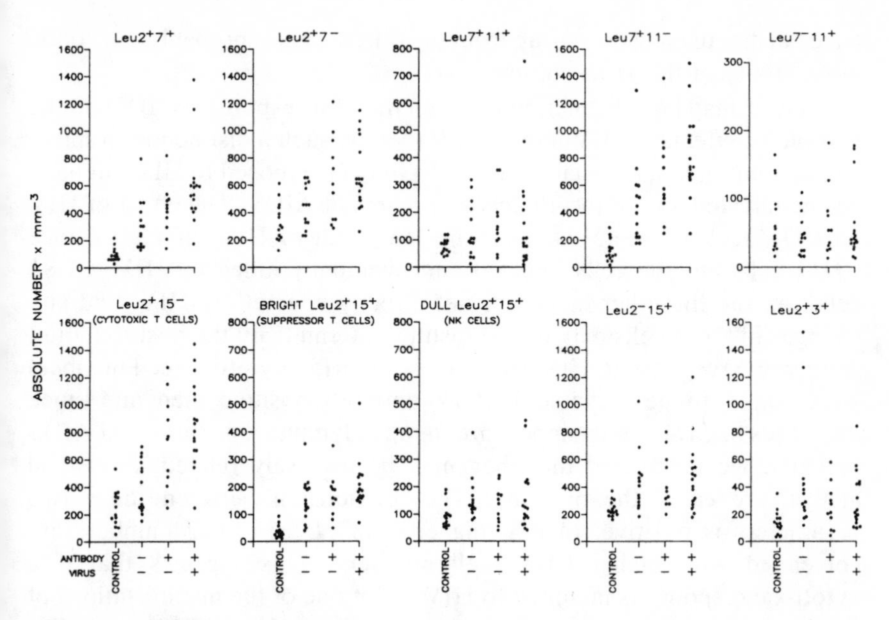

Figure 4. Two-color immunofluorescence on lymphocytes from asymptomatic homosexual men exposed or not exposed to HIV. (Reprinted from *Clinical Immunology and Immunopathology* [23] with permission from Academic Press, Inc.)

formed cells or fibroblasts transfected with vaccinia virus vectors containing HIV proteins. In addition, the cytotoxic responses are not always MHC restricted in the systems studied. MHC-restricted cytotoxic cell responses with naturally infected target cells have not been demonstrated. Even though increases in a phenotypically defined cytotoxic cell population have been identified as being primarily responsible for the increases in $CD8^+$ cells in HIV-infected homosexual men, no correlations have been made between the $CD8^+$ $Leu15^-$ cells and cytotoxic function in HIV-infected patients.

Cells with the greatest natural killer function are phenotypically $Leu7^-$ $Leu11^+$ (15) or dull $CD8^+$ $Leu15^+$. We found no differences in cells with these phenotypes in any of our patient groups, compared with heterosexual controls. Natural killer function in HIV-infected patients, in general, is not abnormal. The only patients who seem to have decreased natural killer function are AIDS patients with overwhelming infections (16).

Although we divided our HIV-antibody-positive men into two groups on the basis of whether they were positive or negative for virus cultured from their peripheral blood lymphocytes, we found no phenotypic differences distinguishing one group from the other. This is probably due to the insensitivity of the virus culture assay.

There may be a distinction between the host response to HIV and the pathologic effects of HIV infection. If there is such a distinction, it might be apparent in homosexual men who have been exposed to HIV but have not manifested any of the clinical symptoms of AIDS. The effect of HIV on $CD4^+$ cell numbers may well be the result of HIV infection, since $CD4^+$ cells are the cells infected and eventually killed by HIV. $CD8^+$ cells, on the the other hand, are not directly affected by HIV, and any changes in these cells may be the result of attempts by the host to mount an immune response to HIV. Since $CD8^+$ $Leu15^-$ cytotoxic cell numbers were found to be elevated in HIV antibody-positive men and since decreases in $CD4^+$ cell numbers are found in lymphadenopathy and AIDS patients, we postulated that they may be inversely related. Statistical analysis revealed the opposite. The relationship between these two parameters is positive; that is, higher $CD8^+$ $Leu15^-$ cell numbers are correlated with higher $CD4^+$ cell numbers. This suggests that if a cytotoxic response is mounted to HIV and if one of the manifestations of that response is an increase in cytotoxic cell numbers ($CD8^+$ $Leu15^-$), then maybe the host is rendered protected and the $CD4^+$ cells are spared from lysis by HIV.

To further examine this, we tested homosexual men who subsequently developed AIDS for the same lymphocyte parameters as the asymptomatic homosexual men (24). The additional patients either were asymptomatic or had chronic lymphadenopathy syndrome (associated with HIV infection) at the time of testing. Of the 39 men who were asymptomatic, 13 subsequently developed AIDS, and of the 21 men with lymphadenopathy, 12 now have AIDS. The time between specimen collection and AIDS diagnosis in the men who developed AIDS was 2 to 38 months (mean, 16 months; median, 17 months). The follow-up study of the men who did not develop AIDS was 7 to 33 months (mean, 21 months; median, 22 months).

Members of all groups had either normal or elevated levels of $CD4^+$ and $CD8^+$ cells, except the patients who later developed AIDS; these patients had elevated numbers of $CD8^+$ cells but decreased numbers of CD4 cells (Fig. 5). Other than decreased $CD4^+$ cell numbers, there were no other lymphocyte subpopulations that distinguished men who are known to have developed AIDS from those who are clinically stable or have yet to progress to developing AIDS (Fig. 6).

If there is an increase in the number of functional cytotoxic cells in HIV-infected patients, it does not appear to protect the host from

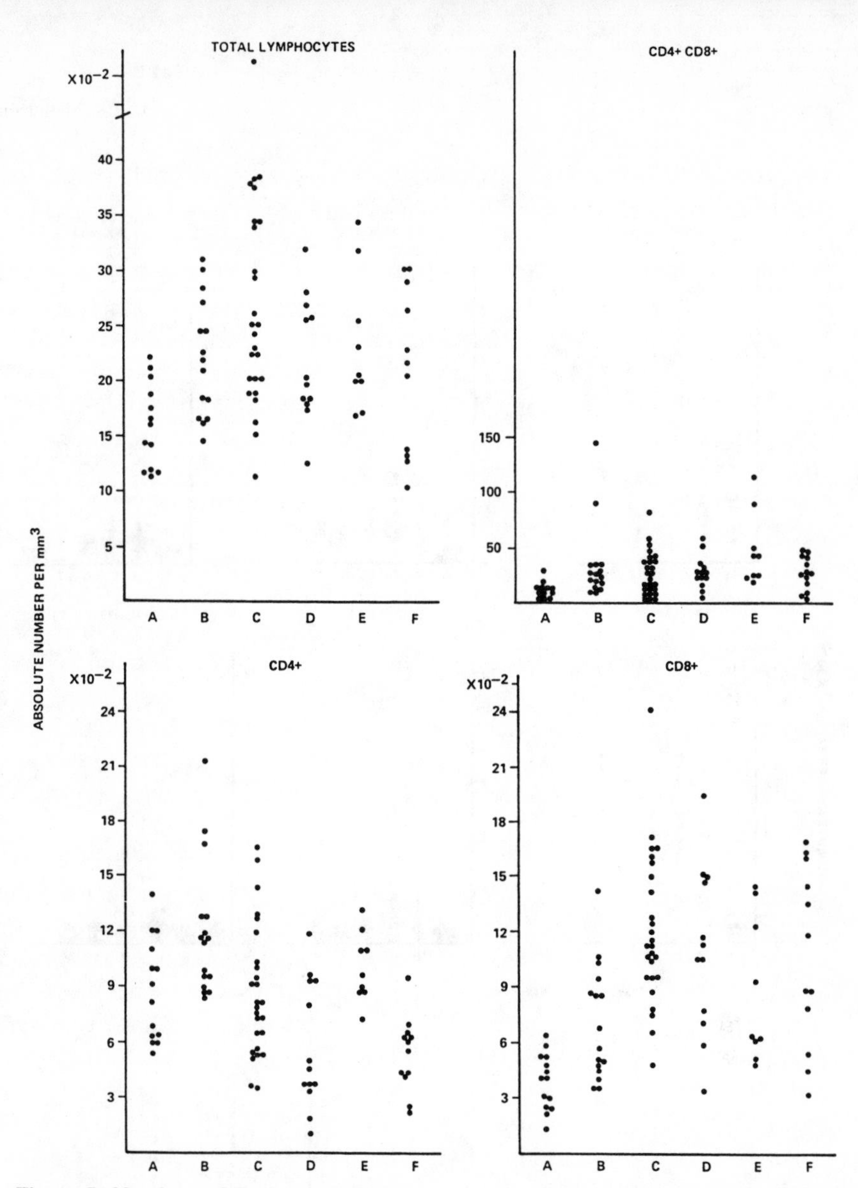

Figure 5. Numbers of CD4 and CD8 cells in homosexual men. Patient groups are as follows:

Characteristic	Group					
	A	B	C	D	E	F
Homosexual	−	+	+	+	+	+
HIV antibody	−	−	+	+	+	+
Lymphadenopathy	−	−	−	−	+	+
Progressed to AIDS	−	−	−	+	−	+

(Reprinted from *Clinical Immunology and Immunopathology* [24] with permission from Academic Press, Inc.)

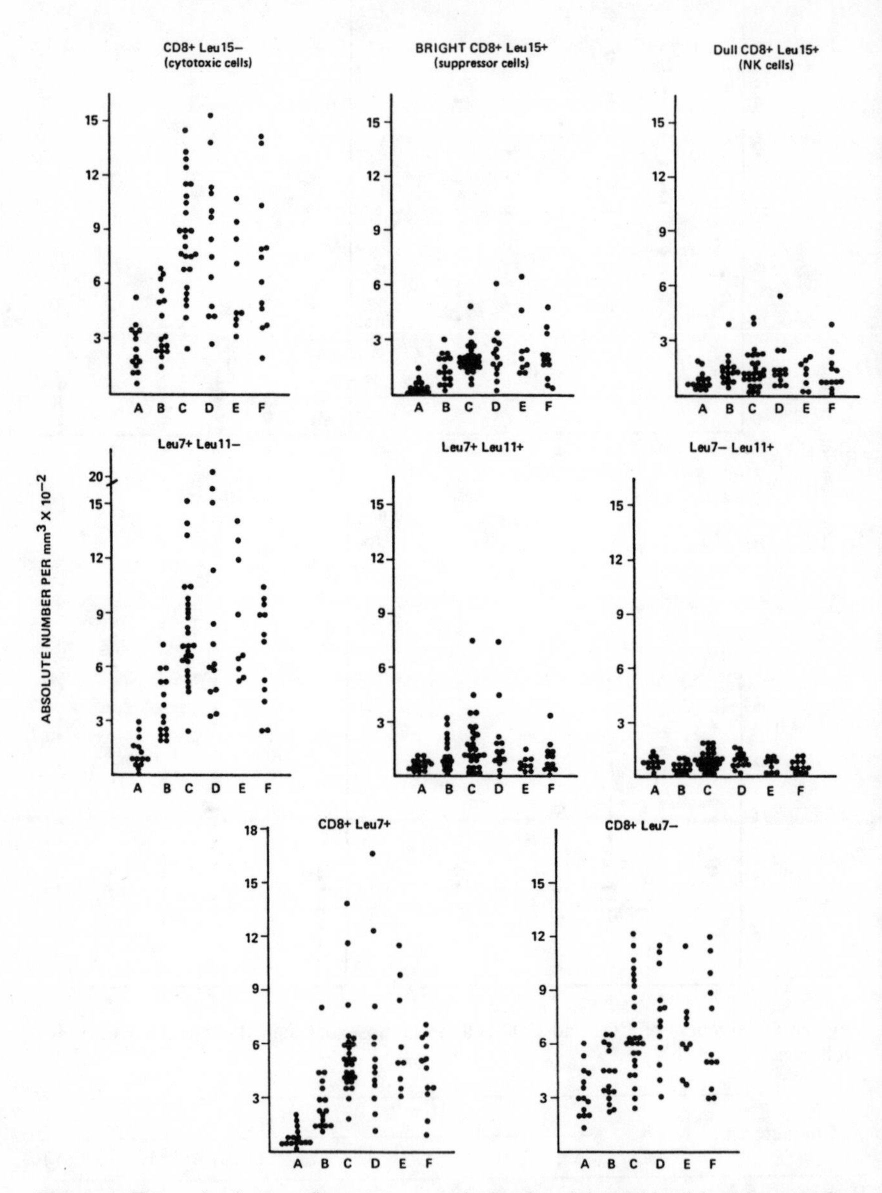

Figure 6. Two-color immunofluorescence analysis of peripheral blood lymphocytes from homosexual men. Groups are as for Fig. 5. (Reprinted from *Clinical Immunology and Immunopathology* [24] with permission from Academic Press, Inc.)

subsequently developing AIDS. $CD4^+$ cell numbers decrease despite elevated cytotoxic cell numbers. Increases in levels of other subsets of the $CD8^+$ cells or in cell populations with natural killer activity may reflect only a general expansion of the cell populations and do not appear to influence the outcome of HIV infection.

Literature Cited

1. **Abo, T., and C. M. Balch.** 1981. Characterization of HNK-1^+ (Leu-7) human lymphocytes. I. Two distinct phenotypes of human NK cells with different cytotoxic capability. *J. Immunol.* **129:**1752–1757.
2. **Ammann, A. J., D. Abrams, M. Conant, D. Chudwin, M. Cowan, P. Volberding, B. Lewis, and C. Casavant.** 1983. Acquired immune dysfunction in homosexual men: immunologic profiles. *Clin. Immunol. Immunopathol.* **27:**315–325.
3. **Clement, L. T., M. K. Dagg, and A. Landay.** 1984. Characterization of human lymphocyte subpopulations: alloreactive cytotoxic T-lymphocyte precursor and effector cells are phenotypically distinct from $Leu2^+$ suppressor cells. *J. Clin. Immunol.* **4:**395–402.
4. **Clement, L. T., C. E. Grossi, and G. L. Gartland.** 1984. Morphologic and phenotypic features of the subpopulation of Leu-2^+ cells that suppresses B cell differentiation. *J. Immunol.* **133:**2461–2468.
5. **Dalgleish, A. G., P. C. L. Beverely, P. R. Clapham, D. H. Crawford, M. F. Greaves, and R. A. Weiss.** 1985. The CD4 (T4) antigen is an essential component of the receptor for the AIDS retrovirus. *Nature* (London) **312:**763–767.
6. **Fahey, J. L., R. Detels, and M. Gottlieb.** 1983. Immune-cell augmentation (with altered T-subset ratio) is common in healthy homosexual men. *N. Engl. J. Med.* **308:**842–843.
7. **Fahey, J. L., H. Prince, M. M. Weaver, J. Groopman, B. Visscher, K. Schwartz, and R. Detels.** 1984. Quantitative changes in the Th or Ts lymphocyte subsets that distinguish AIDS syndromes from other immune subset disorders. *Am J. Med.* **76:**95–100.
8. **Fauci, A. S., H. Macher, D. L. Longo, H. C. Lane, M. Masur, and E. P. Gelman.** 1984. Acquired immunodeficiency syndrome: epidemiologic, clinical, immunologic, and therapeutic considerations. *Ann. Intern. Med.* **100:**92–106.
9. **Fauci, A. S., H. Masur, E. P. Gelmann, P. D. Markham, B. H. Hahn, and H. C. Lane.** 1985. The acquired immunodeficiency syndrome: an update. *Ann. Intern. Med.* **102:**800–813.
10. **Gartner, S., P. Markovits, D. M. Markovitz, M. H. Kaplan, R. C. Gallo, and M. Popovic.** 1986. The role of mononuclear phagocytes in HTLV-III/LAV infection. *Science* **233:**215–219.
11. **Heagy, W., V. E. Kelley, T. B. Strom, K. Mayer, H. M. Shapiro, R. Mandel, and R. Finberg.** 1984. Decreased expression of human class II antigens on monocytes from patients with acquired immune deficiency syndrome. Increased expression with interferon-γ. *J. Clin. Invest.* **74:**2089–2096.
12. **Ho, D. D., T. R. Rota, and M. S. Hirsch.** 1986. Infection of monocyte/macrophages by human T lymphotropic virus type III. *J. Clin. Invest.* **77:**1712–1715.
13. **Klatzmann, D., F. Barre-Sinoussi, M. T. Nugeyre, C. Dauguet, E. Vilmer, C. Griscelli, F. Brun-Vezinet, C. Rouzioux, J. C. Gluckman, J.-C. Chermann, and L. Montagnier.** 1984. Selective tropism of lymphadenopathy associated virus (LAV) for helper-inducer T lymphocytes. *Science* **225:**59–62.
14. **Landay, A., G. L. Gartland, and L. T. Clement.** 1983. Characterization of a phenotyp-

ically distinct subpopulation of Leu-2+ cells that suppresses T cell proliferation responses. *J. Immunol.* **131:**2757–2761.

15. **Lanier, L. L., A. M. Le, J. H. Phillips, N. L. Warner, and G. F. Babcock.** 1983. Subpopulations of human natural killer cells defined by expression of the Leu-7 (HNK-1) and Leu-11 (NK-15) antigens. *J. Immunol.* **131:**1789–1796.
16. **Lew, F., P. Tsang, S. Solomon, I. J. Selikoff, and J. G. Bekesi.** 1984. Natural killer cell function and modulation by α-IFN and IL2 in AIDS patients and prodromal subjects. *J. Clin. Lab. Immunol.* **14:**115–121.
17. **Lewis, D. E., J. M. Puck, G. F. Babcock, and R. R. Rich.** 1985. Disproportionate expansion of a minor T cell subset in patients with lymphadenopathy syndrome and acquired immune deficiency syndrome. *J. Infect. Dis.* **151:**555–559.
18. **McDougal, J. S., S. P. Cort, M. S. Kennedy, C. D. Cabradilla, P. M. Feorino, D. P. Francis, D. Hicks, V. S. Kalyanaraman, and L. S. Martin.** 1985. Immunoassay for the detection and quantitation of infectious human retrovirus, lymphadenopathy-associated virus (LAV). *J. Immunol. Methods* **76:**171–183.
19. **McDougal, J. S., M. S. Kennedy, J. M. Sligh, S. P. Cort, A. Mawle, and J. K. A. Nicholson.** 1986. Binding of HTLV-III/LAV to T4+ T cells by a complex of the 110K viral protein and the T4 molecule. *Science* **231:**382–385.
20. **McDougal, J. S., A. Mawle, S. P. Cort, J. K. A. Nicholson, G. D. Cross, J. A. Scheppler-Campbell, D. Hicks, and J. Sligh.** 1985. Cellular tropism of the human retrovirus HTLV-III/LAV. I. Role of T cell activation and expression of the T4 antigen. *J. Immunol.* **135:**3151–3162.
21. **Melbye, M., R. J. Biggar, P. Ebbesen, C. Neuland, J. J. Goedert, V. Faber, I. Lorenzen, P. Skinhoj, R. C. Gallo, and W. A. Blattner.** 1986. Long-term seropositivity for human T-lymphotropic virus type III in homosexual men without the acquired immunodeficiency syndrome: development of immunologic and clinical abnormalities. *Ann. Intern. Med.* **104:**496–500.
22. **Nicholson, J. K. A., G. D. Cross, C. S. Callaway, and J. S. McDougal.** 1986. In vitro infection of human monocytes with human T-lymphotropic virus type III/lymphadenopathy-associated virus (HTLV-III/LAV). *J. Immunol.* **137:**323–329.
23. **Nicholson, J. K. A., Echenberg, B. M. Jones, H. W. Jaffe, P. M. Feorino, and J. S. McDougal.** 1986. T cytotoxic/suppressor cell phenotypes in a group of asymptomatic homosexual men with and without exposure to HTLV-III/LAV. *Clin. Immunol. Immunopathol.* **40:**505–514.
24. **Nicholson, J. K. A., B. M. Jones, D. F. Echenberg, T. J. Spira, and J. S. McDougal.** 1987. Phenotypic distribution of T cells of patients who have subsequently developed AIDS. *Clin. Immunol. Immunopathol.* **43:**82–87.
25. **Phillips, J. H., and L. L. Lanier.** 1986. Lectin-dependent and anti-CD3 induced cytotoxicity are preferentially mediated by peripheral blood cytotoxic T lymphocytes expressing Leu-7 antigen. *J. Immunol.* **136:**1579–1585.
26. **Plaeger-Marshall, S., C. A. Spina, J. V. Giorgi, R. Mitsuyasu, P. Wolfe, M. Gottlieb, and G. Beall.** 1987. Alterations in cytotoxic and phenotypic subsets of natural killer cells in acquired immune deficiency syndrome (AIDS). *J. Clin. Immunol.* **7:**16–23.
27. **Prince, H. E., J. K. Kressi, C. K. Kasper, S. Kleinman, A. M. Saunders, L. Waldbeser, O. O. Manding, and H. S. Kaplan.** 1985. Distinctive lymphocyte subpopulation abnormalities in patients with congenital coagulation disorders who exhibit lymph node enlargement. *Blood* **66:**64–68.
28. **Schroff, R. W., M. S. Gottlieb, H. E. Prince, L. L. Chai, and J. L. Fahey.** 1983. Immunological studies of homosexual men with immunodeficiency and Kaposi's sarcoma. *Clin. Immunol. Immunopathol.* **27:**300–314.

29. **Walker, B. D., S. Chakrabarti, B. Moss, T. J. Paradis, T. Flynn, A. G. Durno, R. S. Blumberg, J. C. Kaplan, M. S. Hirsch, and R. T. Schooley.** 1987. HIV-specific cytotoxic T lymphocytes in seropositive individuals. *Nature* (London) **328:**345–348.
30. **Wood, G. S., N. L. Warner, and R. A. Warnke.** 1983. Anti-Leu3/T4 antibodies react with cells of monocyte/macrophage and Langerhans lineage. *J. Immunol.* **131:**212–216.
31. **Ziegler-Heitbrock, H. W. L., W. Schramm, D. Satchel, H. Rumpold, D. Kraft, D. Wernicke, K. Von Der Helm, J. Eberle, F. Deinhardt, G. P. Rieber, and G. Rietmuller.** 1985. Expansion of a minor subpopulation of peripheral blood lymphocytes ($T8^+$/ $Leu7^+$) in patients with hemophilia. *Clin. Exp. Immunol.* **61:**633–641.

Chapter 19

Antiviral Therapy for Human Immunodeficiency Virus Infections

Martin S. Hirsch

INTRODUCTION

In the few years that have elapsed since the emergence of acquired immunodeficiency syndrome (AIDS) onto the world scene, surprising progress has been made in the development of antiviral therapeutics directed against the etiologic agent, human immunodeficiency virus type 1 (HIV-1). One agent has already been shown to prolong the lives of patients suffering from AIDS or advanced AIDS-related complex (ARC) (15), and many others are in clinical trials. Furthermore, as knowledge concerning the replication and pathogenesis of HIV-1 increases, new targets for attack will become evident. Antiviral drug combinations will also become more widely used as interactions among agents are clarified.

REPLICATIVE CYCLE OF HIV-1

A number of HIV replicative sites are susceptible to intervention. The envelope glycoprotein of HIV-1 (gp120) binds to the CD4 receptor of helper T lymphocytes, certain monocytes, and other cells (8, 18, 32). HIV-1 is subsequently internalized, presumably by fusion with the cellular membrane, and then is uncoated by cytoplasmic enzymes.

HIV-1 attachment to the CD4 receptor can be blocked in vitro by using monoclonal OKT4 antibodies, suggesting one approach to virus inhibition (32). Blockade by soluble CD4 molecules or synthetic peptides has also been suggested. Another possible approach to blockade of HIV attachment could be the alteration of the chemical composition and

---

Martin S. Hirsch • Massachusetts General Hospital and Harvard Medical School, Boston, Massachusetts 02114.

configuration of the cellular or viral membrane. The resulting structural incongruency between gp120 and the CD4 receptor might inhibit HIV-1 attachment.

HIV-1 RNA-dependent DNA polymerase or reverse transcriptase is introduced into the cytoplasm by HIV-1 and uses the single-stranded dimeric viral 70S RNA genome as a template to transcribe RNA into a minus-strand DNA. Besides reverse transcriptase (RT) activity, HIV DNA polymerase has RNase H activity which subsequently degrades the RNA strand of this DNA-RNA hybrid. Synthesis of a cDNA strand is then initiated. Cytoplasmic double-stranded linear DNA (linear duplex DNA) migrates to the nucleus, where a covalently closed circular DNA is formed. This form of DNA may then become integrated as proviral DNA at random locations in host chromosomes.

Viral RNA is transcribed by host DNA polymerase from the integrated proviral template. After transport to the cytoplasm, viral RNA serves either as mRNA for protein synthesis or as genomic RNA for newly formed viruses.

mRNAs induce *gag* (core proteins), *pol* (RT), and *env* (envelope) protein products. Cleavage of these polyproteins is then followed by modification involving mechanisms such as glycosylation and phosphorylation. The *env*-derived protein products are inserted into the host cell membrane, where further modifications such as glycosylation occur. In addition to the *gag*, *pol*, and *env* genes, HIV-1 contains at least five other genes (3′*orf*, *sor*, *tat*, *art*/*trs*, and *R*). The *tat-III* gene codes for a *trans*-activator protein which boosts HIV-1 gene expression by a positive feedback mechanism (21, 48). The *art*/*trs* gene-encoded protein is thought to relieve negative regulatory effects on the translation of viral mRNA (22, 53). The function of the *R* gene is unknown (58). During the final stage of the HIV replication, all performed elements are assembled and reencapsidated, and the HIV membrane is acquired by budding from the host cell.

APPROACH TO CHEMOTHERAPY OF HIV INFECTION

Inhibition of HIV Attachment and Penetration

AL 721 is a lipid compound with neutral glycerides, phosphatidylcholine, and phosphatidylethanolamine in a 7:2:1 ratio by weight (P. S. Sarin, R. C. Gallo, D. I. Scheer, F. Crews, and A. S. Lippa, Letter, *N. Engl. J. Med.* **313**:1289–1290, 1985). The compound in droplet form extracts cholesterol from cell membranes and may alter the lipid content of viruses that undergo cell membrane budding. It is further presumed that such changes may modify the complementarity of the gp120-CD4 receptor complex, resulting in a reduced attachment of HIV to the cell membrane. It has been shown in preliminary studies (Sarin et al., Letter) that the replication of HIV was reduced when AL 721 was added simultaneously with the virus to cultures of peripheral blood leukocytes or H9 cells. These in vitro studies, which require confirmation, suggest that HIV replication is reduced by over 90% at AL 721 concentrations of 1,000 μg/ml. It is not clear whether AL 721 is absorbed intact from the gastrointestinal tract and circulates in droplet form in vivo. An open study with oral AL 721 in eight patients with AIDS-related complex is currently being evaluated, and other trials are planned.

It has also been reported that an octapeptide (Ala-Ser-Thr-Thr-Asn-Tyr-Thr) is able to block HIV attachment to the CD4 receptor of lymphocytes (45). A 5-amino-acid sequence of this peptide (peptide T) corresponds to a region in the HIV envelope gp120. In studies of HIV replication in vitro, peptide T reduced RT levels by about ninefold. A subsequent study was not able to confirm these observations (J. Sodroski, M. Kowaleski, T. Dorfman, L. Basiripour, C. Rosen, and W. Haseltine, Letter, *Lancet* **i:**1428–1429, 1987). Peptide T was given as an intravenous infusion to four severely ill AIDS patients in Sweden (L. Wetterberg, B. Alexius, J. Saaf, A. Sonnerborg, S. Britton, and C. Pert, Letter, *Lancet* **i:**159, 1987). The substance was well tolerated, but no therapeutic conclusions could be drawn. It is not clear whether sufficient concentrations of such a peptide preparation could be given to block relevant receptor sites without interfering with the function of CD4-bearing cells. Randomized controlled clinical trials of peptide T are planned.

Recently, an exciting new approach to the blockade of HIV infection has emerged which involves the use of soluble CD4 molecules developed by recombinant DNA technology (16). These CD4 molecules inhibit HIV replication in vitro and are currently undergoing animal tests.

Inhibition of HIV RT

HIV RT is inhibited selectively or nonselectively by a variety of drugs. Inhibitors of retrovirus RT could act as enzyme-binding compounds, template-primer-binding compounds, substrate or product analogs, template analogs, or divalent cation-binding agents (10, 11, 47).

Suramin

Suramin is a hexasodium salt derivative of naphthalene trisulfonic acid, which was first produced in 1916 by Bayer as an antiparasitic drug. It is still used in Africa as a treatment of trypanosomiasis and onchocerciasis (38). The compound binds to a wide variety of enzymes and proteins and was shown to inhibit the RT of different animal retroviruses

(9). This inhibitory effect is, however, not specific, and cellular DNA polymerases are inhibited as well (40).

Suramin at >50 μg/ml inhibits HIV RT and HIV replication in vitro. The toxic/therapeutic ratio is small, and normal cell growth is inhibited at concentrations of 100 μg/ml and higher (40). Suramin has to be administered by intravenous infusion and strongly binds to plasma proteins. It is excreted by renal mechanisms and has a half-life in plasma of about 40 days.

A phase 1 clinical trial of 10 patients with AIDS or ARC treated over 5 weeks showed a possible virustatic effect in vivo at suramin concentrations of >100 μg/ml in plasma (4). Several subsequent studies have evaluated the long-term use of the drug. In one study, only 1 of 97 patients showed clinical improvement, and the observed adverse drug effects included fever (41%), skin rash (31%), neutropenia (36%), thrombocytopenia (12%), elevation of creatinine levels in serum (16%), and adrenal insufficiency (28%). Two cases of agranulocytosis and four cases of hepatic failure were probably drug related (34). Because of these results, it is unlikely that suramin will receive further attention as an HIV inhibitor in patients.

AZT

3′-Azido-3′-deoxythymidine (AZT; Zidovudine, Retrovir) is a thymidine analog that was first synthesized by Horwitz et al. in 1964 (30). AZT is active in vitro and in vivo against a number of animal retroviruses (41, 49). It was shown to inhibit HIV replication when present at >0.1 μM (26, 41), whereas inhibition of the multiplication of uninfected lymphocytes was observed when it was present at >1 mM (17). This apparent selectivity is due to the selective interaction of the 5′-triphosphate of AZT (AZTTP) with HIV RT (17). AZT is converted by cellular kinases to its active triphosphate form (AZTTP), and it inhibits HIV RT approximately 100 times more effectively than it inhibits cellular DNA polymerase alpha. AZTTP binds selectively to HIV RT, and the incorporation of azidothymidylate into the growing DNA strand leads to chain termination (17). Furthermore, AZTMP competitively inhibits thymidylate kinase (the enzyme which catalyzes phosphorylation of dTMP to dTDP), resulting in reduced levels of dTTP.

AZT is available in both oral and parenteral forms, and concentrations of 5 μM in plasma can be achieved after a peroral (5 mg/kg of body weight) or intravenous (2.5 mg/kg over 1 h) regimen. The half-life in plasma is approximately 1 h (61). Since AZT is metabolized primarily by glucuronidation, drugs that inhibit this step (e.g., probenecid) can increase the half-life of AZT (60). The drug penetrates the blood-brain barrier effectively and in vitro virucidal doses are achieved within the cerebrospinal fluid (60). AZT was initially evaluated in a phase 1 trial in 19 patients with AIDS or ARC at the National Cancer Institute and Duke University. There were suggestions of clinical and immunologic improvement and a possible virostatic effect in several patients; side effects were minimal in this 6-week clinical trial (61).

From February 1986 to the end of June 1986, 282 patients with AIDS or ARC were enrolled in a randomized double-blind placebo-controlled clinical trial at 12 different centers in the United States (15, 46). Patients with AIDS were enrolled within 4 months of their first attack of *Pneumocystis carinii* pneumonia, and ARC patients had severe clinical signs, e.g., significant weight loss or oral thrush. A total of 145 patients received an oral dose of 250 mg of AZT every 4 h, and 137 patients received placebo. By September 1986, it was apparent that differences in survival had emerged, and the study was terminated by an independent data safety monitoring board. By then, 19 patients (12 with AIDS and 7 with ARC) had died in the placebo group, compared with only 1 who received AZT. Furthermore, there was a significant difference in the incidence of opportunistic infections in patients who received placebo (45 opportunistic infections) compared with AZT recipients (24 opportunistic infections). Patients who received AZT generally gained weight, whereas placebo recipients lost weight. Karnofsky scores of functional capability also improved in AZT recipients, but not in the placebo group. Immunologic changes were strikingly different between the AZT and the placebo groups. Individuals who received AZT generally showed an increase in $CD4^+$ cells, although this effect was lost after 20 weeks for patients with AIDS. Of patients who received AZT, 29% developed skin test reactivity to at least one antigen, but only 9% of placebo recipients lost skin anergy. The exact mechanisms by which these effects occurred have not been fully elucidated, but preliminary data suggest that p24 antigenemia is reduced in AZT recipients (R. E. Chaisson, J.-P. Allain, and P. A. Volberding, Letter, *N. Engl. J. Med.* **315**:1610–1611, 1986). Over a 9-month follow-up period, AZT administration was associated with a four- to sixfold reduction in mortality rate (15).

AZT treatment is not, however, without side effects. During this study megaloblastic anemia occurred frequently in the AZT group and made transfusions necessary for 40 patients (11 in the placebo group). Another side effect which also made AZT dose reduction necessary was neutropenia in approximately 16% of recipients (46). Headache was the most common adverse symptom observed. Nausea, myalgia, and insomnia were also reported more frequently by AZT recipients. These adverse effects were reversible and generally tolerable for patients with AIDS and severe ARC.

Further clinical trials are needed to assess the benefit of AZT in other populations such as asymptomatic HIV carriers and patients with persistent generalized lymphadenopathy, neurologic disease, Kaposi's sarcoma, pediatric AIDS, and less advanced ARC. An independent evaluation of four patients with neurologic syndromes (two with chronic dementia, one with peripheral neuropathy, and one with paraplegia) showed clinical improvement in three of the patients (46). These findings, however, need confirmation in controlled studies.

To facilitate the in vivo evaluation of AZT and other promising new compounds, the National Institutes of Health has developed an extensive network of AIDS treatment evaluation units. Several AIDS treatment evaluation unit multicenter trials are currently under way (28).

Other Nucleoside Analogs

2′,3′-Dideoxycytidine (ddC) is a pyrimidine nucleoside analog that inhibits HIV replication effectively in vitro (39). In certain cell culture systems, ddC reduces HIV replication at several-fold-lower concentrations than a number of other nucleoside analogs do. At ddC concentrations of 0.5 μM, full protection against cytopathogenicity in HIV-infected ATH8 cells was observed. Cellular toxicity was noted at 10-fold-higher concentrations. Our own experiments indicate that the inhibitory concentration of ddC in infected peripheral blood mononuclear cells is generally equal to or slightly lower than that of AZT. Cellular toxicity was insignificant at the doses tested (0.005 to 0.36 μM). The mode of anti-HIV action is probably by the inhibition of RT by the triphosphate of ddC. Phase 1 clinical trials of ascending doses of ddC are currently under way with ARC and AIDS patients. Early results suggest that doses of 0.03 to 0.06 mg/kg of body weight every 4 h are associated with biological activity (reduction of HIV p24 antigen levels in serum), but also with toxic side effects (rash, stomatitis, and peripheral neuropathy).

3′-Azido-2′,3′-dideoxyguanosine at 1 μg/ml was reported to inhibit HIV-induced cytopathogenicity in MT4 cells (25). We studied the effects of this compound on HIV replication in acutely infected peripheral blood leukocytes over a period of 14 days and observed that 3 μg/ml gave complete inhibition of HIV replication. Cellular toxicity was negligible at this concentration.

Recently, 2′,3′-unsaturated analogs of 2′,3′-dideoxynucleosides were synthesized, anticipating a significant therapeutic index against HIV. 2′,3′-Dideoxycytidinene is the unsaturated derivative of ddC; it reduced HIV replication almost as effectively as ddC with low toxicity to host cells

(3). The anti-HIV activity of 2′,3′-dideoxythymidinene (assessed as inhibition of cytopathogenicity in HIV-infected MT4 cells) was about five times greater than the activity of ddC (1). These results must be confirmed by using other cell systems and additional assays such as virus yield.

PFA

Trisodium phosphonoformate (PFA, foscarnet) is a PP_i analog which inhibits DNA polymerases of different herpesviruses and the RNA polymerase of influenza virus (33, 44). It has been used topically as an antiviral agent for the treatment of recurrent herpes labialis and herpes genitalis and systemically in immunosuppressed patients with cytomegalovirus infections.

PFA is known to inhibit RT of several animal retroviruses (13, 35) and inhibits the replication of visna virus, a lentivirus that causes progressive demyelinating illness in sheep (54). We have shown that PFA at >132 μM inhibits HIV RT and HIV replication in lymphoid cell cultures (50). These studies have been confirmed by others (51).

The results of pilot studies in Europe suggest that PFA may be virostatic in vivo and that PFA concentrations active in vitro can be attained in the cerebrospinal fluid. PFA has a short half-life in serum, and continuous infusions (20 mg/kg per h) have been used to achieve concentrations between 75 and 150 μg/ml (250 to 500 μM) in plasma (7, 19). Work is under way to develop orally bioavailable analogs of PFA. Without oral bioavailability, the use of this compound may be limited to specific clinical situations in which intravenous therapy is indicated. However, controlled clinical trials are necessary to demonstrate efficacy.

Antimoniotungstate (HPA-23)

HPA-23 is a mineral condensed polyanion of ammonium-5-tungsto-2-antimoniate. It was reported to show activity against DNA and RNA viruses and partially protected mice from leukemias and sarcomas induced by certain murine retroviruses (31). HPA-23 was shown to inhibit HIV RT in a cell-free assay system (12), but it is unclear how effectively HIV replication in cell cultures is inhibited (2). The drug has a short half-life (<20 min) in serum and must be administered by intravenous infusions. Uncontrolled trials in France have not shown significant clinical benefits (D. Dormont, T. Maillet, H. Di Maria, M. Gardere, F. Barré-Sinoussi, and J. C. Chermann, *Abstr. II Int. Conf. AIDS*, abstr. no. 86, p. 34, 1986). Other heteropolyacid compounds are under study. One, silicotungstate, seems to be as effective as HPA-23 but less toxic in vitro (R. F. Schinazi and C. F. Hill, *Program Abstr. 26th Intersci. Conf.*

Antimicrob. Agents Chemother., abstr. no. 1093, p. 296, 1986); it has not yet been evaluated in patients.

Other RT Inhibitors

A number of additional compounds which inhibit HIV RT have been evaluated in vitro. These include rifabutine (ansamycin) (R. Anand, J. Moore, P. Feorino, J. Curran, and A. Srinivasan, Letter, *Lancet* **i**:97, 1986), which is in early phase 1 clinical trials, as well as various antibiotics (sakyomicin and lucopeptin C) (Y. Inouye, Y. Take, and S. Nakamura, *Program Abstr. 26th Intersci. Conf. Antimicrob. Agents Chemother.*, abstr no. 1094, p. 296, 1986).

Posttranscriptional Processing

Ribavirin

Ribavirin is a synthetic nucleoside derivative of the antibiotic pyrazomycin. It shows broad-spectrum antiviral activity against DNA and RNA viruses (20). The antiviral mechanisms of ribavirin are incompletely understood, but have been best worked out for influenza viruses. Ribavirin resembles guanosine chemically, and its antiviral action can be reversed by adding guanosine to treated cultures. Ribavirin inhibits IMP dehydrogenase, which converts IMP via XMP to GMP. Competition of various enzymes for the resulting decreased pool of GTP may account at least in part for the antiviral effect. Ribavirin further interferes with the guanylation step for 5′ capping of viral mRNA and inhibits viral RNA polymerases which are responsible for mRNA priming and elongation (20).

Ribavirin inhibits animal retroviruses, such as Gross murine leukemia virus, in vitro when present at 1 to 100 μg/ml (52). It was reported to inhibit HIV replication in lymphocyte cultures temporarily when present at 50 to 100 μg/ml (36). Our own studies with acutely infected blood mononuclear cells indicate that ribavirin at 4 to 8 μg/ml may occasionally inhibit HIV replication. Higher concentrations, which are difficult to attain in humans, cause marked toxicity to both infected and uninfected cells (55).

The drug can be given intravenously or perorally and is excreted very slowly by renal mechanisms (half-life in serum, 24 h). The half-life in erythrocytes is about 40 days, since erythrocytes lack specific enzymes to dephosphorylate ribavirin triphosphate effectively (20).

Because ribavirin is a broad-spectrum antiviral drug, several clinical studies have been carried out for a variety of viral infections. Aerosolized ribavirin has been shown to be effective in the treatment of respiratory syncytial virus infections and is licensed in the United States for this purpose. Ribavirin is also effective by peroral or intravenous routes as treatment and postexposure prophylaxis of Lassa fever (37). After promising results in a phase 1 study in which virus-inhibitory ribavirin concentrations were obtained in cerebrospinal fluid (C. Crumpacker, G. Bubley, S. Hussey, L. Schnipper, W. Haegy, R. Finberg, M. F. McLane, J. Allan, and M. Essex, *Abstr. II Int. Conf. AIDS*, p. 34, 1986; C. Crumpacker, G. Bubley, D. Lucey, S. Hussey, and J. Connor, Letter, *Lancet* **ii:**45–46, 1986), a randomized, double-blind placebo-controlled trial was initiated at several centers in the United States (P. W. A. Mansell, P. N. R. Heseltine, R. B. Roberts, G. M. Dickinson, and J. M. Leedom, *Abstr. III Int. Conf. AIDS*, 78.5, 1987). A total of 164 patients with persistent generalized lymphadenopathy, proven HIV culture positivity, and <400 T4 positive cells per mm were enrolled in this 24-week trial. Of 56 patients receiving placebo, 10 (18%) developed AIDS, and of 55 patients treated with 600 mg of ribavirin per day, 6 (11%) developed AIDS. None of 52 patients who were treated with 800 mg of ribavirin daily developed AIDS during the 6 months of study. The significant effect on the development of AIDS could not be explained by a measurable antiviral or immunorestorative effect. The side effects were mild, and only three patients discontinued treatment because of insomnia or nausea and vomiting. The randomization procedures used in this study have been questioned by the Food and Drug Administration. Moreover, a similarly designed study with patients suffering from ARC failed to demonstrate benefit in ribavirin recipients (A. Vernon, R. S. Schulof and The Ribavirin ARC Study Group, *Abstr. III Int. Conf. AIDS*, 78.6, 1987). Further trials of ribavirin are required before conclusions concerning its efficacy can be established.

Other approaches to inhibition of posttranscriptional processing would include the use of agents that block the positive feedback of *tat-III* and *art/trs* gene products which regulate gene expression directed by HIV long terminal repeat sequences (6). No such inhibitors have been described. Synthetic oligonucleotides complementary to viral RNA have also been shown to inhibit HIV replication by hybridization competition (62). It is, however, unclear how the degradation of oligonucleotides by nucleases in serum and tissues can be overcome, allowing such substances to be useful clinically.

Assembly and Release

Interferons are proteins that are produced by virus-infected cells or by cells following challenge by other stimuli including RNA polymers, bacterial lipopolysaccharides, and different mitogenic chemicals. Three major interferon types exist: alpha interferon (IFN-α) (virus-induced leukocyte), beta interferon (IFN-β) (virus-induced fibroblast), and gamma interferon (IFN-γ) (immune). IFN-α and IFN-β have a common cellular receptor, whereas IFN-γ has a separate receptor. After binding to the receptor, IFNs induce the production of other proteins which have antiviral activity. The mechanisms of antiviral action are varied and may differ from virus to virus. In murine retrovirus systems, interferons act mainly at late stage of replication, e.g., assembly and release of mature virus particles.

A number of clinical studies have shown IFN activity against several viruses, such as rhinoviruses, cytomegalovirus, herpes simplex virus, varicella-zoster virus, hepatitis A and B viruses, and papillomavirus. We studied the effect of recombinant IFN-α_A (rIFN-α_A) on the replication of HIV in vitro and found partial inhibition of replication by rIFN-α_A at 4 to 64 U/ml. Doses of >256 U/ml were fully inhibitory (29). Pharmacokinetic data show that rIFN-α_A peak levels range from approximately 200 to 400 U/ml after intramuscular injections of 36×10^6 U (57). Although HIV-inhibitory rIFN-α_A levels can be achieved in vivo, penetration into the cerebrospinal fluid is poor. In a recent study, we further found that rIFN-α_A, rIFN-β, and leukocyte-derived IFN-α show similar concentration-dependent anti-HIV activity. rIFN-γ and lymphocyte-derived IFN-γ were ineffective in acutely infected peripheral blood leukocytes, but showed virus-inhibitory activity in the monocytic cell line U937 (25). The anti-HIV effect of IFN-γ in U937 and MT-4 cells was also observed by others (24, 43).

Double-blind, placebo-controlled trials with rIFN-α_A in patients with AIDS have been carried out to compare 36×10^6 or 3×10^6 U or placebo given three times a week for 12 weeks. No clinical or immunologic benefits were demonstrated (G. H. Friedland, S. H. Landesman, C. C. Crumpacker, and the Interferon Working Group, *Abstr. III Int. Conf. AIDS*, TH4.6, p. 156, 1987).

Certain polymers of double-stranded RNA have long been known to have antiviral and immunomodulatory activities, at least in part through the induction of interferon. One of these, a mismatched double-stranded RNA called ampligen [poly(I) $\cdot$ poly($C_{12}U$)], has shown in vitro activity against HIV (42). Ampligen has also been reported to reduce HIV load and augment immune responses in 10 patients with AIDS, ARC, or lymphadenopathy syndrome (5). Controlled clinical trials of ampligen are planned. Other studies of IFN-α are also in progress.

Another approach might be the use of specific inhibitors of glycosylation of envelope glycoproteins, e.g., castanospermine (1,6,7,8-tetrahydroxyoctahydroindolizine, a plant alkaloid that inhibits glucosidase I).

Walker et al. have reported that castanospermine inhibited the replication of HIV-1 and inhibited syncytium formation induced by the envelope glycoprotein of the virus (56). The effects of castanospermine may be related to modifications of the envelope glycoprotein, which alter steps following binding of HIV to the CD4 receptor. We have tested the glycosylation inhibitor swainsonine and have found no effect on HIV replication (M. W. Vogt and M. S. Hirsch, unpublished data).

Combinations of Antiviral Drugs

Combination chemotherapy is used widely in the therapy of bacterial and fungal infections. Although combinations have not been used extensively in the therapy of virus infections, this approach offers potential benefits such as synergistic antiviral interactions, lower toxicity, and delay in development of drug resistance.

We have found synergy in vitro with the following drug combinations: PFA plus rIFN-α_A (26), PFA plus ribavirin (Vogt and Hirsch, unpublished data), AZT plus rIFN-α_A (27), AZT plus lymphoblastoid IFN (Vogt and Hirsch, unpublished data), ddC plus rIFN-α_A (5), and AZT plus castanospermine (V. Johnson, B. Walker, and M. Hirsch, unpublished data). It has further been reported that AZT plus acyclovir act synergistically (60); the mechanism of this interaction is unclear, since acyclovir does not have an anti-HIV activity of its own. Although in vitro findings must be interpreted with caution, clinical trials of AZT plus IFN-α, as well as AZT plus acyclovir, are under way.

Synergy can, however, not always be expected when anti-HIV drugs are combined. In a variety of different experimental systems, we have demonstrated that ribavirin antagonizes the anti-HIV effect of AZT (55). The responsible mechanism appears to be inhibition of AZT phosphorylation by metabolic changes induced by ribavirin. It is therefore important to study interactions of drugs that use similar metabolic pathways and that may be used in combination, such as ddC with other nucleoside analogs.

Another approach which has not been studied extensively so far is the use of immunomodulators such as IFN-γ or interleukin-2 in combination with anti-HIV compounds. There are now dozens of immune modulators in clinical trial, and studies of combinations with antiviral agents are under way. The combination of the immune modulator granulocyte-macrophage colony-stimulating factor plus AZT has recently been shown to have synergistic anti-HIV activity in vitro (23). Clinical trials of this combination are planned.

SUMMARY AND FUTURE PROSPECTS

Over the past 3 years the number of agents with in vitro anti-HIV-1 activity has grown at a rapid rate. Federal and industry commitment to rapid drug discovery and development programs has increased dramatically. Although the activity of several anti-HIV agents was discovered as the result of previous experience with other retroviral systems, recent insights into the replication of HIV have allowed a more directed search for new anti-HIV compounds. Developments such as the use of synthetic peptides or oligonucleotides are examples. Computer-assisted evaluation of virus structure-function relationships may help elucidate other possible targets for new drugs. Further research is needed to develop drugs that will act against latently integrated proviral DNA. Moreover, the emergence of new human retroviruses capable of causing similar syndromes, e.g., HIV-2, makes the task still more complex (7). New drugs should undergo rapid, but thorough, in vitro and animal testing, and promising compounds should be made available for phase 1 clinical studies as soon as possible after development. Maximal efficiency and organization are essential, given the large number of currently infected individuals and ominous predictions for the future. These efforts can be sustained only if the current level of commitment by both governmental and industrial sources is maintained or increased.

Literature Cited

1. **Baba, M., R. Pauwels, P. Herdewijn, E. De Clercq, J. Desmyter, and M. Vandeputte.** 1987. Both 2′,3′-dideoxythymidine and its 2′,3′-unsaturated derivative (2′,3′-dideoxythymidinene) are potent and selective inhibitors of human immunodeficiency virus replication in vitro. *Biochem. Biophys. Res. Commun.* **142:**128–134.
2. **Balzarini, J., H. Mitsuya, E. De Clercq, and S. Broder.** 1986. Comparative inhibitory effects of suramin and other selected compounds on the infectivity and replication of human T-cell lymphotropic virus (HTLV-III)/lymphadenopathy associated virus (LAV). *Int. J. Cancer.* **37:**451–457.
3. **Balzarini, J., R. Pauwels, P. Herdewijn, E. De Clercq, D. A. Cooney, G.-J. Kang, M. Dalal, D. G. Johns, and S. Broder.** 1986. Potent and selective anti-HTLV-III/LAV activity of 2′,3′-dideoxycytidinene, the 2′,3′-unsaturated derivative of 2′,3′-dideoxycytidine. *Biochem. Biophys. Res. Commun.* **140:**735–742.
4. **Broder, S., R. Yarchoan, J. M. Collins, H. C. Lane, P. D. Markham, R. W. Klecker, R. R. Redfield, H. Mitsuya, D. F. Hoth, E. Gelmann, J. E. Groopman, L. Resnick, R. C. Gallo, C. E. Myers, and A. S. Fauci.** 1985. Effects of suramin on HTLV-III/LAV infection presenting as Kaposi's sarcoma or AIDS related complex: clinical pharmacology and suppression of virus replication in vivo. *Lancet* **ii:**627–630.
5. **Carter, W. A., et al.** 1987. Clinical, immunological, and virological effects of ampligen, a mismatched double-stranded RNA, in patients with AIDS or AIDS-related complex. *Lancet* **i:**1286–1292.
6. **Chen, I. S. Y.** 1986. Regulation of AIDS virus expression. *Cell* **47:**1–2.
7. **Clavel, F., K. Mansinho, S. Chamaret, D. Geutarat, V. Favier, J. Nina, M. Santos-**

Ferreira, J. Champalimaud, and L. Montagnier. 1987. Human immunodeficiency virus type 2 infection associated with AIDS in West Africa. *N. Engl. J. Med.* **316:**1180–1185.

8. **Dalgleish, A. G., P. C. L. Beverley, P. R. Clapham, D. H. Crawford, M. F. Greaves, and R. A. Weiss.** 1984. The CD4 (T4) antigen is an essential component of the receptor of the AIDS retrovirus. *Nature* (London) **312:**763–766.
9. **De Clercq, E.** 1979. Suramin: a potent inhibitor of the reverse transcriptase of RNA tumor viruses. *Cancer Lett.* **8:**9–22.
10. **De Clercq, E.** 1986. Chemotherapeutic approaches to the treatment of the acquired immune deficiency syndrome (AIDS). J. Med. Chem. **29:**1561–1569.
11. **De Clercq, E.** 1987. Suramin in the treatment of AIDS: mechanism of action. *Antiviral Res.* **7:**1–10.
12. **Dormont, D., B. Spire, F. Barre-Sinoussi, L. Montagnier, and J. C. Chermann.** 1985. Inhibition of RNA-dependent DNA polymerases of AIDS and SAIDS retroviruses by HPA-23 (ammonium-21-tungsto-9-antimoniate). *Ann. Inst. Pasteur Virol.* **136E:**75–83.
13. **Eriksson, B., G. Stening, and B. Oberg.** 1982. Inhibition of reverse transcriptase activity of avian myeloblastosis virus by pyrophosphate analogues. *Antiviral Res.* **2:**81–95.
14. **Farthing, C. F., A. G. Dalgleish, A. Clark, M. McClure, A. Chanas, and B. G. Gazzard.** 1987. Phosphonoformate (foscarnet): a pilot study in AIDS and AIDS related complex. *AIDS* **1:**21–25.
15. **Fischl, M. A., D. D. Richman, M. H. Grieco, and the AZT Collaborative Working Group.** 1987. The efficacy of azidothymidine (AZT) in the treatment of patients with AIDS and AIDS-related complex—a double-blind placebo-controlled trial. *N. Engl. J. Med.* **317:**185–191.
16. **Fisher, R. A., J. M. Bertonis, W. Meier, V. A. Johnson, D. S. Costopoulos, T. Liu, R. Tizarol, B. D. Walker, M. S. Hirsch, R. T. Schooley, and R. A. Flavell.** 1988. HIV infection is blocked in vitro by recombinant soluble CD4. *Nature* (London) **331:**76–78.
17. **Furman, P. A., J. A. Fyfe, M. H. St. Clair, K. Weinhold, J. L. Rideout, G. A. Freeman, S. Nusinoff Lehrman, D. P. Bolognesi, S. Broder, H. Mitsuya, and D. W. Barry.** 1986. Phosphorylation of 3′-azido-3′-deoxythymidine and selective interaction of the 5′-triphosphate with human immunodeficiency virus reverse transcriptase. *Proc. Natl. Acad. Sci. USA* **83:**8333–8337.
18. **Gartner, S., P. Markovits, D. M. Markovitz, M. H. Kaplan, R. C. Gallo, and M. Popovic.** 1986. The role of mononuclear phagocytes in HTLV-III/LAV infection. *Science* **233:**215–219.
19. **Gaub, J., C. Pedersen, A. G. Poulsen, L. R. Mathiesen, K. Ulrich, B. O. Lindhardt, V. Faber, J. Gerstoft, B. Hofmann, J. O. Lernestedt, C. M. Nielsen, J. O. Nielsen, and P. Platz.** 1987. The effect of foscarnet (phosphonoformate) on human immunodeficiency virus isolation, T cell subsets, and lymphocyte function in AIDS patients. *AIDS* **1:**27–33.
20. **Gilbert, B. E., and V. Knight.** 1986. Biochemistry and clinical applications of ribavirin. *Antimicrob. Agents Chemother.* **30:**201–205.
21. **Goh, W. C., C. Rosen, J. Sodroski, D. D. Ho, and W. A. Haseltine.** 1986. Identification of a protein encoded by the *trans*-activator gene *tatIII* of human T-cell lymphotropic retrovirus type III. *J. Virol.* **59:**181–184.
22. **Goh, W. C., J. G. Sodroski, C. A. Rosen, and W. A. Haseltine.** 1987. Expression of the *art* gene protein of human T-lymphotropic virus type III (HTLV-III/LAV) in bacteria. *J. Virol.* **61:**633–637.
23. **Hammer, S. M., and J. M. Gillis.** 1987. Synergistic activity of granulocyte-macrophage colony-stimulating factor and 3′-azido-3′-deoxythymidine against human immunodeficiency virus in vitro. *Antimicrob. Agents Chemother.* **31:**1046–1050.

24. **Hammer, S. M., J. M. Gillis, J. E. Groopman, and R. M. Rose.** 1986. In vitro modification of human immunodeficiency virus infection by granulocyte-macrophage colony-stimulating factor and gamma interferon. *Proc. Natl. Acad. Sci. USA* **83:**8734–8738.
25. **Hartshorn, K. L., D. Neumeyer, M. W. Vogt, R. T. Schooley, and M. S. Hirsch.** 1987. Activity of interferons alpha, beta, gamma against human immunodeficiency virus replication in vitro. *AIDS Res. Hum. Retroviruses* **3:**125–133.
26. **Hartshorn, K. L., E. G. Sandstrom, D. Neumeyer, T. J. Paradis, T. C. Chou, R. T. Schooley, and M. S. Hirsch.** 1986. Synergistic inhibition of human T-cell lymphotropic virus type III replication in vitro by phosphonoformate and recombinant alpha-A interferon. *Antimicrob. Agents Chemother.* **30:**189–191.
27. **Hartshorn, K. L., M. W. Vogt, T.-C. Chou, R. S. Blumberg, R. Byington, R. T. Schooley, and M. S. Hirsch.** 1987. Synergistic inhibition of human immunodeficiency virus in vitro by azidothymidine and recombinant alpha A interferon. *Antimicrob. Agents Chemother.* **31:**168–172.
28. **Hirsch, M. S., and J. C. Kaplan.** 1987. Treatment of human immunodeficiency virus infections. *Antimicrob. Agents Chemother.* **31:**839–843.
29. **Ho, D. D., K. L. Hartshorn, T. R. Rota, C. A. Andrews, J. C. Kaplan, R. T. Schooley, and M. S. Hirsch.** 1985. Recombinant human interferon alfa-A suppresses HTLV-III replication in vitro. *Lancet* **i:**602–604.
30. **Horwitz, J. P., J. Chua, and M. Noel.** 1964. Nucleosides. V. The mononesylates of 1-(2′-deoxy-beta-D-lyxofuranosyl)thymidine. *J. Org. Chem.* **29:**2076–2078.
31. **Jasmin, C., J.-C. Chermann, G. Herve, A. Teze, P. Souchay, C. Boy-Loustau, N. Raybaud, F. Sinoussi, and M. Raynaud.** 1974. In vivo inhibition of murine leukemia and sarcoma viruses by the heteropolyanion 5-tungsto-2-antimoniate. *J. Natl. Cancer Inst.* **53:**469–474.
32. **Klatzmann, D., E. Champagne, S. Chamaret, J. Gruest, D. Geutard, T. Hercend, J. C. Gluckman, and L. Montagnier.** 1984. T-lymphocyte T4 molecule behaves as the receptor for human retrovirus LAV. *Nature* (London) **312:**767–768.
33. **Larsson, A., and B. Oberg.** 1981. Selective inhibition of herpesvirus DNA synthesis by foscarnet. *Antiviral Res.* **1:**55–62.
34. **Levine, A. M., P. S. Gill, J. Cohen, J. G. Hawkins, S. C. Formenti, S. Aguilar, P. R. Meyer, M. Krailo, J. Parker, and S. Rasheed.** 1986. Suramin antiviral therapy in the acquired immunodeficiency syndrome. *Ann. Intern. Med.* **105:**32–37.
35. **Margalith, M., H. Falk, and A. Panet.** 1982. Differential inhibition of DNA polymerase and RNase H activities of the reverse transcriptase by phosphonoformate. *Mol. Cell. Biochem.* **43:**97–103.
36. **McCormick, J. B., J. P. Getchell, S. W. Mitchell, and D. R. Hicks.** 1984. Ribavirin suppresses replication of lymphadenopathy associated virus in cultures of human adult T lymphocytes. *Lancet* **ii:**1367–1369.
37. **McCormick, J. B., I. King, P. A. Webb, C. L. Scribner, R. B. Craven, K. M. Johnson, L. H. Elliott, and R. Belmont-Williams.** 1986. Lassa fever. Effective therapy with ribavirin. *N. Engl. J. Med.* **314:**20–26.
38. **Meshnick, S. R.** 1984. The chemotherapy of African trypanosomiasis, p. 165–199. *In* J. M. Mansfield (ed.), *Parasitic Diseases*, vol. 2. Marcel Dekker, Inc., New York.
39. **Mitsuya, H., and S. Broder.** 1986. Inhibition of the in vitro infectivity and cytopathic effect of human T-lymphotropic virus type III/lymphadenopathy-associated virus (HTLV-III/LAV) by 2′,3′-dideoxynucleosides. *Proc. Natl. Acad. Sci. USA* **83:**1911–1915.
40. **Mitsuya, H., M. Popovic, R. Yarchoan, S. Matsushita, R. C. Gallo, and S. Broder.** 1984.

Suramin protection of T-cells in vitro against infectivity and cytopathic effect of HTLV-III. *Science* **226:**172–174.
41. **Mitsuya, H., K. J. Weinhold, P. A. Furman, M. H. St. Clair, S. Nusinoff-Lehrman, R. C. Gallo, D. Bolognesi, D. W. Barry, and S. Broder.** 1985. 3′-Azido-3′-deoxythymidine (BW A509U): an antiviral agent that inhibits the infectivity and cytopathic effect of human T-lymphotropic virus type III/lymphadenopathy-associated virus in vitro. *Proc. Natl. Acad. Sci. USA* **82:**7096–7100.
42. **Montefiori, D. C., and W. M. Mitchell.** 1987. Antiviral activity of mismatched double-stranded RNA against human immunodeficiency virus in vitro. *Proc. Natl. Acad. Sci. USA* **84:**2985–2989.
43. **Nakashima, H., T. Yoshida, S. Harada, and N. Yamamoto.** 1986. Recombinant human interferon gamma suppresses HTLV-III replication in vitro. *Int. J. Cancer* **38:**433–436.
44. **Oberg, B.** 1983. Antiviral effects of phosphonoformate (PFA, foscarnet sodium). *Pharmacol. Ther.* **19:**387–415.
45. **Pert, C. B., J. M. Hill, M. R. Ruff, R. M. Berman, W. G. Robey, I. O. Arthur, F. W. Ruscetti, and W. L. Farrar.** 1986. Octapeptides deduced from the neuropeptide receptor-like pattern of antigen T4 in brain potently inhibit human immunodeficiency virus receptor binding and T-cell infectivity. *Proc. Natl. Acad. Sci. USA* **83:**9254–9258.
46. **Richman, D. D., M. A. Fischl, M. H. Grieco, and the AZT Collaborative Working Group.** 1987. The toxicity of azidothymidine (AZT) in the treatment of patients with AIDS and AIDS-related complex—a double blind, placebo-controlled trial. *N. Engl. J. Med.* **317:**192–197.
47. **Ronen, D., and Y. Teitz.** 1984. Inhibition of the synthesis of Moloney leukemia virus structural proteins by N-methylisatin-beta-4′,4′-diethylthiosemicarbazone. *Antimicrob. Agents Chemother.* **26:**913–916.
48. **Rosen, C. A., J. G. Sodroski, W. C. Goh, A. I. Dayton, J. Lippke, and W. A. Haseltine.** 1986. Post-transcriptional regulation accounts for the transactivation of the human T-lymphotropic virus type III. *Nature* (London) **319:**555–559.
49. **Rupprecht, R. M., I. G. O'Brien, I. D. Rossoni, and S. Nusinoff-Lehrman.** 1986. Suppression of mouse viremia and retroviral disease by 3′-azido-3′-deoxythymidine. *Nature* (London) **323:**467–469.
50. **Sandstrom, E. G., J. C. Kaplan, R. E. Byington, and M. S. Hirsch.** 1985. Inhibition of human T-cell lymphotropic virus type III in vitro by phosphonoformate. *Lancet* **i:**1480–1482.
51. **Sarin, P. S., Y. Taguchi, D. Sun, A. Thornton, R. C. Gallo, and B. Oberg.** 1985. Inhibition of HTLV-III/LAV replication by foscarnet. *Biochem. Pharmacol.* **34:**4075–4079.
52. **Shannon, W. M.** 1977. Selective inhibition of RNA tumor virus replication in vitro and evaluation of candidate antiviral agents in vivo. *Ann. N.Y. Acad. Sci.* **284:**472–507.
53. **Sodroski, J., W. C. Goh, C. Rosen, A. Dayton, E. Terwilliger, and W. Haseltine.** 1986. A second post-transcriptional trans-activator gene required for HTLV-III replication. *Nature* (London) **321:**412–417.
54. **Sundquist, B., and E. Larner.** 1979. Phosphonoformate inhibition of visna virus replication. *J. Virol.* **30:**847–851.
55. **Vogt, M. W., K. L. Hartshorn, P. A. Furman, T.-C. Chou, J. A. Fyfe, L. A. Coleman, C. Crumpacker, R. T. Schooley, and M. S. Hirsch.** 1987. Ribavirin antagonism of anti-HIV effect of azidothymidine by phosphorylation inhibition. *Science* **235:**1376–1379.
56. **Walker, B. D., M. Kowalski, W. C. Goh, K. Kozarsky, M. Krieger, C. Rosen, L. Rohrschneider, W. A. Haseltine, and J. Sodroski.** 1987. Inhibition of human immunode-

ficiency virus syncytium formation and replication by castanospermine. *Proc. Natl. Acad. Sci. USA* **84:**8120–8124.

57. **Wills, R. J., S. Dennis, H. E. Spiegel, D. M. Gibson, and P. I. Nadler.** 1984. Interferon kinetics and adverse reactions after intravenous, intramuscular, and subcutaneous injection. *Clin. Pharmacol. Ther.* **35:**722–727.
58. **Wong-Staal, F., P. K. Chanda, and J. Gheayeb.** 1987. Human immunodeficiency virus: the eighth gene. *AIDS Res. Hum. Retroviruses* **3:**33–39.
59. **Yarchoan, R., G. Berg, P. Browers, M. A. Fischel, A. R. Spitzer, A. Wichman, J. Grafman, R. V. Thomas, B. Safai, A. Brunetti, C. F. Perno, P. J. Schmidt, S. M. Larson, C. E. Myers, and S. Broder.** 1987. Response of human-immunodeficiency virus associated neurological disease to 3′-azido-3′-deoxythymidine. *Lancet* **i:**132–135.
60. **Yarchoan, R., and S. Broder.** 1987. Development of antiretroviral therapy for the acquired immunodeficiency syndrome and related disorders. *N. Engl. J. Med.* **316:**557–564.
61. **Yarchoan, R., R. W. Klecker, K. J. Weinhold, P. D. Markham, H. K. Lyerly, D. T. Durack, E. Gelmann, S. Nusinoff Lehrmann, R. M. Blum, D. W. Barry, G. M. Shearer, M. A. Fischl, H. Mitsuya, R. C. Gallo, J. M. Collins, D. P. Bolognesi, C. E. Myers, and S. Broder.** 1986. Administration of 3′-azido-3′-deoxythymidine, an inhibitor of HTLV-III replication, to patients with AIDS and AIDS-related-complex. *Lancet* **i:**575–580.
62. **Zamecnik, P. C., J. Goodchild, Y. Taguchi, and P. S. Sarin.** 1986. Inhibition of replication and expression of human T-cell lymphotropic virus type III in cultured cells by exogenous synthetic oligonucleotides complementary to viral RNA. *Proc. Natl. Acad. Sci. USA* **83:**4143–4146.

Part IV.

EPSTEIN-BARR VIRUS INFECTIONS

Chapter 20

Chronic Epstein-Barr Virus Infection: a Critical Assessment

Stephen E. Straus

INTRODUCTION

During the past decade a series of studies reported evidence that Epstein-Barr virus (EBV) is associated with chronic illness (1, 5, 8, 14, 17, 27, 31, 41, 45, 52, 55, 59). Although in itself not surprising, because of the propensity of all of the other well-characterized human herpesviruses to cause recurrent and chronic infections, the hypothesis that EBV causes such infections has been difficult to prove and, in general, has been subject to considerable controversy. This review summarizes the evidence favoring and disputing the "chronic EBV disease" hypothesis.

EBV INFECTION

Biology of EBV

Except for another recently described B-lymphotropic virus about which little is known, EBV is unique among human herpesviruses in its restricted tropism and biological properties (35, 43). EBV infects B lymphocytes, salivary glands, and oral and genital epithelium (49, 50). Only two genotypic classes of EBV are recognized. Virus from the saliva of seropositive persons typically transforms cord blood B lymphocytes. Unusual strains with a small but perhaps critical difference in DNA sequence tend to lyse infected B cells and are not readily detected by classic cord blood transformation assays.

B lymphocytes, the major focus of studies of host responses to EBV, may not be the ideal subject of such investigations, because active virus

---

Stephen E. Straus • Medical Virology Section, National Institute of Allergy and Infectious Diseases, Bethesda, Maryland 20892.

replication is largely restricted to salivary tissues. Nonetheless, we have learned that the regulation of EBV infection, at least at the lymphocyte level, requires a concerted and sustained cellular immune response (19, 57). Deficiencies in immune-system containment of EBV-infected B cells permit their proliferation and, in extreme instances, their dangerous infiltration of organs (18, 29, 32, 42, 47, 59). Immune deficiency presumably also releases constraints on EBV residing in salivary tissues because rates of oropharyngeal excretion of EBV rise in that setting.

Epidemiology and Transmission

Little is known of the spread of EBV, since there are no adequate animal models and transmission to volunteers was found to be difficult and is no longer considered an ethically appropriate undertaking (15). It is believed that the virus is acquired primarily via infected saliva, but the recent demonstration of EBV in cervical epithelium and the suggestion of its presence in genital ulcers compel us to consider that other routes of transmission may be important in some people (49, 50). Transfused blood is an uncommon vehicle for spreading EBV, and it is reasonable, although unproven, that transplanted organs would also convey the virus (41).

EBV is usually acquired during childhood, but in developed nations one-half to two-thirds of initial infections are delayed until adolescence or adulthood, when they are far more likely to be symptomatic. Eventually, 95 to 100% of people become infected. Fetal and perinatal infections have been documented but are in general considered to be of little consequence (16).

Diagnosis

Acute infectious mononucleosis, most often caused by EBV, is diagnosed clinically and confirmed by detecting substantial numbers of circulating atypical lymphocytes and elevated titers of heterophile antibodies. Heterophile antibodies are directed at animal erythrocyte antigens and have no specificity for EBV proteins or those of other known causes of the syndrome. Their presence reflects a general state of virus-induced humoral immune activation.

On the other hand, acute EBV infection, be it symptomatic or asymptomatic, is accompanied by virus-specific serologic responses that are diagnostic (Fig. 1) (23). The acute development of immunoglobulin M (IgM) antibodies to the viral capsid antigen (VCA), significant increases in IgG anti-VCA levels, and the later appearance of antibodies to EBV nuclear antigens (EBNA) all indicate primary infection. Most symptomatic primary EBV infections also result in the emergence of antibodies to viral early antigens (EA). These antibodies are operationally distin-

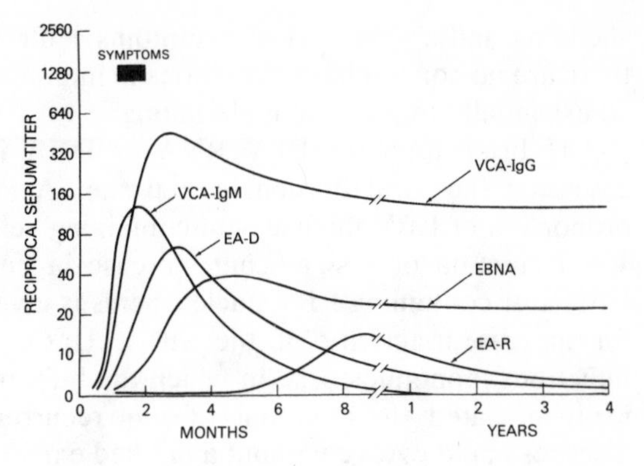

Figure 1. Typical serologic responses to EBV antigens following primary symptomatic infection. (Reprinted from *Annual Review of Medicine* [28] with permission from Annual Reviews, Inc.)

guished as diffuse (EA-D) or restricted (EA-R) in pattern, depending upon their distribution in fluorescent antibody-stained lymphocytes in which they are expressed. Anti-EA-D appears first, wanes after several months, and is then replaced by anti-EA-R responses, which themselves persist at only low levels or become undetectable in 1 to 4 years (25).

During symptomatic infections, virus can be recovered with high frequency from saliva and can be detected by anticomplementary immunofluorescence for EBNA in very small proportions (<0.1%) of circulating or lymph node lymphocytes.

A major difficulty in discerning whether EBV causes recurrent or chronic disease is the lack of corresponding diagnostic tools. Serologic tests are of limited value for this purpose. IgG antibodies to VCA and EBNA persist for life. The levels of these EBV antibodies might not undergo significant elevations with reactivation of infection. This would not be surprising, in light of the failure of serologic responses to reflect local reactivation of some other herpesviruses. For example, antibody titers to herpes simplex virus do not change significantly with recurrent oral or genital herpes.

Virus reactivation during pregnancy and in the compromised host is associated with increases in levels of antibodies to viral EAs, particularly to EA-R, but the frequency with which they persist in normal individuals and the usual lack of symptoms attributable to reactivation in any setting cause these indications to be inadequately specific (16, 21, 25, 26, 53, 54).

Virus shedding is also a nonspecific indicator of recurrent or chronic disease. Detection of EBV in saliva of normal seropositive persons is common, and increases in the titer or incidence of shedding correlate inversely with cellular immune competence (37, 51). It is therefore difficult to presume that there is a relationship between increased virus shedding and a given set of symptoms, unless one could ascertain that there are no concomitant alterations in immune function, which in itself is an essentially impossible undertaking.

Hybridization for EBV RNA or DNA is one sensitive means of assessing the viral burden in a tissue. As with EBNA staining, the proportion of EBV nucleic acid-containing cells is normally very low (3, 40). Detection of a significant increase in numbers of cells expressing EBNA or containing EBV nucleic acids is a valid and reasonable marker for increased activity of the virus. Unfortunately, just as in acute infectious mononucleosis, in which the bulk of pathology is immunologically mediated, the symptoms of some recurrent or chronic EBV-induced diseases could evolve without a marked expansion of the level of EBV in tissue.

EBV-RELATED DISEASES

Acute Infectious Mononucleosis

Acute infectious mononucleosis is characterized by fever, lymphadenopathy, pharyngitis, atypical lymphocytosis, and heterophile antibodies. EBV causes over 90% of cases in which all five of these signs and findings are present. Acute infectious mononucleosis is associated with numerous other less common symptoms and findings, including fatigue. Typically, the fatigue is debilitating for only a few weeks or months.

The unique aspect of EBV-induced mononucleosis is the degree to which disease results from the exuberant, perhaps excessive, immune responses to the virus in individuals in a remarkably narrow age group. Many of the lymphocytes that circulate and infiltrate lymphoid organs are reactive T cells (48). Selected symptoms and features of acute infectious mononucleosis are ameliorated by treatment with corticosteroids, an approach that would be predictably detrimental to the course of other infections in which virus-mediated injury is dominant.

Fatal Infectious Mononucleosis

Rare cases have been described in which neutropenia, thrombocytopenia, or anemia exceed in severity or duration the mild forms in which they commonly complicate acute infectious mononucleosis (38, 46). These patients succumb after weeks or months of illness, primarily to hemorrhage and secondary infection. One cohort of subjects at risk for fatal mononucleosis is that of young boys with an X-linked inherited disorder (42). Appearing normal in all regards until they become infected with EBV, 60 to 70% die of an explosive B-cell proliferation, hemocytopenias, and opportunistic infection. The survivors experience chronic immune problems and delayed lymphomas. The immune-system problems which predispose to fatal mononucleosis are diverse, but probably include defects in interferon and natural killer cell responses to EBV (59).

The diagnosis of these fatal disorders may be readily made, because many patients still possess anti-IgM anti-VCA at the time of death and their tissues contain increased numbers of EBV-bearing B lymphocytes.

Burkitt's Lymphoma and Nasopharyngeal Carcinoma

The EBV-related malignancies Burkitt's lymphoma and nasopharyngeal carcinoma are epidemiologically important in Central Africa and Southeast Asia (11, 39). A discussion of the potential role of EBV in their causation is beyond this review. Briefly, EBV may be functioning in these syndromes directly as an oncogenic virus, as a nonspecific stimulator of cell proliferation, or as a cofactor to other carcinogens or endogenous regulatory elements such as the c-*myc* sequences (7, 10).

Other B-Cell Lymphomas

Transplant recipients and patients with the acquired immunodeficiency syndrome are at increased risk for polyclonal and monoclonal EBV-associated B-cell lymphomas (18, 29, 32, 47). Proliferation of EBV-infected B cells becomes clinically evident when they are released sufficiently from normal immune-system containment. At some point, individual clones of cells become committed to unbridled proliferation. Demonstration of extremely elevated EBV antibody titers and increased amounts of viral nucleic acids and EBNA in tissues is diagnostic, but one or more of these features may not be seen in all patients. In the setting of organ transplantation, withdrawal of immune-system suppressants allows regression of the polyclonal tumors. Unfortunately, acquired immunodeficiency syndrome patients do not have this luxury.

Chronic EBV Infection

There are a handful of reports of individual cases or small series of patients with protracted illness attributed to EBV (1, 5, 8, 14, 17, 24, 41, 45, 52, 55). Some illnesses represent chronic sequelae of the X-linked lymphoproliferative syndrome or of EBV-associated aplastic anemia. Others appear to represent a spectrum of severe and occasionally fatal conditions unrelated to the above in which the cardinal features include acquired humoral and cellular immune dysfunction, fever, weight loss, interstitial pneumonitis, and, occasionally, chronic hepatitis or uveitis. In some patients, titers of IgG antibodies to VCA or EA are extraordinarily elevated, being >1:5,120 and >1:640, respectively. In some, antibodies to one (EBNA-1) or all EBNA proteins are lacking (22, 36).

In each instance the imprecision or lack of availability of certain diagnostic tools makes it difficult to prove that EBV was at fault rather than being nonspecifically reactivated secondary to immune defects of other cause. The types of observation which favor EBV as the etiologic agent include detection of increased EBV content in affected tissues and demonstration that the EBV itself is hard to recover and possesses the lytic rather than immortalizing phenotype and genotype (1, 3, 19, 45).

Chronic Fatigue Syndrome

Much of the controversy surrounding the hypothesis that EBV induces chronic infection relates not to studies of the rare severe illnesses mentioned above, but to a series of reports since 1982 in which groups of far healthier adults with debilitating fatigue and other constitutional symptoms were noted to have unusual profiles of EBV antibodies and a variety of subtle immunologic problems (5, 14, 27, 44, 52, 56). Legions of these tired patients have subsequently seized upon the reports as providing answers to their long-unheeded complaints.

There is evidence that such individuals do suffer organic illness. In some cases the patients fall ill along with relatives, fellow employees, or neighbors. Numerous reports of such epidemics of neuromyasthenia, or severe fatigue, have appeared since 1934, and although community hysteria has been considered, many of the data favor an infectious agent or agents (12, 20, 24, 33, 34). In some outbreaks an incubation period is estimated to be only 2 to 3 weeks, which is far shorter than the 6 to 12 weeks typical for acute infectious mononucleosis.

The most recent and best-publicized epidemic of this type was first noted by private practitioners in Incline Village, Nevada, in early 1985. To date, some 300 persons from that and surrounding communities have been recorded to experience weeks to months of severe fatigue, difficulty with concentration, feverishness, diffuse aches, and a range of neurologic problems. Many are still ill as of this writing (P. Cheney, personal communication).

Investigators from Harvard University, in collaboration with local physicians and epidemiologists from the Centers for Disease Control, conducted independent assessments of the character of the syndrome and of its serologic features. Scientists from the Centers for Disease Control reported higher levels of antibodies to EBV in a small group of patients than in controls (24). Importantly, they noted increased levels of antibodies to other viruses as well, including cytomegalovirus, herpes simplex virus, and measles virus, suggesting a broad polyclonal response of uncertain cause.

Many individuals considered to have the features of the chronic fatigue syndrome are not part of epidemics, but report nonetheless that their illnesses followed an infectious-type illness pattern with either coryzal, influenzal, gastrointestinal, lymphadenopathic, or neurologic features. Salit chose to identify such individuals as having "sporadic post-infectious neuromyasthenia" (44). Other than their epidemiology, these illnesses differ little from those of epidemic neuromyasthenia.

In both the epidemic and sporadic settings, two-thirds or more of patients are women; their mean age is in the mid 30s, and most are Caucasian and were previously very active, even driven. Objective physical findings or abnormalities of routine hematologic or chemical assessments are few. In fact, gross lymphadenopathy, neurologic findings, or abnormalities of routine laboratory tests strongly suggest alternative diagnoses (52, 56).

Patients with illnesses of sporadic onset often exhibit subtle immune abnormalities including mild partial hypogammaglobulinemia, elevated levels of immune complexes, and elevated levels of the interferon-induced enzyme 2′,5′-oligoadenylate synthetase (8, 14, 17, 27, 31, 52). Reduced levels of gamma interferon and interleukin-2 synthesis in vitro, inadequate outgrowth inhibition of EBV-infected B cells by T cells from patients, and increased levels of suppressor cell activity have also been reported (30, 52, 58).

The patients with the chronic fatigue syndrome are quite heterogeneous in terms of their historical and laboratory features (Tables 1 and 2). In a minority, the chronic illnesses began during an episode of classic acute infectious mononucleosis. Many, but not all, have higher levels of IgG anti-VCA and EA than seen in most normal individuals (14, 27, 52, 55). Henle et al. recently noted that anti-EBNA-2 titers equal or exceed those of anti-EBNA-1 for many of these patients, a pattern characteristic of normal individuals 3 to 12 months after acute infectious mononucleosis (22). As many as 15% of the patients totally lack anti-EBNA or anti-EBNA-1. Interestingly, about 3% of patients with chronic neuromyasthenia lack all antibodies to EBV, confirming that in at least these few individuals, factors other than EBV can trigger this syndrome.

In studies of approximately 100 chronically fatigued patients at the National Institutes of Health, we found there to be little correlation between immune abnormalities or EBV titer and severity or type of clinical complaint (S. Straus and J. Dale, unpublished data). Among a cohort of 27 patients studied prospectively by the National Institutes of Health for at least 2 years and chosen in part because of their unusual EBV antibody profiles, antibodies to cytomegalovirus were also significantly elevated, as in the Centers for Disease Control study of Nevada patients (24; S. Straus and J. Dale, unpublished observations).

Table 1. Common Recurring or Persisting Complaints in Patients with the Chronic Fatigue Syndrome

Feature	Avg rate (%)
Persisting severe fatigue	90–100
Sore throat	80–100
Difficulty concentrating	80–90
Increased headache	70–90
Tender lymph nodes	75–85
Muscle aches	70–80
Joint aches	70–80
Feverishness	70–80
Difficulty sleeping	50–70
Allergies, sensitivities	40–60
Abdominal pain	30–50
Depression	30–40
Light-headedness	20–30

The clinical heterogeneity, the diversity of immune findings, and the serologic responses to several antigens argue in favor of multiple etiologies for the syndrome (4, 6, 9, 24, 44). EBV may be the most likely inciting agent in patients in whom the chronic illness was initiated as acute infectious mononucleosis (56). The similarity of complaints, despite the possibility of multiple etiologies, suggests a common underlying immune pathogenesis. Perhaps the disorder involves interleukins or other lympho-

Table 2. Results of Common Laboratory Tests for Patients with the Chronic Fatigue Syndrome

Test	Result
Blood and differential counts	Normal
Chemical profile	Normal
Urinalysis	Normal
Sedimentation rate	Normal
Thyroid hormones	Normal
Delayed-type hypersensitivity	Normal
T helper/suppressor ratio	Normal, some high
Heterophile	Negative, rare positive
Quantitative immunoglobulin	Normal, some low
Serum complement	Normal
Circulating immune complexes	Normal, many high
Antinuclear antibodies	Absent, few low positive
IgG antibodies to EBV-VCA, EA	Normal, many high
IgG antibodies to EBNA	Normal, few absent
Antibodies to other viruses	Normal, many high

kines, which are known in experimental therapeutic trials to instigate many of the symptoms reported by neuromyasthenic patients (28).

Treatment

The therapeutic approach to all EBV-related disorders is a supportive one, depending heavily on education, guidance, counseling, and common sense. Antiviral drugs, most notably acyclovir, have been evaluated to some extent in many EBV-related conditions. Although a reduction of virus shedding in saliva was seen during intravenous acyclovir therapy of patients with acute infectious mononucleosis, no clinical benefits have been associated with that therapy (2). Significant improvement was attributed to acyclovir in a few patients with EBV-associated polyclonal B-cell lymphomas and with severe chronic EBV infections (18, 45).

The desperation of patients with chronic neuromyasthenia has compelled many of them to seek a range of traditional and nontraditional therapies, about which anecdotes abound. Anti-inflammatory agents to reduce fever and discomfort and mild soporifics for sleep do seem to be beneficial. Professional counseling and appropriate use of psychotropic agents are also appropriate for some patients. No merit, however, can yet be ascribed to the numerous anecdotes of improvements caused by nystatin, ranitidine, isoprinosine, ribavirin, amantadine, transfer factor, or acupuncture, to name but a few.

In one trial, intramuscular immunoglobulin therapy was thought to benefit patients (13). A more extensive controlled trial is needed to validate those findings. Preliminary results of a placebo-controlled trial of intravenous and peroral acyclovir treatment conducted at the National Institutes of Health indicate a high rate of response to placebo and no greater responses among acyclovir recipients (S. E. Straus, J. K. Dale, G. Armstrong, O. Preble, T. Lawley, and W. Henle, *Clin. Res.* **35:**618A, 1987).

Given the controversial nature of the chronic fatigue syndrome and the lack of adequate measurements of the subjective complaints, it is incumbent upon the physician to exclude other diagnoses or intercurrent illnesses and to guide the patient away from harmful or destructive practices and toward a more balanced acceptance of his limitations.

FUTURE DIRECTIONS

Normal humoral and cellular immune responses to EBV and its constituent proteins and the state and expression of EBV genes in latency must be better defined to identify sensitive barometers of virus reactivation. Some of these will help discern which viral reactivation events are clinically significant. There must also be a fuller description of the metabolic and immunologic features of chronic fatigue. It is hoped that such studies will indicate the patients for whom antiviral drugs may be useful and the ones for whom immunological interventions should be considered.

Literature Cited

1. **Alfieri, C., F. Ghibu, and J. H. Joncas.** 1984. Lytic, nontransforming Epstein-Barr virus (EBV) from a patient with chronic active EBV infection. *Can. Med. Assoc. J.* **131:**1249–1252.
2. **Andersson, J., S. Britton, I. Ernberg, U. Andersson, W. Henle, B. Skoldenberg, and A. Tisell.** 1986. Effect of acyclovir on infectious mononucleosis: a double-blind, placebo-controlled study. *J. Infect. Dis.* **153:**283–290.
3. **Andiman, W., L. Gradoville, L. Heston, R. Heydorff, M. E. Savage, G. Kitchingman, D. Shedd, and G. Miller.** 1983. Use of cloned probes to detect Epstein-Barr viral DNA in tissues of patients with neoplastic and lymphoproliferative diseases. *J. Infect. Dis.* **148:**967–977.
4. **Arnold, D. L., G. K. Radda, P. J. Bore, P. Styles, and D. J. Taylor.** 1984. Excessive intracellular acidosis of skeletal muscle on exercise in a patient with a post-viral exhaustion/fatigue syndrome. *Lancet* **i:**1367–1369.
5. **Ballow, J., J. Seeley, D. Purtilo, S. St. Onge, K. Sakamoto, and F. R. Rickles.** 1982. Familial chronic mononucleosis. *Ann. Intern. Med.* **97:**821–825.
6. **Behan, P. O., W. M. H. Behan, and E. J. Bell.** 1985. The postviral fatigue syndrome—an analysis of the findings in 50 cases. *J. Infect.* **10:**211–222.
7. **Bornkamm, G. W., M. V. Knebel-Doeberitz, and G. M. Lenoir.** 1984. No evidence for differences in the Epstein-Barr virus genome carried in Burkitt lymphoma cells and nonmalignant lymphoblastoid cells from the same patients. *Proc. Natl. Acad. Sci. USA* **81:**4930–4934.
8. **Borysiewicz, L. K., S. J. Haworth, J. Cohen, J. Mundin, A. Rickinson, and J. G. P. Sissons.** 1986. Epstein Barr virus-specific immune defects in patients with persistent symptoms following infectious mononucleosis. *Q. J. Med.* **58:**111–121.
9. **Buchwald, D., J. L. Sullivan, and A. L. Komaroff.** 1987. Frequency of "chronic active Epstein-Barr virus infection" in a general medical practice. *J. Am. Med. Assoc.* **257:**2303–2307.
10. **Croce, C. M.** 1985. Chromosomal translocations, oncogenes, and B-cell tumors. *Hosp. Pract.* **1:**41–47.
11. **de Thé, G., A. Geser, N. E. Day, P. M. Tukei, E. H. Williams, D. P. Beri, P. G. Smith, A. G. Dean, G. W. Bornkamm, P. Feorino, and W. Henle.** 1978. Epidemiological evidence for causal relationship between Epstein-Barr virus and Burkitt's lymphoma from Ugandan prospective study. *Nature* (London) **274:**756–761.
12. **Dillon, M. J., W. C. Marshall, J. A. Dudgeon, and A. J. Steigman.** 1974. Epidemic neuromyasthenia: outbreak among nurses at a children's hospital. *Br. Med. J.* **1:**301–305.
13. **DuBois, R. E.** 1986. Gamma globulin therapy for chronic mononucleosis syndrome. *AIDS Res.* **2:**S191–S195.
14. **DuBois, R. E., J. K. Seeley, I. Brus, K. Sakamoto, M. Ballow, S. Harada, T. A. Bechtold, G. Pearson, and D. T. Purtilo.** 1984. Chronic mononucleosis syndrome. *South. Med. J.* **77:**1376–1382.

15. **Evans, A. S.** 1947. Experimental attempts to transmit infectious mononucleosis to man. *Yale J. Biol. Med.* **20:**19–26.
16. **Fleisher, G., and R. Bologonese.** 1984. Epstein-Barr virus infections in pregnancy: a prospective study. *J. Pediatr.* **104:**374–379.
17. **Hamblin, T. J., J. Hussain, A. N. Akbar, Y. C. Tang, J. L. Smith, and D. B. Jones.** 1983. Immunological reason for chronic ill health after infectious mononucleosis. *Br. Med. J.* **287:**85–88.
18. **Hanto, D. W., G. Frizzera, K. J. Gajl-Peczalska, K. Sakamoto, D. T. Purtilo, H. H. Balfour, R. L. Simmons, and J. S. Najarian.** 1982. Epstein-Barr virus-induced B-cell lymphoma after renal transplantation. *N. Engl. J. Med.* **306:**913–918.
19. **Haynes, B. F., R. T. Schooley, C. R. Payling-Wright, J. E. Grouse, R. Dolin, and A. S. Fauci.** 1979. Emergence of suppressor cells of immunoglobulin synthesis during acute Epstein-Barr virus-induced infectious mononucleosis. *J. Immunol.* **123:**2095–2101.
20. **Henderson, D. A., and A. Shelokov.** 1959. Epidemic neuromyasthenia—clinical syndrome? *N. Engl. J. Med.* **260:**757–764.
21. **Henle, W., and G. Henle.** 1981. Epstein-Barr virus-specific serology in immunologically compromised individuals. *Cancer Res.* **41:**4222–4225.
22. **Henle, W., G. Henle, J. Andersson, I. Ernberg, G. Klein, C. A. Horwitz, G. Marklund, L. Rymo, C. Wellinder, and S. E. Straus.** 1987. Antibody responses to Epstein-Barr virus-determined nuclear antigen (EBNA)-1 and EBNA-2 in acute and chronic Epstein-Barr virus infection. *Proc. Natl. Acad. Sci. USA* **84:**570–574.
23. **Henle, W., G. Henle, and C. A. Horwitz.** 1974. Epstein-Barr virus-specific diagnostic tests in infectious mononucleosis. *Hum. Pathol.* **5:**551–565.
24. **Holmes, G. P., J. E. Kaplan, J. A. Stewart, B. Hunt, P. F. Pinsky, and L. B. Schonberger.** 1987. A cluster of patients with a chronic mononucleosis-like syndrome: is Epstein-Barr virus the cause? *J. Am. Med. Assoc.* **257:**2297–2302.
25. **Horwitz, C. A., G. Henle, H. Rudnick, and E. Latts.** 1985. Long-term serological follow-up of patients with Epstein-Barr virus after recovery from infectious mononucleosis. *J. Infect. Dis.* **151:**1150–1153.
26. **Horwitz, C. A., W. Henle, G. Henle, and H. Schmitz.** 1975. Clinical evaluation of patients with infectious mononucleosis and development of antibodies to the R component of the Epstein-Barr virus-induced early antigen complex. *Am. J. Med.* **58:**330–338.
27. **Jones, J. F., C. G. Ray, L. L. Minnich, M. J. Hicks, R. Kibler, and D. O. Lucas.** 1985. Evidence for active Epstein-Barr virus infection in patients with persistent, unexplained illnesses: elevated antibody anti-early antigen antibodies. *Ann. Intern. Med.* **102:**1–7.
28. **Jones, J. F., and S. E. Straus.** 1987. Chronic Epstein-Barr virus infection. *Annu. Rev. Med.* **38:**195–209.
29. **Katz, B. Z., W. A. Andiman, R. Eastman, K. Martin, and G. Miller.** 1986. Infection with two genotypes of Epstein-Barr virus in an infant with AIDS and lymphoma of the central nervous system. *J. Infect. Dis.* **153:**601–604.
30. **Kibler, R., D. O. Lucas, M. J. Hicks, B. T. Poulos, and J. F. Jones.** 1985. Immune function in chronic active Epstein-Barr virus infection. *J. Clin. Immunobiol.* **5:**46–54.
31. **Kuis, W., J. J. Roord, B. J. M. Zegers, A. B. Rickinson, J. G. Kapsenberg, H. The, and J. W. Stoop.** 1985. Heterogeneity of immune defects in three children with a chronic active Epstein-Barr virus infection. *J. Clin. Immunol.* **5:**377–385.
32. **Martin, P. J., H. M. Shulman, W. H. Schuback, J. A. Hansen, A. Fefer, G. Miller, and E. D. Thomas.** 1984. Fatal Epstein-Barr-virus-associated proliferation of donor B cells after treatment of acute graft-versus-host disease with a murine anti-T-cell antibody. *Ann. Intern. Med.* **101:**310–315.

33. **McEvedy, C. P., and A. W. Beard.** 1970. Royal Free epidemic of 1955: a reconsideration. *Br. Med. J.* **1:**7–11.
34. **McEvedy, C. P., and A. W. Beard.** 1970. Concept of benign myalgic encephalomyelitis. *Br. Med. J.* **1:**11–15.
35. **Miller, G.** 1985. Epstein-Barr virus, p. 563–589. *In* B. N. Fields (ed.), *Virology*. Raven Press, New York.
36. **Miller, G., E. Grogan, D. K. Fischer, J. C. Niederman, R. T. Schooley, W. Henle, G. Lenois, and C.-R. Liu.** 1985. Antibody responses to two Epstein-Barr virus nuclear antigens defined by gene transfer. *N. Engl. J. Med.* **312:**750–755.
37. **Miller, G., J. C. Niederman, and L. Andrews.** 1973. Prolonged oropharyngeal excretion of Epstein-Barr virus after infectious mononucleosis. *N. Engl. J. Med.* **288:**229–232.
38. **Neel, E. U.** 1976. Infectious mononucleosis: death due to agranulocytosis and pneumonia. *J. Am. Med. Assoc.* **236:**1493–1494.
39. **Nonoyama, M., and J. S. Pagano.** 1973. Homology between Epstein-Barr virus DNA and viral DNA from Burkitt's lymphoma and nasopharyngeal carcinoma determined by DNA-DNA reassociation kinetics. *Nature* (London) **242:**44–47.
40. **Pagano, J. S., C.-H. Huang, and Y.-T. Huang.** 1976. Epstein-Barr virus genome in infectious mononucleosis. *Nature* (London) **263:**787–789.
41. **Purtilo, D. T., L. A. Paquin, K. Sakamoto, L. M. Hutt, J. P. S. Yang, S. Sparling, N. Beberman, and R. A. McAuley.** 1980. Persistent transfusion-associated infectious mononucleosis with transient acquired immunodeficiency. *Am. J. Med.* **68:**437–440.
42. **Purtilo, D. T., K. Sakamoto, V. Barnabei, J. Seeley, T. Bechtold, G. Rogers, J. Yetz, and S. Harada.** 1982. Epstein-Barr virus-induced diseases in boys with the X-linked lymphoproliferative syndrome (XLP). *Am. J. Med.* **73:**49–56.
43. **Salahuddin, S. Z., D. V. Ablashi, P. D. Markham, S. F. Josephs, S. Sturzenegger, M. Kaplan, G. Halligan, P. Biberfeld, F. Wong-Staal, B. Kramarsky, and R. C. Gallo.** 1986. Isolation of a new virus, HBLV, in patients with lymphoproliferative disorders. *Science* **234:**596–601.
44. **Salit, I. E.** 1985. Sporadic postinfectious neuromyasthenia. *Can. Med. Assoc. J.* **133:**659–663.
45. **Schooley, R. T., R. W. Carey, G. Miller, W. Henle, R. Eastman, E. J. Mark, K. Kenyon, E. O. Wheeler, and R. H. Robin.** 1986. Chronic Epstein-Barr virus infection associated with fever and interstitial pneumonitis: clinical and serologic features and response to antiviral chemotherapy. *Ann. Intern. Med.* **104:**636–643.
46. **Shadduck, R. K., A. Winkelstein, Z. Zeigler, J. Lichter, M. Goldstein, M. Michaels, and B. Rabin.** 1979. Aplastic anemia following infectious mononucleosis: possible immune etiology. *Exp. Hematol.* **7:**264–271.
47. **Shearer, W. T., J. Ritz, M. J. Finegold, C. Guerra, H. M. Rosenblatt, D. E. Lewis, M. S. Pollack, L. H. Taber, C. V. Sumaya, F. C. Grumet, M. L. Cleary, R. Warnke, and J. Sklar.** 1985. Epstein-Barr virus-associated B-cell proliferations of diverse clonal origins after bone marrow transplantation in a 12-year-old patient with severe combined immunodeficiency. *N. Engl. J. Med.* **312:**1151–1159.
48. **Sheldon, P. J., E. H. Hemsted, M. Papamichail, and E. J. Holborow.** 1973. Thymic origin of atypical lymphoid cells in infectious mononucleosis. *Lancet* **i:**1153–1154.
49. **Sixbey, J. W., S. M. Lemon, and J. S. Pagano.** 1986. A second site of Epstein-Barr virus shedding: the uterine cervix. *Lancet* **ii:**1122–1124.
50. **Sixbey, J. W., J. G. Nedrud, N. Raab-Traub, R. A. Hanes, and J. S. Pagano.** 1984. Epstein-Barr virus replication in oropharyngeal epithelial cells. *N. Engl. J. Med.* **310:**1225–1230.
51. **Strauch, G., L. Andrews, G. Miller, and N. Siegel.** 1974. Oropharyngeal excretion of

Epstein-Barr virus by renal transplant recipients and other patients treated with immunosuppressive drugs. *Lancet* **i**:234–237.
52. **Straus, S. E., G. Tosato, G. Armstrong, T. Lawley, O. T. Preble, W. Henle, R. Davey, G. Pearson, J. Epstein, I. Brus, and R. M. Blaese.** 1985. Persisting illness and fatigue in adults with evidence of Epstein-Barr virus infection. *Ann. Intern. Med.* **102**:7–16.
53. **Sumaya, C. V.** 1977. Endogenous reactivation of Epstein-Barr virus infections. *J. Infect. Dis.* **135**:374–379.
54. **Sumaya, C. V., R. N. Boswell, Y. Ench, D. L. Kisner, E. M. Hersh, J. M. Reuben, and P. W. A. Mansell.** 1986. Enhanced serological and virological findings of Epstein-Barr virus in patients with AIDS and AIDS-related complex. *J. Infect. Dis.* **154**:864–870.
55. **Tobi, M., Z. Ravid, V. Feldman-Weiss, E. Ben-Chetrit, A. Morag, I. Chowers, Y. Michaeli, M. Shalit, and H. Knobler.** 1982. Prolonged atypical illness associated with serological evidence of persistent Epstein-Barr virus infection. *Lancet* **i**:61–64.
56. **Tobi, M., and S. E. Straus.** 1985. Chronic Epstein-Barr virus disease: a workshop held by the National Institute of Allergy and Infectious Diseases. *Ann. Intern. Med.* **103**:951–953.
57. **Tosato, G., I. T. Magrath, and R. M. Blaese.** 1982. T cell-mediated immunoregulation of Epstein-Barr virus- (EBV) induced B lymphocyte activation in EBV-seropositive and EBV-seronegative individuals. *J. Immunol.* **128**:575–579.
58. **Tosato, G., S. Straus, W. Henle, S. E. Pike, and R. M. Blaese.** 1985. Characteristic T cell dysfunction in patients with chronic active Epstein-Barr virus infection (chronic infectious mononucleosis). *J. Immunol.* **134**:3082–3088.
59. **Virelizier, J.-L., G. Lenoir, and C. Griscelli.** 1978. Persistent Epstein-Barr virus infection in a child with hypergammaglobulinemia and immunoblastic proliferation associated with a selective defect in immune interferon secretion. *Lancet* **ii**:231–234.

Chapter 21

Novel Forms of Epstein-Barr Virus Persistence

A. B. Rickinson

INTRODUCTION

The Epstein-Barr virus (EBV) presents one of the most interesting and accessible models for the study of virus persistence and pathogenesis in humans. EBV combines the classic features of a latent infection (26), common to all members of the herpesvirus group, with a capacity for cell growth transformation (12, 28) which distinguishes this agent from the other human herpesviruses and which marks it out as a potential human tumor virus. The established view of EBV as a strictly B-lymphotropic agent is having to be modified in the light of evidence indicating that pharyngeal epithelium can sustain a productive EBV infection in vivo (8, 39). In considering the virus-host interaction, therefore, we are dealing with at least two lineages of potential target cells, B lymphocytes and epithelia. The relationship between these two types of target as possible reservoirs of EBV persistence in vivo is not yet understood; indeed, the mechanism of virus persistence in the immunocompetent host remains one of the most interesting unsolved questions of EBV biology. My purpose here is to consider how some of the more recent findings about EBV might influence current thinking on this issue.

EBV INFECTION OF NORMAL B CELLS IN VITRO

Cells of the B-lymphoid lineage are still the only targets in which experimental EBV infection is readily demonstrable in vitro. The B-cell infection is initiated through binding of the virus, via its envelope

A. B. Rickinson • Department of Cancer Studies, University of Birmingham, Birmingham B15 2TJ, United Kingdom.

glycoprotein gp340, to the 140-kilodalton C3d receptor molecule CR2 on the B-cell surface (25, 43). Expression of CR2 is largely, if not entirely, restricted to the B-lymphocyte lineage, and thus the observed B lymphotropism of the virus reflects the cellular distribution of this receptor molecule. Staining with monoclonal antibodies to CR2 suggests that the receptor is first expressed at about the time when the developing B cell begins to express surface immunoglobulin M and that subsequently CR2 expression is retained until that cell enters terminal plasmacytoid differentiation (16, 44). However, several groups have shown in in vitro experiments with fetal bone marrow that EBV can infect cells at earlier stages of B-cell ontogeny, either during or even prior to immunoglobulin gene rearrangement (4, 9, 18). Whether this represents interaction of the virus with the CR2 molecule, perhaps expressed at very low levels on the membranes of precursor B cells, or with some alternative receptor structure remains to be seen. In this context, monoclonal antibody staining suggests that a CR2-related molecule is also expressed on the basal and suprabasal cells of stratified squamous epithelium (38a, 54), but there is as yet no conclusive proof that this provides the route whereby EBV gains access to epithelial cells.

Perhaps the most interesting property of EBV in vitro is its capacity to bring about the growth transformation of human B cells into permanent lymphoblastoid cell lines (LCLs). Such transformed cells carry multiple episomal copies of the EBV genome and express a limited number of viral gene products, the so-called latent viral proteins, through which the virus-induced transformation of cell growth is achieved (3). Many, although perhaps not all, of the latent viral proteins have now been identified. They include several nuclear antigens: EBV nuclear antigen 1 (EBNA-1), encoded by the BKRF1 reading frame of the viral genome (41); EBNA-2, encoded by BYRF1 (14); the EBNA-3 family of up to three proteins, thought to be encoded by adjacent reading frames BLRF3-BERF1, BERF2a-BERF2b, and BERF3-BERF4 (15, 17); and EBNA leader protein (EBNA-LP), encoded by highly spliced exons from *Bam*HI W and Y regions of the viral genome (40, 50). In addition, a well-characterized latent membrane protein (LMP) encoded by BNLF1 is also constitutively expressed in all EBV-transformed LCLs (13). The biological functions of these individual proteins are not as yet known, except for EBNA-1, one of whose functions is to bind to a specific region of the viral genome (the origin of plasmid replication, *ori-p*) and in this way to maintain the genome in episomal plasmid form (53). How these latent virus proteins cooperate to achieve cell growth transformation is currently the focus of intense investigation, but it seems clear that either directly or indirectly, they must be able to deregulate the physiological pathways which are normally involved in moving a B cell from the resting to the proliferative state. Thus, many aspects of the EBV-induced activation of resting B cells into cell cycle mirror precisely what is seen when a B cell encounters either its specific antigen or a potent mitogen such as *Staphylococcus aureus* protein A (48). In particular, both methods of activation induce the expression on the B-cell surface of a series of cellular activation antigens, the best characterized of which is the 45-kilodalton CD23 molecule (35, 47). This protein now appears to play a crucial role in growth signal transduction at the cell surface; there is evidence to suggest, first, that it serves as a receptor for a T-cell-derived B-cell growth factor (7) and, second, that a cleaved form of CD23 shed into supernatant medium from the B-cell surface itself has direct B-cell growth factor activity (42). The recent demonstration that one of the EBV latent proteins, EBNA-2, selectively up-regulates the cellular expression of CD23 (49) is the first example of how the virus can interact with cellular pathways to achieve growth transformation. Constitutive expression of EBNA-2, and, thus, constitutive activation of the CD23 pathway, could therefore underlie the autocrine stimulation of growth which has been reported for EBV-transformed LCLs (6). The principal difference between the EBV-induced and the antigen-induced systems of B-cell activation is that the virally infected cells remain locked in the proliferating lymphoblastoid phase and tend not to complete the normal cycle of B-cell differentiation to plasma cells. Significantly, in the few cells which can be observed in LCLs to have moved out of cycle and toward the plasmacytoid state, there appears to be a down-regulation of EBV gene expression (51). This is an interesting observation, because it indicates just how sensitive the virus-cell interaction can be to changes in the state of differentiation of the infected cell.

To pursue this theme, it is clear that the sensitivity of B cells to EBV-induced transformation in vitro is itself greatly dependent upon the precise identity of the target cell. Not all mature resting B cells in peripheral blood appear to be responsive to the growth-transforming effects of EBV infection. Certain cells show evidence of virus gene expression, in that one or more proteins of the EBNA series are detectable postinfection, but the cells do not go on to up-regulate CD23 or to achieve entry into cycle (46). The reasons for this are unclear, but it may be that the physiological state of the target cell determines whether the cascade of events leading to growth transformation can be initiated. In this context, it is known that B cells which have been activated from the resting state before exposure to EBV are comparatively insusceptible to virus-induced transformation to LCLs (11).

These observations serve to emphasize just how limited is our understanding of the various EBV–B-cell interactions which might exist in vivo, where an even greater heterogenity of target cells is available to the virus than is commonly studied in in vitro systems. This is a point worth bearing in mind in the ensuing discussion.

INTERACTION OF EBV WITH ENDEMIC BL CELLS

The strong association between EBV and the endemic or high-incidence form of Burkitt's lymphoma (BL) is well established; in more than 95% cases of this disease the virus genome is present in the tumor cells, again in episomal form (21). The consistency with which the virus is seen in this particular malignancy and its absence from many other types of B-cell lymphoma, even when such tumor cells possess virus receptors, make this unlikely to be a casual association and argue strongly in favor of an etiological role for EBV in the context of BL. The proven ability of EBV to induce sustained B-cell proliferation not only in vitro but also in vivo in experimentally inoculated tamarins (1b) is further circumstantial evidence of a causative role in BL. However, the development of this tumor clearly requires other events in addition to EBV infection within a particular target cell; one of the key events common to all cases of BL is a chromosomal translocation whereby the c-*myc* gene on chromosome 8 is juxtaposed with one of the immunoglobulin gene loci on chromosome 14, 2, or 22 (2). The translocation per se and/or subsequent mutation in the rearranged c-*myc* locus result in constitutive activation of c-*myc* expression from the rearranged allele, with apparent down-regulation of the normal allele on the uninvolved chromosome 8 (27). It is still not known how EBV infection, c-*myc* gene activation, and perhaps other genetic changes cooperate to achieve malignant transformation, or, indeed, in what order these events occur at the level of the target cell (2, 20).

Recently (38), we have been examining the type of virus-cell interaction which exists in BL cells and comparing this with the type of interaction which EBV consistently establishes when it transforms a normal B cell in vitro to produce an LCL. The panel of BL cell lines used for such a study had recently been derived from EBV-positive tumors arising mainly from the BL-endemic areas of Africa and New Guinea. During cell line establishment and subsequent passage in vitro, it was noted that many BL lines changed their cell growth and cell surface phenotype from the pattern typical of the original biopsy cells, i.e., growth of single cells and absence of B-cell activation antigens, toward a more LCL-like pattern, i.e., growth in tight clumps and expression of B-cell activation antigens (37). This was not due to overgrowth of the culture by normal EBV-carrying B cells contaminating the tumor cell suspension, but was the result of phenotypic change occurring within the BL cell population itself; thus, all of the derived cell lines carried the relevant chromosomal marker indicative of their malignant origin. Individual BL tumors showed different degrees of progression toward the LCL-like phenotype, so that a panel of lines established from different biopsies displayed marked phenotypic differences between individual lines, even though all tumors were identical at the biopsy stage (31). Recognition of these in vitro changes allowed us to identify a minority of EBV-positive BL cell lines which remained phenotypically stable on serial passage, i.e., which showed no LCL-like progression. Such lines were clearly the best in vitro correlates of the in vivo tumor, and their analysis has proved particularly instructive.

The study has revealed a new form of EBV-cell interaction, present in phenotypically stable BL cell lines, which is quite unlike that seen in LCLs. This new interaction is characterized by expression of EBNA-1, with no detectable expression of at least two other latent viral proteins, EBNA-2 and LMP (38). Most interestingly, BL cell lines which subsequently moved toward a more LCL-like phenotype did so with accompanying up-regulation of both EBNA-2 and LMP. In more recent experiments, expression of EBNA-LP has been shown to be absent in phenotypically stable BL cell lines, but to be up-regulated in lines which progress to a more LCL-like phenotype (5). Preliminary observations indicate that expression of the EBNA-3 family of latent proteins in BL cells may also be phenotype dependent in a similar way (M. Rowe, personal communication).

Although there has been limited opportunity to examine BL biopsies directly, results to date suggest that the unusually restricted pattern of EBV latent-gene expression described above is indeed true of the BL tumor cells arising in vivo (38). Viewed in a wider context, this observation could help to explain the peculiar advantage which EBV-positive BL tumor cells appear to enjoy over EBV-infected normal B cells in their capacity to evade the powerful virus-specific cytotoxic T-cell surveillance which EBV-infected individuals (including BL patients) demonstrably possess (32, 33). One obvious possibility is that one or more of those EBV latent proteins whose expression is down-regulated in BL cells provide the dominant target antigens for EBV-specific cytotoxic T-cell surveillance. Indeed, there is now evidence suggesting a role both for EBNA-2 and for LMP as target antigens in this context (23a, 24a, 45; R. J. Murray, D. Wang, L. S. Young, F. Wang, M. Rowe, E. Kieff, and A. B. Rickinson, submitted for publication).

EBV INFECTION OF VIRUS-NEGATIVE BL CELLS IN VITRO

Sporadic BL, a malignancy which is both histologically and karyotypically indistinguishable from the endemic disease, occurs at very low incidence worldwide, and this form of the tumor is often not associated with EBV (21). It must therefore be concluded that in certain circumstances EBV can greatly facilitate the development of the Burkitt tumor, but that involvement of the virus is not obligatory. Cell lines established from cases of EBV-negative sporadic BL have in fact proved to be useful experimental tools in the laboratory, since they are often susceptible to EBV infection in vitro. Significantly, the EBV-negative BL cell lines themselves do not show the kind of phenotypic progression toward an LCL-like phenotype which is so often seen with endemic (EBV-positive) BL lines. However, after in vitro infection with a transforming virus strain (such as B95.8), EBV-converted lines arise and are reproducibly much more LCL-like both in growth pattern and in cell surface phenotype than is the original EBV-negative parent (36).

This is strong evidence that the progression of BL cells to an LCL-like phenotype in vitro is actually driven by a resident EBV genome. Indeed, the types of changes in cell growth pattern and surface phenotype which are induced by EBV in this situation are exactly those which are induced by the virus in normal B cells during EBV-induced growth transformation (35, 47, 48). Given the difficulties of working with freshly isolated normal B cells as targets, infection of EBV-negative BL cells with whole virus and/or transfection of these cells with EBV latent genes cloned in appropriate expression vectors therefore offers an important new approach to dissecting the roles of individual virus genes in the transformation process (49).

One particular application of this approach which is relevant to the present discussion centers on the use of a deletion mutant of EBV, the P3HR1 virus strain, which lacks the genomic sequence encoding EBNA-2 (1) and which has no in vitro growth-transforming ability for normal B cells (23). P3HR1-induced converts of sporadic BL cell lines can be produced by in vitro infection and compared directly with parallel B95.8-induced converts of the kind described above. Recently, we tested five sets of paired converts, each set being derived from a different EBV-negative parent line, for their pattern of EBV latent gene expression. All B95.8-induced converts expressed the full spectrum of latent proteins (EBNA-1, EBNA-2, the EBNA-3 family, EBNA-LP, and LMP); in contrast, all P3HR1-induced converts expressed EBNA-1, the EBNA-3 family, and EBNA-LP, but were negative not only for EBNA-2 (because of the deletion) but also for LMP (24a). Moreover, the P3HR1-induced converts remained close to the original parental cell growth and cell surface phenotype and, as such, were clearly distinct from the much more LCL-like B95.8-induced converts (1a). This again reflects the way in which the resident pattern of EBV latent gene expression can influence the cellular phenotype.

One must, of course, exercise caution in interpreting experiments with deletion mutants, but these findings do illustrate the fact that different forms of latent EBV infection can exist within B cells. To an extent, the P3HR1-induced converts resemble endemic BL cell lines which retained a highly restricted form of EBV latent gene expression and which were phenotypically stable in vitro; conversely, the B95.8-induced converts resemble endemic BL cell lines which began to express the full spectum of virus latent genes and which, in consequence, moved to a more LCL-like phenotype with serial passage. Again, it is clear that these two distinct types of EBV latent infection are perceived differently by the T-cell system; thus, all the above B95.8-induced converts were recognized and killed by EBV-specific cytotoxic T cells, whereas the P3HR1-induced converts, lacking EBNA-2 and LMP, were not (24a).

IMPLICATIONS FOR EBV PERSISTENCE IN VIVO

One of the paradoxes of the EBV carrier state is that virus-infected B cells can regularly be detected in the lymphoid systems of infected individuals many years after the primary infection, even though such individuals possess strong EBV-specific cytotoxic T-cell surveillance, which should be capable of eliminating all virus-infected B cells in vivo (29). We attempted to explain this paradox by postulating that EBV-infected cells were indeed removed by T-cell surveillance, but that their numbers were continually being replenished by B cells freshly infected during their transit through the epithelial reservoir of chronic EBV replication which persists in the pharynx (30, 52). In this way, the apparent virus carrier status of the B-lymphoid system could be maintained, even though virus-infected B-cell numbers were continually being eroded by the T-cell response.

One of the assumptions we made in drawing up the model in its original form was that EBV infection of a B cell necessarily leads to the expression of the full spectrum of EBV latent proteins, to lymphoblastoid transformation, and hence to recognition and elimination of that cell by the immune response (24). Recent observations, outlined above, now indicate that EBV can enter into a different type of interaction with certain target B cells. In this situation, there is a much more restricted pattern of latent gene expression, such that the infected cell does not become sensitive to EBV-specific cytotoxic T-cell surveillance. Are these observations relevant to the normal mechanisms of EBV persistence in vivo?

First, it is possible that the unusual type of infection described above is unique to BL cells. The argument would be that BL cells represent a highly selected population, in which the very rare circumstances of a chromosomal translocation leading to c-*myc* gene activation have liberated EBV-induced B-cell growth from its usual dependence upon the full spectrum of EBV latent proteins. Selection pressure from the prevailing T-cell response would therefore favor outgrowth of a rare subclone in which expression of the now redundant EBV proteins had been lost. In this scenario, the type of infection seen in BL cells is therefore a result of the intense selection through which the tumor evolves in vivo (34).

However, a second, more interesting, possibility is that the virus-cell interaction apparent in BL cells reflects a form of infection which normal cells can sustain at particular stages of their in vivo life history. The inference from the work outlined above is that the expression of certain EBV latent genes, in particular EBNA-2 and LMP, is crucially important in determining the type of infection established by the virus in a target cell. Perhaps there are stages of B-cell differentiation, for instance, in B cells already activated through contact with their specific antigen, in which expression of these crucial viral proteins is suppressed. In in vitro infection experiments, therefore, one might only perceive this altered form of virus-cell interaction in a negative sense, that is, as a failure of the infected cell to show virus-induced growth transformation to an LCL.

To pursue the argument further, one interesting candidate for a normal cell population capable of sustaining the above type of EBV latency would be that subset of B cells from which the Burkitt tumor is itself derived. Recently, we identified a population of actively cycling cells within the germinal centers of lymphoid tissue which show a remarkable phenotypic similarity to BL biopsy cells (10). These germinal-center cells express two important markers characteristic of BL cells, the common acute lymphoblastic leukemia antigen cALLA and the glycolipid antigen BLA; they are also positive for conventional pan-B markers such as CD19 and CD20, yet, like BL, they do not express any of the activation antigens usually associated with the B-lymphoblastoid state. Relatively little is known about the detailed biology of germinal centers, except that these structures are oligoclonal foci of antigen-driven B-cell proliferation and are involved in the generation of B-cell memory (19). One interesting suggestion is that they are actually the sites at which somatic mutation of immunoglobulin genes occurs, thus allowing selection of rare mutant clones producing higher-affinity antibody for the antigen in question (22). Germinal-center cells are clearly at a unique stage of B-cell differentiation and as such represent a potentially interesting target population for EBV. At present, however, there is no direct information about their susceptibility to EBV infection either in vivo or in vitro or about the type of infection they might sustain.

Until these issues have been examined experimentally, one must keep an open mind about possible sites of EBV persistence within the B-lymphoid system. There may be situations in which normal B cells can harbor EBV for long periods in a form of latency which mimics that recently recognized in BL cells. Such cells might begin to express the full spectrum of EBV latent proteins only when they or their progeny reach a stage of differentiation in which the earlier repression of virus gene expression is relaxed; only then would such cells become susceptible to immune T-cell surveillance. If this proves to be the case, long-term persistence of EBV may yet be possible within the B-lymphoid system, and our original view of epithelial cells as the sole reservoir of virus persistence may have to be revised.

Literature Cited

1. **Bornkamm, G. W., J. Hudewitz, U. K. Freese, and U. Zimber.** 1982. Deletion of the non-transforming EBV strain P3HR-1 causes fusion of the large internal repeat to the DSL region. *J. Virol.* **43:**952–968.

1a. **Calender, A., M. Billaud, J. P. Aubry, J. Banchereau, M. Vuillaume, and G. M. Lenoir.** 1987. Epstein-Barr virus (EBV) induces expression of B cell activation markers in in vitro infection of EBV-negative B-lymphoma cells. *Proc. Natl. Acad. Sci. USA* **84:**8060–8064.

1b. **Cleary, M. L., M. A. Epstein, S. Finerty, R. F. Dorfman, G. W. Bornkamm, J. K. Kirkwood, A. J. Morgan, and J. Sklar.** 1985. Individual tumors of multifocal EB virus-induced malignant lymphomas in tamarins arise from different B cell clones. *Science* **228:**722–724.

2. **Cory, S.** 1986. Activation of cellular oncogenes in hemopoietic cells by chromosome translocation. *Adv. Cancer Res.* **47:**189–234.

3. **Dambaugh, T., H. Hennessy, S. Fennewald, and E. Kieff.** 1986. The virus genome and its expression in latent infection, p. 13–45. *In* M. A. Epstein and B. G. Achong (ed.), *The Epstein-Barr Virus: Recent Advances*. William Heinemann, London.

4. **Ernberg, I., K. Falk, and M. Hansson.** 1987. Progenitor and pre-B lymphocytes transformed by Epstein-Barr virus. *Int. J. Cancer* **39:**190–197.

5. **Finke, J., M. Rowe, B. Kallin, I. Ernberg, A. Rosen, J. Dillner, and G. Klein.** 1987. Monoclonal and polyclonal antibodies against Epstein-Barr virus nuclear antigen 5 detect multiple protein species in Burkitt's lymphoma and lymphoblastoid cell lines. *J. Virol.* **61:**3870–3878.

6. **Gordon, J., S. C. Ley, M. D. Melamed, L. S. English, and N. C. Hughes-Jones.** 1984. Immortalized B lymphocytes produce B-cell growth factor. *Nature* (London) **310:**145–147.

7. **Gordon, J., A. J. Webb, L. Walker, G. R. Guy, and M. Rowe.** 1986. Evidence for an association between CD23 and the receptor for a low molecular weight B cell growth factor. *Eur. J. Immunol.* **16:**1627–1630.

8. **Greenspan, J. S., D. Greenspan, E. T. Lennette, D. I. Abrams, M. A. Conant, V. Petersen, and U. K. Freese.** 1985. Replication of Epstein-Barr virus within the epithelial cells of oral "hairy" leukoplakia, a AIDS-associated lesion. *N. Engl. J. Med.* **313:**1564–1571.

9. **Gregory, C. D., C. Kirchgens, C. F. Edwards, L. S. Young, M. Rowe, A. Forster, T. H. Rabbitts, and A. B. Rickinson.** 1987. Epstein-Barr virus-transformed human precursor B cell lines: altered growth phenotype of lines with germlike or rearranged but non-expressed heavy chain genes. *Eur. J. Immunol.* **17:**1199–1207.
10. **Gregory, C. D., T. Tursz, C. F. Edwards, C. Tetaud, M. Talbot, B. Caillou, A. B. Rickinson, and M. Lipinski.** 1987. Identification of a subset of normal B cells with a Burkitt's lymphoma (BL)-like phenotype. *J. Immunol.* **139:**313–318.
11. **Henderson, E., G. Miller, J. Robinson, and L. Heston.** 1977. Efficiency of transformation of lymphocytes by Epstein-Barr virus. *Virology* **76:**152–162.
12. **Henle, W., V. Diehl, G. Kohn, H. Zur Hausen, and G. Henle.** 1967. Herpes-type virus and chromosome marker in normal leukocytes after growth with irradiated Burkitt cells. *Science* **157:**1064–1065.
13. **Hennessy, K., S. Fennewald, M. Hummel, T. Cole, and E. Kieff.** 1984. A membrane protein encoded by Epstein-Barr virus in latent growth-transforming infection. *Proc. Natl. Acad. Sci. USA* **81:**7207–7211.
14. **Hennessy, K., and E. Kieff.** 1985. A second nuclear protein is encoded by Epstein-Barr virus in latent infection. *Science* **227:**1238–1240.
15. **Hennessy, K., F. Wang, E. Woodland Bushman, and E. Kieff.** 1986. Definitive identification of a member of the Epstein-Barr virus nuclear protein 3 family. *Proc. Natl. Acad. Sci. USA* **83:**5693–5697.
16. **Iida, K., L. Nadler, and V. Nussenzweig.** 1983. Identification of the membrane receptor for the complement fragment C3d by means of a monoclonal antibody. *J. Exp. Med.* **158:**1021–1033.
17. **Kallin, B., J. Dillner, I. Ernberg, B. Ehlin-Henriksson, A. Rosen, W. Henle, G. Henle, and G. Klein.** 1986. Four virally determined nuclear antigens are expressed in Epstein-Barr virus-transformed cells. *Proc. Natl. Acad. Sci. USA* **83:**1499–1503.
18. **Katamine, S., M. Otsu, K. Tada, S. Tsuchiya, T. Sato, N. Ishida, T. Honjo, and Y. Ono.** 1984. Epstein-Barr virus transforms precursor B cells even before immunoglobulin gene rearrangements. *Nature* (London) **309:**369–372.
19. **Klaus, G. G. B., J. H. Humphrey, A. Kunkl, and D. W. Dongworth.** 1980. The follicular dentritic cell: its role in antigen presentation in the generation of immunological memory. *Immunol. Rev.* **53:**3–28.
20. **Lenoir, G., and G. Bornkamm.** 1987. Burkitt's lymphoma, a human cancer model for the study of the multistep development of cancer: proposal for a new scenario. *Viral Oncol.* **6:**173–206.
21. **Lenoir, G. M.** 1986. Role of the virus, chromosomal translocations and cellular oncogenes in the aetiology of Burkitt's lymphoma, p. 183–205. *In* M. A. Epstein and B. G. Achong (ed.), *The Epstein-Barr Virus: Recent Advances*. William Heinemann, London.
22. **MacLennan, I. C. M., and D. Gray** 1986. Antigen-driven selection of virgin and memory B cells. *Immunol. Rev.* **91:**61–85.
23. **Miller, G., J. Robinson, L. Heston, and M. Lipman.** 1974. Differences between laboratory strains of Epstein-Barr virus based on immortalization, abortive infection and interference. *Proc. Natl. Acad. Sci. USA* **71:**4006–4010.

23a. **Moss, D. J., I. S. Misko, S. R. Burrows, K. Burman, R. McCarthy, and J. B. Sculley.** 1988. Cytotoxic T-cell clones discriminate between A- and B-type Epstein-Barr virus transformants. *Nature* (London) **331:**719–721.

24. **Moss, D. J., A. B. Rickinson, L. E. Wallace, and M. A. Epstein.** 1981. Sequential appearance of Epstein-Barr virus nuclear and lymphocyte-detected membrane antigens in B cell transformation. *Nature* (London) **291:**664–666.

24a. **Murray, R. J., L. S. Young, A. Calender, C. D. Gregory, M. Rowe, G. M. Lenoir, and**

A. B. Rickinson. 1988. Different patterns of Epstein-Barr virus gene expression and of cytotoxic T-cell recognition in B-cell lines infected with transforming (B95.8) or nontransforming (P3HR1) virus strains. *J. Virol.* **62:**894–901.

25. **Nemerow, G. R., C. Mold, V. K. Schwend, V. Tollefson, and N. R. Cooper.** 1987. Identification of gp350 as the viral glycoprotein mediating attachment of Epstein-Barr virus to the Epstein-Barr virus C3d receptor of B cells: sequence homology of gp350 and C3 complement fragment C3d. *J. Virol.* **61:**1416–1420.
26. **Nilsson, K., G. Klein, W. Henle, and G. Henle.** 1971. The establishment of lymphoblastoid cell lines from adult and from foetal human lymphoid tissue and its dependence on EBV. *Int. J. Cancer* **8:**443–450.
27. **Nishikura, K., A. Ar-Rushidi, J. Erikson, R. Watt, G. Rovera, and C. M. Croce.** 1983. Differential expression of the normal and of the translocated human c-myc oncogenes in B cells. *Proc. Natl. Acad. Sci. USA* **80:**4822–4826.
28. **Pope, J. H., M. K. Horne, and W. Scott.** 1968. Transformation of foetal human leukocytes in vitro by filtrates of a human leukaemic cell line containing herpes-like virus. *Int. J. Cancer* **3:**857–866.
29. **Rickinson, A. B., D. J. Moss, J. H. Pope, and N. Ahlberg.** 1980. Long-term T-cell-mediated immunity to Epstein-Barr virus in man. IV. Development of T cell memory in convalescent infectious mononucleosis patients. *Int. J. Cancer* **25:**59–65.
30. **Rickinson, A. B., Q. Y. Yao, and L. E. Wallace.** 1985. The Epstein-Barr virus as a model of virus-host interactions. *Br. Med. Bull.* **41:**75–79.
31. **Rooney, C. M., C. D. Gregory, M. Rowe, S. Finerty, C. Edwards, H. Rupani, and A. B. Rickinson.** 1986. Endemic Burkitt's lymphoma: phenotypic analysis of Burkitt's lymphoma biopsy cells and of the derived tumor cell lines. *J. Natl. Cancer Inst.* **77:**681–687.
32. **Rooney, C. M., A. B. Rickinson, D. J. Moss, G. M. Lenoir, and M. A. Epstein.** 1985. Cell-mediated immunosurveillance mechanisms and the pathogenesis of Burkitt's lymphoma, p. 249–264. *In* G. M. Lenoir, G. T. O'Conor, and C. L. M. Olweny (ed.), *Burkitt's Lymphoma: a Human Cancer Model.* International Agency for Research on Cancer, Lyon, France.
33. **Rooney, C. M., M. Rowe, L. E. Wallace, and A. B. Rickinson.** 1985. Epstein-Barr virus-positive Burkitt's lymphoma cells not recognised by virus-specific T cell surveillance. *Nature* (London) **317:**629–631.
34. **Rowe, D. T., M. Rowe, G. I. Evan, L. E. Wallace, P. J. Farrell, and A. B. Rickinson.** 1986. Restricted expression of EBV latent genes and T-lymphocyte-detected membrane antigen in Burkitt's lymphoma cells. *EMBO J.* **5:**2599–2607.
35. **Rowe, M., J. E. K. Hildreth, A. B. Rickinson, and M. A. Epstein.** 1982. Monoclonal antibodies to Epstein-Barr virus-induced, transformation-associated cell surface antigens: binding patterns and effect upon virus-specific T-cell cytotoxicity. *Int. J. Cancer* **29:**373–381.
36. **Rowe, M., C. M. Rooney, C. F. Edwards, G. M. Lenoir, and A. B. Rickinson.** 1986. Epstein-Barr virus status and tumour cell phenotype in sporadic Burkitt's lymphoma. *Int. J. Cancer* **37:**363–372.
37. **Rowe, M., C. M. Rooney, A. B. Rickinson, G. M. Lenoir, H. Rupani, D. J. Moss, H. Stein, and M. A. Epstein.** 1985. Distinctions between endemic and sporadic forms of Epstein-Barr virus-positive Burkitt's lymphoma. *Int. J. Cancer* **35:**435–442.
38. **Rowe, M., D. T. Rowe, C. D. Gregory, L. S. Young, P. J. Farrell, H. Rupani, and A. B. Rickinson.** 1987. Differences in B cell growth phenotype reflect novel patterns of Epstein-Barr virus latent gene expression in Burkitt's lymphoma cells. *EMBO J.* **6:**2743–2751.

38a. **Sixbey, J. W., D. S. Davis, L. S. Young, L. Hutt-Fletcher, T. F. Tedder, and A. B.**

Rickinson. 1987. Human epithelial cell expression of an Epstein-Barr virus receptor. *J. Gen. Virol.* **68:**805–811.

39. **Sixbey, J. W., J. G. Nedrud, N. Raab-Traub, R. A. Hanes, and J. S. Pagano.** 1984. Epstein-Barr virus replication in oropharyngeal epithelial cells. *N. Engl. J. Med.* **310:**1225–1230.
40. **Speck, S. H., A. Pfitzner, and J. L. Strominger.** 1987. An Epstein-Barr virus transcript from a latently infected, growth-transformed B-cell line encodes a highly repetitive polypeptide. *Proc. Natl. Acad. Sci. USA* **83:**9298–9302.
41. **Summers, W. P., E. A. Grogan, D. Shedd, M. Robert, C.-R. Liu, and G. Miller.** 1982. Stable expression in mouse cells of nuclear neoantigen after transfer of a 3.4 megadalton cloned fragment of Epstein-Barr virus DNA. *Proc. Natl. Acad. Sci. USA* **79:**5688–5692.
42. **Swendeman, S., and D. A. Thorley-Lawson.** 1987. The activation antigen BLAST-2, when shed, is an autocrine BCGF for normal and transformed B cells. *EMBO J.* **6:**1637–1642.
43. **Tanner, J., J. Weis, D. Fearon, Y. Whang, and E. Kieff.** 1987. Epstein-Barr virus gp 350/220 binding to the B lymphocyte C3d receptor mediates adsorption, capping and endocytosis. *Cell* **50:**203–213.
44. **Tedder, T. F., L. T. Clement, and M. D. Cooper.** 1984. Expression of C3d receptors during human B cell differentiation: immunofluorescence analysis with the HB-5 monoclonal antibody. *J. Immunol.* **133:**678–683.
45. **Thorley-Lawson, D. A., and E. S. Israelsohn.** 1987. Generation of specific cytotoxic T cells with a fragment of the Epstein-Barr virus-encoded p63/latent membrane protein. *Proc. Natl. Acad. Sci. USA* **84:**5384–5389.
46. **Thorley-Lawson, D. A., and K. P. Mann.** 1985. Early events in Epstein-Barr virus infection provide a model for B cell activation. *J. Exp. Med.* **161:**45–59.
47. **Thorley-Lawson, D. A., L. M. Nadler, A. K. Bhan, and R. T. Schooley.** 1985. BLAST-2 (EBVCS), an early cell surface marker of human B cell activation, is superinduced by Epstein-Barr virus. *J. Immunol.* **134:**3007–3012.
48. **Walker, L., G. R. Guy, G. Brown, M. Rowe, A. E. Milner, and J. Gordon.** 1986. Characterisation of novel activation states that precede the entry of G_0 B cells into cycle. *Immunology* **58:**583–589.
49. **Wang, F., C. D. Gregory, M. Rowe, A. B. Rickinson, D. Wang, M. Birkenbach, H. Kikutani, T. Kishimoto, and E. Kieff.** 1987. Epstein-Barr virus nuclear protein 2 specifically induces expression of the B cell activation antigen. *Proc. Natl. Acad. Sci. USA* **84:**3452–3456.
50. **Wang, F., L. Petti, D. Braun, S. Seung, and E. Kieff.** 1987. A bicistronic Epstein-Barr virus mRNA encodes two nuclear proteins in latently infected growth-transformed lymphocytes. *J. Virol.* **61:**945–954.
51. **Wendel-Hansen, V., A. Rosen, and G. Klein.** 1987. EBV-transformed lymphoblastoid cell lines down-regulate EBNA in parallel with secretory differentiation. *Int. J. Cancer* **39:**404–408.
52. **Yao, Q. Y., A. B. Rickinson, and M. A. Epstein.** 1985. A re-examination of the Epstein-Barr virus carrier state in healthy seropositive individuals. *Int. J. Cancer* **35:**35–42.
53. **Yates, J., N. Warren, and B. Sugden.** 1985. Stable replication of plasmids derived from Epstein-Barr virus in various mammalian cells. *Nature* (London) **313:**812–815.
54. **Young, L. S., D. Clark, J. W. Sixbey, and A. B. Rickinson.** 1986. Epstein-Barr virus receptors on human pharyngeal epithelia. *Lancet* **i:**240–242.

Chapter 22

Suppressor T-Cell Regulation of Latent Epstein-Barr Virus Infection: Characterization of a Suppressor T-Cell Line

Giovanna Tosato and Sandra E. Pike

Epstein-Barr virus (EBV) causes acute infectious mononucleosis in humans and is associated with two forms of human cancer, Burkitt's lymphoma and nasopharyngeal carcinoma. In addition, EBV infects most adult individuals asymptomatically (for a review, see reference 16). During both primary infection and latency, EBV can generally be recovered from the peripheral blood, where it infects a small proportion of the circulating B cells (1). These B cells naturally infected with EBV give rise spontaneously to long-term B-cell lines that express the EBV-encoded nuclear antigen EBNA, have a B-cell phenotype, and secrete immunoglobulin (19). Similarly, human B cells become experimentally infected with EBV, and as a consequence of this infection, a proportion of the B cells give rise to long-term EBNA-positive B-cell lines (3–5, 14).

It is known that several regulatory mechanisms contribute to the immune control of B cells naturally infected with EBV and prevent their unregulated expansion in vivo (16). These include natural killer cells (10), virus-specific cytotoxic T cells (11–13), and suppressor T cells (18, 20).

In previous studies we have cloned suppressor T lymphocytes from the peripheral blood of two patients with acute EBV-induced infectious mononucleosis, and by this technique we have physically separated suppressor T cells from cytotoxic T cells (20). These suppressor/noncy-

Giovanna Tosato and Sandra E. Pike • Laboratory of Molecular Immunology, Division of Biochemistry and Biophysics, Office of Biologics Research and Review, Food and Drug Administration, Bethesda, Maryland 20892.

totoxic T-cell clones inhibited pokeweed mitogen (PWM)-induced immunoglobulin production by autologous as well as by random allogeneic mononuclear cells. In addition, the T-cell clones inhibited EBV-induced immunoglobulin production and immortalization by autologous B cells newly infected with EBV. These results demonstrated that during acute EBV-induced infectious mononucleosis, regulatory T cells become activated and inhibit EBV-infected B cells by a mechanism other than cytotoxicity.

In the studies discussed here, we have addressed the possibility that inhibitory T cells without cytotoxic activity contribute to the control of latent EBV infection in EBV-seropositive normals. To this end, we sought to select a T-cell line from a normal EBV-seropositive individual that would suppress EBV-infected cells without killing them.

In initial experiments we were unsuccessful in growing suppressor T-cell clones from normal EBV-seropositive individuals, using limiting-dilution cloning of peripheral blood T cells in culture medium containing interleukin-2 (IL-2) (20). This is probably due to a relative infrequency of these regulatory T cells in normal blood. We therefore attempted to enrich for EBV-related regulatory T cells by multiple stimulations of peripheral blood T cells from a normal EBV-seropositive individual with autologous irradiated (10,000 rads) B cells immortalized by EBV (VD-B) at a T-to-B-cell ratio of 3:1 to 4:1. After each stimulation, lasting 4 to 7 days, the T cells were washed free of stimulator B cells and cultured for 7 to 10 days in culture medium containing IL-2 and irradiated (2,000 rads) autologous mononuclear cells (10^6/ml) as feeders. After each of the initial four stimulations, a reduction in the total T-cell number was observed (Fig. 1). Thereafter, however, we observed an increase in cell number after each exposure of the T cells to the autologous EBV-infected B-cell line. By this technique of stimulation with autologous EBV-transformed B cells, followed by rest in the presence of autologous mononuclear cells and IL-2, we were able to maintain this T-cell line in continuous culture for longer than 1 year. The experiments described below were performed with this T-cell line (named VD-T), which had been in continuous culture for 4 to 6 months.

Analysis of the phenotype of the cell line at 4 months after its initiation, and periodically thereafter, demonstrated that 90 to 100% of the cells were T3$^+$ and T8$^+$ (T4$^-$, Leu 7$^-$). As expected from its growth pattern, this T-cell line showed enhanced [^{3}H]thymidine incorporation when cultured in the presence of autologous EBV-infected B cells which had been irradiated. Similarly, VD-T proliferated in response to allogeneic EBV-transformed B cells which expressed the histocompatibility antigen A2. In contrast, VD-T failed to proliferate in response to

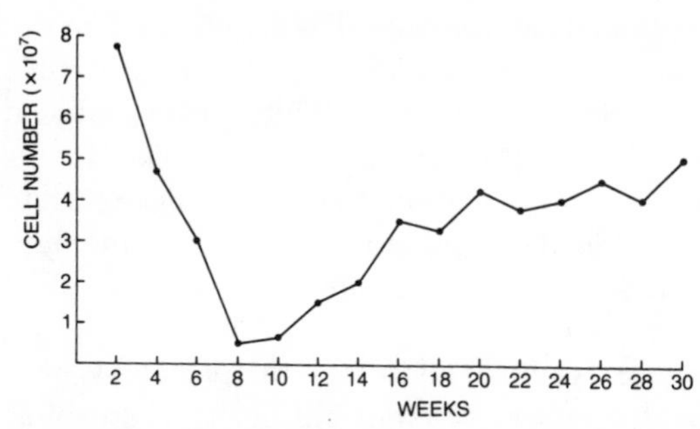

Figure 1. Selection and growth characteristics of an EBV-specific long-term T-cell line. T cells (10^8 in all) were cultured in the presence of autologous irradiated (10,000 rads) lymphoblastoid B cells in 24-well culture plates (Limbro, 2×10^6 T cells per well with 0.5×10^6 B cells per well in 2 ml of culture medium). After 5 to 7 days of culture the cells were removed from the wells, washed in RPMI 1640 medium, and density fractionated to remove nonviable cells. After counting, the cells were recultured for 7 to 10 days in 24-well culture plates (2×10^6 cells per well) in the presence of irradiated (3,000 rads) autologous mononuclear cells (10^6/well) and IL-2 (10 U/ml). Alternating cycles of culture in the presence of autologous lymphoblastoid B cells (stimulation; 5 to 7 days), followed by culture in the presence of autologous mononuclear cells (rest; 7 to 10 days), were used for the lifetime of the cell line.

autologous B cells activated by *Staphylococcus aureus* (Fig. 2). Thus, VD-T could specifically recognize EBV-infected cells and appeared to be restricted by class I histocompatibility antigens.

In contrast to the restrictive pattern of proliferation, the VD-T cell line inhibited immunoglobulin production in a nonrestrictive manner.

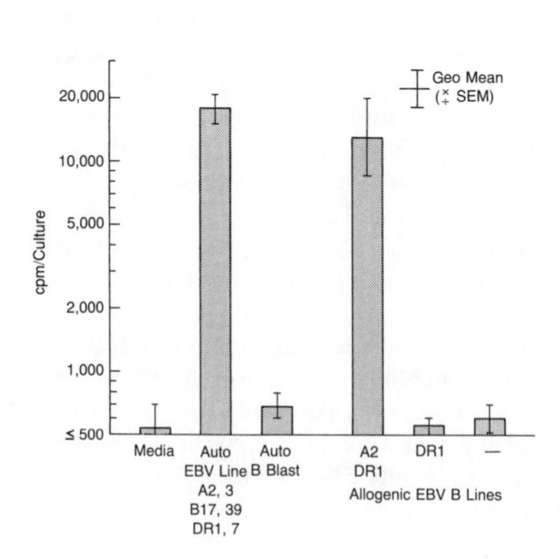

Figure 2. The VD-T cell line proliferates specifically in response to histocompatibility antigen A2-positive, EBV-infected B cells. Cells from cell line VD-T (20×10^3 cells per well) were cultured for 3 days in flat-bottom microtiter plates either alone or in the presence of stimulator B cells (50×10^3 per well). These included autologous EBV-immortalized B cells (Auto EBV Line) (10,000 rads), autologous *S. aureus* activated B-cell blasts (Auto B Blast) (5,000 rads), and allogeneic EBV-induced lymphoblastoid cells (10,000 rads). These allogeneic lines shared either histocompatibility antigens A2 plus DR1, only DR1, or no histocompatibility antigens with the responding T cells, as indicated.

Table 1. VD-T Cell Line Inhibits Immunoglobulin Production[a]

Responding cells	Stimulus	No. of PFC/well at VD-T cell concn:			
		0 cells/well	10^3 cells/well	10^4 cells/well	10^5 cells/well
VD mononuclear cells	PWM	3,200	120	25	20
VD-B cells	EBV	2,600	530	140	150
Allogeneic mononuclear cells					
1	PWM	2,200	80	6	5
2	PWM	3,400	11	18	10
3	PWM	2,750	25	26	52
4	PWM	1,800	0	10	15

[a] T-cell suppression of PWM-induced immunoglobulin production was determined by comparing the number of immunoglobulin-secreting cells produced by PWM-activated mononuclear cells either cultured alone or mixed in coculture with allogeneic T cells. Normal indicator mononuclear cells were cultured for 5 to 7 days in 96-well round-bottom microtiter plates (Costar) at a cell density of 100×10^3 to 125×10^3 cells per well without or with PWM (Gibco Laboratories) at a final dilution of 1:80. T-cell suppression of EBV-induced immunoglobulin production was evaluated by comparing the number of immunoglobulin-secreting cells produced by EBV-infected B cells either cultured alone or mixed in coculture with autologous T cells. B cells were cultured for 8 to 14 days in the presence of EBV (B95-8 strain) at a cell density of 50,000 B cells per well in 96-well round-bottom plates as the indicator response. Effector T cells were added in coculture as described above. At the end of culture, the number of immunoglobulin-secreting cells (plaque-forming cells; PFC) was assayed with a modified reverse-hemolytic plaque assay (20).

Both PWM- and EBV-induced responses by the autologous mononuclear cells were suppressed by VD-T (Table 1). Also, PWM-induced immunoglobulin production by random allogeneic mononuclear cells was inhibited by the T-cell line.

We then tested whether the VD-T cell line would kill either K562, the natural killer-sensitive line, or the autologous, EBV-induced lymphoblastoid cell line. In contrast to an alloreactive T-cell line used as a positive control, the T-cell line VD-T failed to kill the autologous lymphoblastoid cell line VD-B even at high effector-to-target ratios. Similarly, VD-T showed little or no cytotoxic activity against K562. Normal mononuclear cells served as a positive control for K562 killing (Fig. 3).

In addition to its inhibitory effects on immunoglobulin production induced by PWM or EBV, the VD-T cell line inhibited the proliferation of autologous B cells immortalized by EBV. Growth was observed in 100% of wells containing 300 VD-B lymphoblastoid B cells per well (Fig. 4). The addition of various numbers of the suppressor T cells resulted in growth in only a proportion of the wells. No growth was observed when 50,000 suppressor T cells were added, and only 50% of the wells showed growth when 6,000 such T cells were added. Thus, this T-cell line, a suppressor but not cytotoxic, can control the growth of an established lymphoblastoid cell line.

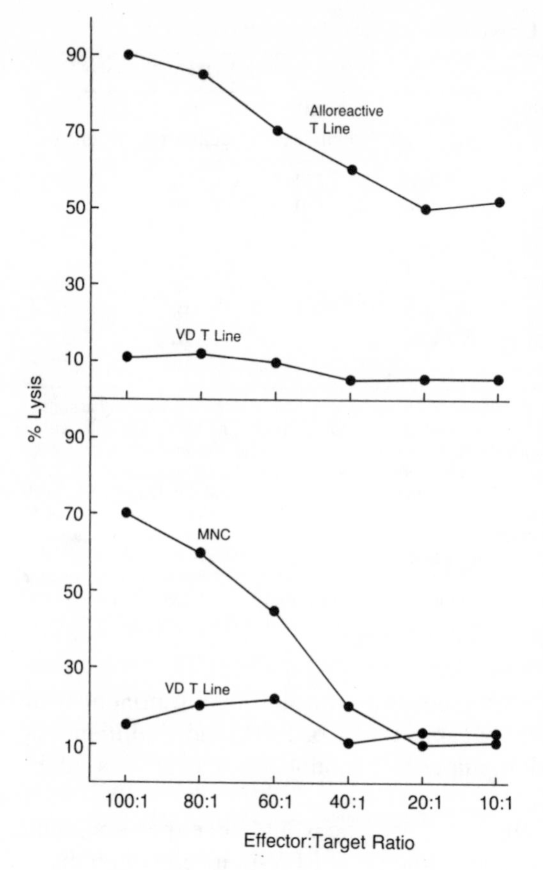

Figure 3. The VD-T cell line fails to kill the natural killer-sensitive cell line K562 and the autologous EBV-induced B-cell line. Cytotoxicity was measured using a 4.5-h chromium release assay. Target cells for cytotoxicity included EBV-infected B cells (A) and the natural killer-sensitive K562 cell line, obtained from the American Type Culture Collection, Rockville, Md. (B). Each assay contained targets incubated with medium alone in the absence of effectors (spontaneous release) and targets incubated in 10% Triton X-100 (maximum release). Percent target lysis was calculated as: % target lysis = 100 × [(experimental release − spontaneous release)/(maximum release − spontaneous release)]. Experimental release equals the counts per minute in the supernatant of effectors and targets. In all the experiments the spontaneous release was <30% of the maximum release.

Taken together, the results presented above confirm the existence of suppressor and noncytotoxic T cells which inhibit the growth of EBV-infected B cells. Cells with this property were previously isolated from the blood of patients with acute EBV-induced infectious mononucleosis (20). We have now been able to isolate them from the blood of a normal EBV-seropositive individual by stimulation of peripheral blood T cells with autologous B cells infected with EBV, followed by the addition of IL-2. By this technique, we have selected for a T-cell subset capable of proliferating in response to EBV-activated B cells and have probably lost most other T-cell clones. Interestingly, the cell line was mainly composed of T cells with a suppressor/cytotoxic phenotype, suggesting that, at least in certain individuals, proliferation in response to EBV-infected B cells is carried out predominantly by T8-positive T cells. Proliferation by the

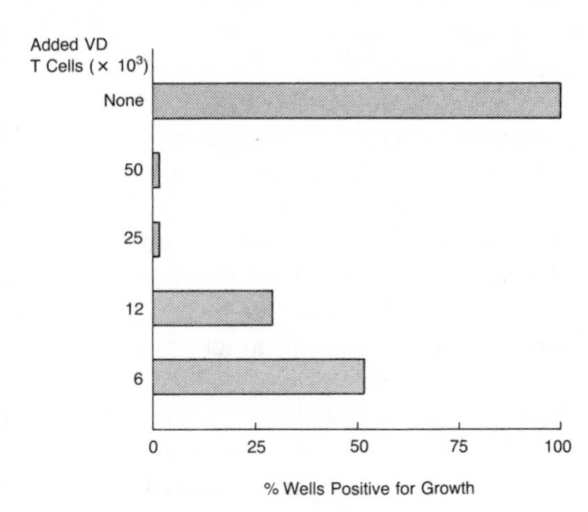

Figure 4. The VD-T cell line inhibits the growth of the autologous EBV-induced B-cell line. B cells from the established lymphoblastoid cell line VD-B (300 cells per well) were cultured in 96-well round-bottom microtiter plates with or without the addition of cells from the VD-T cell line at various concentrations in 10% fetal calf serum medium. Plates were visually examined weekly with phase-contrast microscopy for evidence of outgrowth, and each well was scored after 3 weeks of culture. Wells were considered positive for B-cell outgrowth if clumps of large progressively growing cells were recognized, as described (17).

T8-positive cell line was restricted to EBV-positive B cells expressing the histocompatibility antigen A2. This restricted pattern of recognition suggests that the T-cell line recognized an EBV-encoded antigen. Among EBV-induced proteins, the latent infection membrane protein represents a candidate, because it is the only known EBV-encoded product expressed on the cell surface (2, 6, 8, 9). Another possibility is that the T-cell line recognizes a B-cell activation antigen hyperexpressed on EBV-infected cells. Among these, CD23 is a good candidate because it is expressed at high levels on EBV-activated B cells, but only at low levels on B cells activated by other means (7, 15).

In addition to its ability to proliferate in response to EBV-infected A2-positive B cells, the T-cell line inhibited immunoglobulin production and proliferation by cocultured B cells in a dose-dependent manner. However, in contrast to the EBV specificity of its proliferation, inhibition of both immunoglobulin production and proliferation was EBV nonspecific and nonrestricted by histocompatibility antigens. Thus, VD-T inhibited proliferation and immunoglobulin production induced not only by EBV but also by PWM. In addition, VD-T inhibited autologous as well as random allogeneic responses.

A property the cell line consistently failed to express was cytotoxicity. This suggests that both inhibition of immunoglobulin production and proliferation by VD-T were not attributable to its killing the inhibited targets. It is possible that the cell line produces inhibitory factors, even though culture line supernatants were not inhibitory. Alternatively, VD-T could consume growth factors which are required for the growth of the cocultured cells.

Previous studies from our laboratory had suggested the existence of a suppressor-T-cell regulation of EBV-infected B cells in EBV-seropositive normal individuals (19). We have now grown a suppressor/noncytotoxic T-cell line from the blood of a normal EBV-seropositive person and have thus physically separated suppressor T cells from cytotoxic T cells. Characteristic features of this cell line include specific proliferation in response to EBV-infected B cells and virus-nonspecific suppression of both immunoglobulin production and proliferation.

The existence in vivo of T cells with functional properties similar to those of the cell line described here may provide an alternative explanation for the persistence of EBV-infected B cells after primary EBV infection (12). These virus-infected cells, under suppressor-T-cell control, would be prevented from expanding and would persist in relatively constant low numbers in the circulation.

Inhibition of growth and immunoglobulin production by the VD-T cell line was virus nonspecific. If this cell line represents an in vitro expansion of regulatory cells present in vivo, one would expect that suppression in vivo would also be virus nonspecific. We have seen, however, that the VD-T line only proliferates in response to EBV-infected B cells. For this reason, sufficiently high concentrations of specific regulatory cells in vivo would only be reached in physical proximity to EBV-infected cells. As a consequence of their physical proximity to the regulatory T cells, EBV-infected B cells would be inhibited in preference to other cells. In this formulation, specificity of regulatory effector functions would be assured by a specific control of the growth of the regulatory suppressor T cells.

Much needs to be learned about immune-T-cell regulation of EBV-infected B cells. We have discussed here the isolation and characterization of a suppressor/noncytotoxic T-cell line which inhibits the growth of EBV-infected B cells. The existence of cells with similar properties in vivo would provide an explanation for the continuous presence of EBV-infected B cells in the blood of most EBV-seropositive normal individuals.

CONCLUSIONS

Suppression and cytotoxicity are believed to play an important role in the regulation of EBV infection in humans. To analyze the role of suppressor T cells in the control of latent EBV infection, we sought to grow and characterize these cells. By repetitive stimulation of peripheral blood T cells from a normal EBV-seropositive individual with autologous EBV-infected B cells and culture in the presence of IL-2, a long-term T-cell line was obtained which proliferated specifically in response to EBV-infected B cells expressing the histocompatibility antigen A2. In addition to proliferating specifically in response to EBV-infected B cells, this cell line inhibited both proliferation and immunoglobulin production in autologous cells which were activated with either PWM or EBV. This suppressive T-cell line had no cytotoxic activity for either the natural killer-sensitive K562 cell line or the autologous EBV-induced lymphoblastoid cell line. These results suggest that, during latent EBV infection, suppressor T cells can be identified which proliferate specifically in response to EBV-infected cells and are capable of virus nonspecifically inhibiting B-cell activation in the absence of cytotoxicity.

Literature Cited

1. **Diehl, V., G. Henle, W. Henle, and G. Khon.** 1968. Demonstration of a herpes group virus in cultures of peripheral leukocytes from patients with infectious mononucleosis. *J. Virol.* **2:**663–669.
2. **Fennewald, S., V. van Santen, and E. Kieff.** 1984. Nucleotide sequence of an mRNA transcribed in latent growth-transforming virus infection indicates that it may encode a membrane protein. *J. Virol.* **51:**411–419.
3. **Gerber, P., and B. H. Hoyer.** 1971. Induction of cellular DNA synthesis in human leukocytes by Epstein-Barr virus. *Nature* (London) **231:**46–47.
4. **Henle, G., W. Henle, and V. Diehl.** 1968. Relation of Burkitt's tumor-associated herpes-type virus to infectious mononucleosis. *Proc. Natl. Acad. Sci. USA* **59:**94–101.
5. **Henle, W., V. Diehl, G. Kohn, H. zur Hausen, and G. Henle.** 1967. Herpes type virus and chromosome marker in normal leukocytes after growth with irradiated Burkitt cells. *Science* **157:**1064–1065.
6. **Hennessy, K., S. Fennewald, M. Hummel, T. Cole, and E. Kieff.** 1984. A membrane protein encoded by Epstein-Barr virus in latent growth-transforming infection. *Proc. Natl. Acad. Sci. USA* **81:**7207–7211.
7. **Kintner, C., and B. Sugden.** 1981. Identification of antigenic determinants unique to the surfaces of cells transformed by Epstein-Barr virus. *Nature* (London) **294:**458–461.
8. **Liebowitz, D., D. Wang, and E. Kieff.** 1986. Orientation and patching of the latent infection membrane protein encoded by Epstein-Barr virus. *J. Virol.* **58:**233–237.
9. **Mann, K., D. Staunton, and D. A. Thorley-Lawson.** 1985. Epstein-Barr virus-encoded protein found in plasma membranes of transformed cells. *J. Virol.* **55:**710–720.
10. **Masucci, M. G., M. T. Bejarano, G. Masucci, and E. Klein.** 1983. Large granular lymphocytes inhibit *in vitro* growth of autologous Epstein-Barr virus infected B cells. *Cell. Immunol.* **76:**311–321.
11. **Moss, D. J., A. B. Rickinson, and J. H. Pope.** 1979. Long-term T cell-mediated immunity to Epstein-Barr virus in man. III. Activation of cytotoxic T cells in virus-infected leukocyte cultures. *Int. J. Cancer* **23:**618–625.
12. **Moss, D. J., A. B. Richinson, L. E. Wallace, and M. A. Epstein.** 1981. Sequential appearance of Epstein-Barr virus nuclear and lymphocyte-detected membrane antigens in B cell transportation. *Nature* (London) **291:**664–666.

13. **Rickinson, A. B., D. J. Moss, and J. H. Pope.** 1979. Long-term T cell immunity to Epstein-Barr virus in man. II. Components necessary for regression in virus infected leukocyte cultures. *Int. J. Cancer* **23:**610–617.
14. **Rosen, A., P. Gergely, M. Jondal, G. Klein, and S. Britton.** 1977. Polyclonal immunoglobulin production after Epstein-Barr virus infection of human lymphocytes *in vitro*. *Nature* (London) **267:**52–54.
15. **Thorley-Lawson, D. A., L. M. Nadler, A. K. Bhan, and R. T. Schooley.** 1985. Blast-2 (EBVCS), an early cell surface marker of human B cell activation, is superinduced by Epstein-Barr virus. *J. Immunol.* **134:**3007–3012.
16. **Tosato, G., and R. M. Blaese.** 1985. Epstein-Barr virus infection and immunoregulation in man. *Adv. Immunol.* **37:**99–149.
17. **Tosato, G., R. M. Blaese, and R. Yarchoan.** 1985. Relationship between immunoglobulin production and immortalization by Epstein-Barr virus. *J. Immunol.* **135:**959–964.
18. **Tosato, G., I. T. Magrath, and R. M. Blaese.** 1982. T cell-mediated immunoregulation of Epstein-Barr virus-induced B lymphocyte activation in EBV-seropositive and EBV-seronegative individuals. *J. Immunol.* **128:**575–579.
19. **Tosato, G., A. D. Steinberg, R. Yarchoan, C. A. Heilman, S. E. Pike, V. DeSeau, and R. M. Blaese.** 1984. Abnormally elevated frequency of Epstein-Barr virus-infected B cells in the blood of patients with rheumatoid arthritis. *J. Clin. Invest.* **73:**1789–1795.
20. **Wang, F., R. M. Blaese, K. C. Zoon, and G. Tosato.** 1987. Suppressor T cell clones from patients with acute Epstein-Barr virus-induced infectious mononucleosis. *J. Clin. Invest.* **79:**7–14.

INDEX